The Resilience of the Roman Empire

Regional case studies on the relationship between population and food resources

Edited by
Dimitri Van Limbergen,
Sadi Maréchal and Wim De Clercq

BAR INTERNATIONAL SERIES **3000** | 2020

Published in 2020 by
BAR Publishing, Oxford

BAR International Series 3000

The Resilience of the Roman Empire

ISBN 978 1 4073 5770 6 hardback
ISBN 978 1 4073 5694 5 paperback
ISBN 978 1 4073 5695 2 e-format

DOI https://doi.org/10.30861/9781407356945

A catalogue record for this book is available from the British Library

© the editors and authors severally 2020

COVER IMAGE *Vallus, Moissonneuse des Trévires, Buzenol-Montauban (B), Collection Musées gaumais, Virton (B)*

The Authors' moral rights under the 1988 UK Copyright,
Designs and Patents Act are hereby expressly asserted.

All rights reserved. No part of this work may be copied, reproduced, stored, sold, distributed, scanned, saved in any form of digital format or transmitted in any form digitally, without the written permission of the Publisher.

Links to third party websites are provided by BAR Publishing in good faith and for information only. BAR Publishing disclaims any responsibility for the materials contained in any third party website referenced in this work.

BAR titles are available from:

BAR Publishing
122 Banbury Rd, Oxford, OX2 7BP, UK
EMAIL info@barpublishing.com
PHONE +44 (0)1865 310431
FAX   +44 (0)1865 316916
       www.barpublishing.com

Celebrating the publication of the 3000th title in the BAR International Series

As we reach 3000 titles in the BAR International Series, I look back and wonder what the founders, my father Anthony Hands and David Walker, would have thought of this unique milestone in academic archaeology. It was nearly fifty years ago that they published the first volume in the series, printed on a commercial press in a room at our home, collated (to deafening rock music) by local teenagers using our table-tennis tables alongside kitchen and dining tables, then stapled and guillotined on industrial machines in the laundry room. It truly was a home industry. This was a necessity back then when the difficulties of publishing excavation reports and scholarly research were often prohibitive, and as archaeologists themselves, David and Anthony wanted to find a solution to the problem. And so, BAR was born.

From the beginning, it was the founders' commitment 'to create a worldwide databank in archaeological research that is relevant in 100 years' time'. They actively sought important work from all continents of the world creating, over the years, the largest series of academic archaeology in the world. We have been proud this year to launch the BAR Digital Collection for Libraries, allowing the complete BAR Series (International and British) to be available for research in a whole new way. A specialised in-depth search is being developed which will allow the complete collection to be searched, and there is general excitement and anticipation to see what the analysis of this vast body of information brings to light. Of course, these discoveries will unfold over the next fifty years and longer as the Collection itself grows to BAR 5000 and beyond.

It is with great respect, fondness and gratitude that we thank all our authors, editors, reviewers, readers and customers who have been essential to this Series for five decades – and who continue to make BAR such a unique publishing environment.

Annabel Hands, Director, BAR Publishing
Oxford
September 2020

Of Related Interest

Economy and Cultural Change in the Pre-Roman Iron Age in Northern Central Europe
Joanna Ewa Markiewicz

Oxford, BAR Publishing, 2019   BAR International Series **2933**

Economia e Territorio
L'Adriatico centrale tra tarda Antichità e alto Medioevo
Edited by Enrico Cirelli, Enrico Giorgi and Giuseppe Lepore

Oxford, BAR Publishing, 2019   BAR International Series **2926**

Il paesaggio trasformato
La pianura a sud di Padova tra Romanizzazione e Tarda Antichità
Michele Matteazzi

Oxford, BAR Publishing, 2019   BAR International Series **2921**

Romans in the Middle and Lower Danube Valley, 1st century BC–5th century AD
Case Studies in Archaeology, Epigraphy and History
Edited by Eric C. De Sena and Calin Timoc

Oxford, BAR Publishing, 2017   BAR International Series **2882**

The Archaeology of the Roman Rural Economy in the Central Balkan Provinces
Rural settlements and store buildings
Olivera Ilić

Oxford, BAR Publishing, 2017   BAR International Series **2849**

Grumentum and Roman Cities in Southern Italy/Grumentum e le città romane nell'Italia meridionale
Rural settlements and store buildings
Edited by Attilio Mastrocinque, Chiara Maria Marchetti and Rossana Scavone

Oxford, BAR Publishing, 2016   BAR International Series **2830**

Trade and Economic Contacts Between the Volga and Kama Rivers Region and the Classical World
Andrey Bezrukov

Oxford, BAR Publishing, 2015   BAR International Series **2727**

Rain Harvesting in the Rainforest
The Ancient Maya Agricultural Landscape of Calakmul, Campeche, Mexico
Helga Geovannini Acuña

Oxford, BAR Publishing, 2008   BAR International Series **1879**

For more information, or to purchase these titles, please visit www.barpublishing.com

Contents

1. Introduction: Food for a growing Empire: reframing an old debate 1
   Dimitri Van Limbergen, Sadi Maréchal and Wim De Clercq

2. The expansion of agricultural land into marginal areas in northern Gaul 9
   Pierre Ouzoulias

3. Farming for a growing population: developments in agriculture in the provinces of *Germania* 31
   Maaike Groot

4. Viticulture and demography in the Laetanian region (*Hispania Citerior Tarraconensis*),
   1st c. BC – 3rd c. AD 47
   Antoni Martín i Oliveras, Víctor Revilla Calvo, César Carreras Monfort and José Remesal Rodríguez

5. Growing grapes in populous landscapes: demography, food, land and vine agroforestry in central
   Adriatic Italy 71
   Dimitri Van Limbergen

6. Population decline and wine industry: societal transformation on Late Antique Delos (Greece) 109
   Emlyn K. Dodd

7. Cities and sustenance in Roman Asia Minor 129
   Rinse Willet

Introduction
Food for a growing Empire: reframing an old debate

Dimitri Van Limbergen, Sadi Maréchal and Wim De Clercq

Ghent University

Abstract: This introduction explains the premise, content and aims of the volume, while it also frames the historiographical debate on land and people in the Roman world. Finally, it briefly touches upon possible research directions in the future regarding this theme.

Keywords: Roman economy, demography, agriculture, land, resilience, regionality, globalisation, economic growth

Introduction

At a certain point in their existence, all successful pre-industrial societies became faced with a fundamental challenge inherent to peasant-based economies: feeding a growing population while coping with the limits imposed by the natural environment and the available farming and processing techniques. Still, as medieval and later European history has repeatedly shown us (Grigg 1980), such problems did not automatically – or at least not immediately – lead to catastrophic Malthusian scenarios. Instead, it has rather reminded us how the threat of overpopulation stimulated agrarian communities to adopt a wide variety of strategies that – despite the inevitable decline of overall living standards – enabled them to maintain the balance between population and resources. Some of these solutions might be defined as "Boserupian" responses – that is, seeing population growth as a major driving force behind agricultural intensification and agro-technological innovation – and include the reduction of the natural fallow (ultimately leading to the omission of the technique in favour of rotation or multiple cropping systems) and changes in the type of crops grown (high- vs. low-yielding crops or crops with low vs. high agronomic needs). In other cases the flag does not cover the cargo, and societies often resorted to using a combination of demographic and agrarian adjustments, as there are birth control, migration (seasonal or permanent), changes in the organization of labour, or the expansion of the cultivated area into newly acquired or marginal territory, often at the expense of grazing- and woodland.

The link between demography and agriculture in economies preceding the industrialization era is a widely acknowledged feature in historical studies focusing on 13th-19th century Europe. Recent work by scholars such as Bruce W. Frier, Neville Morley and Walter Scheidel has stressed the need for a deeper integration of population issues in Roman scholarship, as well as the profound lack thereof in socio-economic studies on classical antiquity (Frier 2001; Morley 2011; Scheidel 2001). However, archaeology and ancient history have already provided us with potential clues that may hint at the existence of population pressure in certain places and periods of the Mediterranean under Roman hegemony. One may recall here some of the survey evidence for the Italian peninsula, which suggests the bringing into cultivation of new and often marginal land in areas such as southern Etruria (Tuscany) (Potter 1979), central Picenum (Marche) (Van Limbergen et al. 2017) and the Po plain (e.g. Traina 1983; Bottazzi et al. 1990) in the Late Republic and the Early Empire. Other possible signs include the many remains of land reclamations, such as drainage works in marshy lands (Quilici and Quilici Gigli 1995; De Haas 2010; Frassine 2013; Pelgrom 2018; Walsh et al. 2014) or terrace constructions on hillslopes (Foxhall 1996). And what about some of the comments made by the ancient agronomists, like Columella who in the 1st century AD discusses the practice of turning woodland and pasturage into arable land; an action that in the end resulted in diminishing returns (Col. *Rust.* 3.11.3; 2.1.3-5; 2.8; 17.3)? Or what to say with regard to some of the epigraphic sources, which contain references to competition over (marginal) land in the Early Imperial period in Italy? (Paci 1996/1997; Campagnoli and Giorgi 2003)? Are we seeing here the effects of (over)population on (limited) land availability?

Whatever the case, the potential impact of demographic developments on agrarian structures deserves a more prominent place in explanatory models of the Roman economy. This volume wishes to address this hiatus, and brings together a group of international scholars to discuss the relationship between population dynamics and regional development in the Roman world from the perspective of archaeology. By adopting a comparative approach, the focus of the volume lies on exploring the various ways in which regional communities actively responded to population growth – or decline for that matter – in order to keep going on the land available to them. The theoretical framework – or at least starting point – for our case studies are the agricultural intensification models developed by Thomas Malthus and Ester Boserup. In order to advance the debate on the validity of these models for identifying the societal and economic pathways of the Roman world, we incorporate the concepts of resilience and diversity into our approach, and shift our attention from the longue-durée to how people managed to sustain themselves along the way, that is, over shorter periods of time.

When trying to decipher the dynamic between demography and land use, archaeologists and historians have indeed often resorted to the works of Thomas Malthus and Ester Boserup. In essence the Malthusian theory on the relationship between population and ecology in pre-industrial economies posits that a population continues to grow until it begins to approach – and eventually surpasses – the carrying capacity of the land that is available to them. As such, this process leads to so-called 'positive checks'; that is, any kind of event that increased the death rate within a society, e.g. famine, war, and epidemic diseases, hence leading to a reduction of the population, and thus to demographic decline and socio-economic collapse (Malthus 1798). This process has become known as the 'Malthusian trap'.

The Boserup model, on the other hand, turns this view around. Instead of considering population size as an outcome dependent on – and thus restricted by – the available farm land, it sees population as an independent variable able to trigger agro-technological progress. In Boserup's view, this progress or change in land use as a response to population growth followed an extensive-intensive trajectory, along which people over and over again adapted more intensive labour- and capital demanding agricultural strategies and technologies, aimed in the first place at increasing the cropping frequency – and thus the carrying capacity – of the land (Boserup 1965).

The main reasons for the success – and at the same time the problematic nature – of these explanatory models in archaeology and ancient history are their simplicity and universality. Indeed, the Malthusian population dynamic is highly attractive because of the inescapable logic behind it, that is, the fact that there are physical limits to the amount of cultivable land on the earth. The Boserupian reasoning too has that very same appeal, as it starts from a unitary course, applicable to all pre-modern societies, which sees communities responding to population pressure in an identical way, by intensifying their cropping systems. But these are also the principal reasons why both models have (rightly) been criticized.

The most obvious and important critique on Malthus is that he assumed that soil capacity was a static and thus unchangeable element. So he did not consider the effects of an evolving agriculture, both in technological advancements that improved the fertility, productivity and thus efficiency of land use, and in interventions that increased the quantity of cultivable land, such as land clearance, irrigation, drainage and land reclamation; all actions that enabled to raise the level of carrying capacity, and thus to maintain the balance between population and resources. To this, we might add that the total carrying capacity of the land can also change 'naturally', independent from human intervention, because of climatic effects (Lo Cascio and Malanima 2005).

The Boserup model did acknowledge the power of human agency to alter the productive capacity of its environment, but its main weakness was the unilateral and universal nature of these interventions (Boserup 1965). Indeed, the model displays a significant lack in variability and diversity when it comes to human productive and intensification strategies. There are many ways in which an agricultural regime may be intensified, that is, not only through proper intensification with an increase in labour and capital, but also through a range of specialisation and diversification strategies. In other words, no single path towards intensification exists. So the first model considered population size as limited by man's inability to alter and adapt its environment, while the second model reversed man's position and saw population as a key driver for environmental adaptations, even if from a too narrow point of view.

By mostly adopting this essence of Malthus and Boserup, however, scholars have too often stressed the apparent juxtaposition of both models, while they are – in a sense – compatible. Indeed, Malthus did acknowledge that population growth might stimulate people to intensify their production, while Boserup realised that responses other than the intensification of production could also be the outcome of population growth. So we might in fact imagine a range of intermediate scenario's in which, for example, a Boserupian response to population growth might prevent the Malthusian trap to set itself in motion.

The aim of this volume is thus not to discard the very basics of the theories of Malthus and Boserup, but rather to deconstruct too strict Malthusian (cf. Erdkamp 2016) or Boserupian scenarios, and as such introduce novel and more layered ways of thinking by exploring resilience and variability in human responses to population (growth). The last decade has seen a firm resurgence of sustainability and resilience studies in archaeo-historical scholarship. These have greatly enhanced our knowledge on how case-specific interactions of endogenous and exogenous factors either helped or hindered people in dealing with such unfavourable circumstances in medieval and early modern Europe, Classic and post-Classic America, and some parts of pre-modern Asia, Africa and Australia (e.g. Costanza et al. 2007; Curtis 2014; Faulseit 2015; Fisher et al. 2009). There is a lacuna, however, for the Mediterranean area in Roman times (ca. 500 BC – AD 500). Indeed, it remains very much an open question how Roman civilization managed – at least for a while – to respond actively to population (growth) in its various sub-regions, and thus to remain stable over a long period of time. In order to start formulating some answers, this volume follows a regional trajectory by systematically going through some of its more important subareas in a series of six case studies (Figure 1.1). The focus is hereby on the Western Roman Empire between the 1st and the 3rd century AD. We believe that this approach can successfully offer interesting new points of reflection, as it will show how local (re)actions to demographic changes were in part determined by local environmental and climate within the vast Roman Empire.

Contents of the volume

The first chapter by Pierre Ouzoulias serves as both a broader introduction to the theories of Malthus and Boserup, and a critical reappraisal of recent syntheses on ancient demography. It then applies the Boserup paradigm to explore the expansion of agriculture in four marginal areas of northern Roman Gaul (Figure 1.1, region 1). Despite natural adversities, all regions show an intensification of both agriculture and habitation during the High Empire. In the Haye forest, a network of small and medium-sized villas got the most out of the poor soils through a system of walled and terraced fields. Similarly, in the Châtillon forest, the very thin layer of top soil covering a bedrock was intensely worked and even fertilized. The archaeological data from the Vosges foothills points to a remarkable connectivity of what appears to be a secluded area with the larger Roman cultural sphere. In the Brie Boisée district, it was not the introduction of new technologies or techniques, but a more intense and more productive application of traditional agricultural methods that improved the output of the land. Ouzoulias interestingly links this more intense cultivation of marginal lands to the growing urbanisation and the accompanying rise in demography in this part of the Empire. Indigenous family units running modest farms still formed the agricultural basis of this land, but the incorporation in a larger network incited them to increase the scale and maximize the results of their agricultural system. Such small farms are also the first victims when the times of boom are over and demand is on the wane.

The second chapter by Maaike Groot shifts the attention more to the northeast, and further explores the ideas of Ester Boserup for understanding agricultural developments in the provinces of *Germania Inferior* and *Germania Superior*, in particular to entangle the relationship between demography and animal husbandry (Figure 1.1, region 2). The archaeological evidence shows (at least) three ways how cattle breeding could cope with a rising demand as a result of increasing urbanisation. One response was to breed larger cattle. Three strategies seem to have been adopted: improve the cattle's nutrition, selective breeding

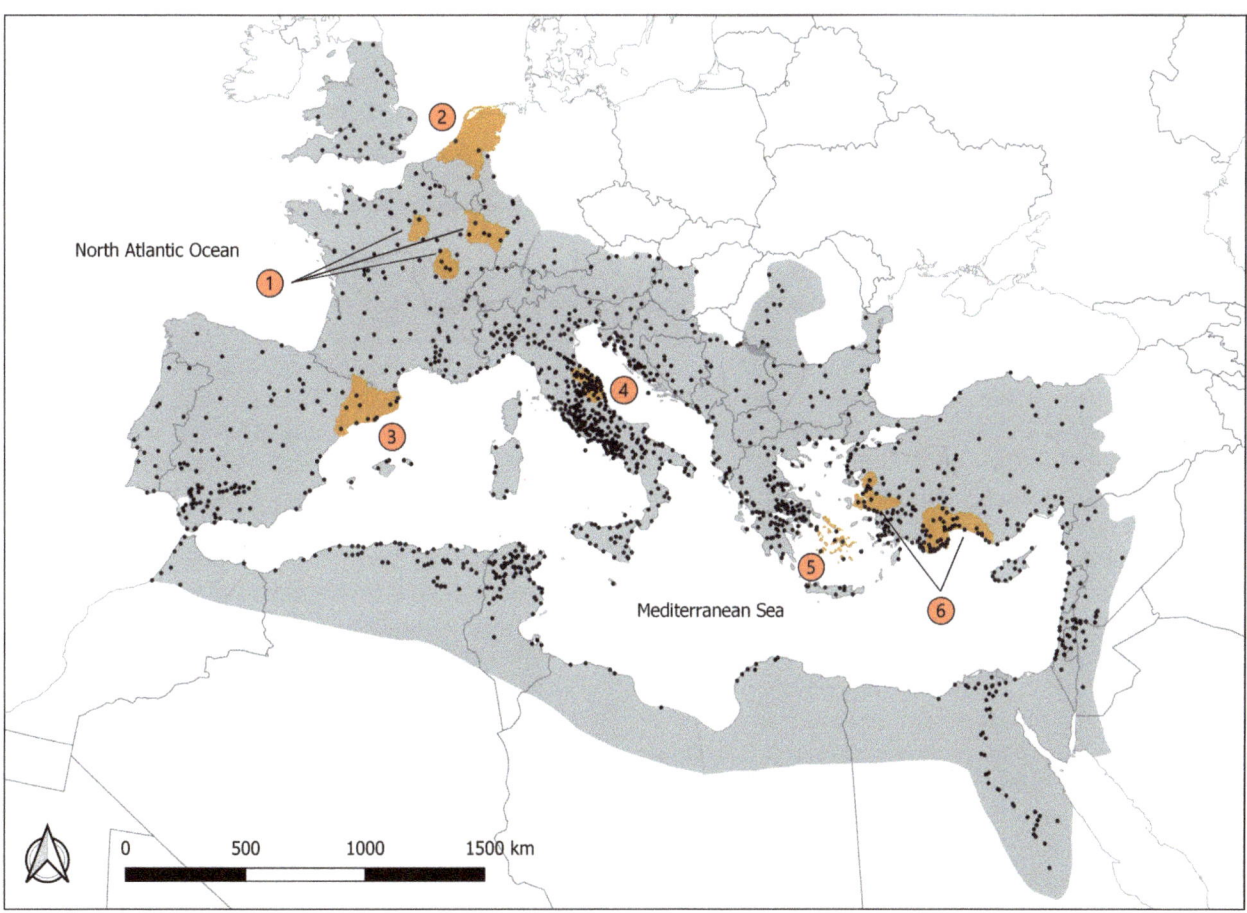

Figure 1.1. Localisation of the six study areas (yellow) within the Roman Empire (limits in AD 117), with indication of Roman towns (based on Hanson 2016).

of larger animals, and importing larger species for crossbreeding with local stock. A second response was to increase the stock by conquering new lands: marginal areas were put to good use. At last, a third response consisted of a specialization of cattle breeding. Evidence points to farms that focussed on wool production or the breeding of horses. In urban contexts, this specialization can be distinguished in the development of new crafts (glue production, bone working, tanning, etc.). Urbanisation and population growth and the resulting 'pressure' on farmers was not necessarily a bad thing, concludes Groot, as the new relationship between city and countryside also created new opportunities, new products and sometimes more wealth.

With the third chapter by Antoni Martin I Oliveras et al., we move to the heartland of the Mediterranean and the Roman wine industry with a case study of the Laetanian region in coastal Roman Spain (Figure 1.1, region 3). The intensification and specialization of viticulture, most notably perceivable by an increase of rural estates and amphora workshops, is linked to a demographic rise during the late Republic and Early Imperial period. The article investigates the complex economic interplay between production, trade and consumption from the regional level, to the inter-regional and eventually extra-regional empire-wide market. Hopkin's tax-and-trade model is used as a starting point for this analysis. The use of mathematics, statistics and linear programming models allows the authors to analyse, interpret, and make predictions and reconstructions about the evolution of an ancient economic system.

The fourth chapter by Dimitri Van Limbergen explores the potential of the *arbustum* as an ingenious response to land constraints in central Adriatic Italy in Early and Mid-Imperial times (ca. 25 BC-AD 200) (Figure 1.1, region 4). The *arbustum* was a plantation with vines trained on rows of host trees placed within crop fields. These fields were usually reserved for grain, legumes and vegetables, but sometimes they were also used for animal rearing. This type of silvo-arable agroforestry is a long-standing tradition in Italy (with later variants playing a central role in commercial viticulture up until the mid-20th century) (Sereni 19), but scholarly discussion on the *arbustum* has largely revolved around its place within subsistence agriculture and small-scale viticulture (Tchernia 1986). While the origin of this cultivation technique undeniably lies within this context, the system clearly broke through into Roman commercial farming as well. In fact, already in the 2nd century BC, Cato recommends the *arbustum* to farmers who grow vines for the urban market. About two centuries later, both Pliny the Elder and Columella consider the *arbustum* fully part of the Italian wine landscape. Furthermore, the literary evidence suggests that Italian farmers began systematizing and perfecting the *arbustum* from the mid-1st century BC onwards, with the practice reaching its most organized and widespread form in the course of the 1st and 2nd century AD. Recent archaeological and historical research in central Adriatic Italy has identified this period as a time of significant urban and rural demographic vitality. Taking as a starting point the possible causal link between these two processes, this chapter represents a first attempt to study this ancient agroforestry system, and in particular to analyze its qualities as a sustainable agricultural strategy in this part of Roman Italy. As such, it aims at integrating vine agroforestry into our narratives of viticulture and wine production in the Late Republic and the Early/High Empire.

The relationship between population and local viticulture is also discussed in the next chapter by Emlyn Dodd, this time for Delos (Figure 1.1, region 5). This chapter is atypical, however, in the way that it 1) discusses the potential effects of population decline (and not growth) on the local wine industry, 2) does so not for the mainland, but for an island in the Mediterranean, and 3) focuses on Late Antiquity (4th-6th century AD) rather than the core period of this volume, that is the 1st to 3rd century AD. Still, its inclusion in this volume represents a valuable opportunity to explore the dynamic between population and land use in a unique geographical setting, and within the distinct framework of island archaeology, characterised by a high degree of interconnectivity and the common formation of productive niches. In particular, the author combines an original archaeological dataset (i.e. wine production installations) with socio-cultural and socio-economic theory to sketch a picture of unexpected resilience in a time of allegedly 'negative developments' (population decrease). He so provides a seminal example of how a reduced population might 'respond' agriculturally in a positive way, thus contradicting typical Malthusian or Boserupian scenarios. At the same time, his study serves as a firm reminder not to reconstruct societal evolution *a priori* in too strict terms of prosperity and crisis.

The sixth and final chapter by Rinse Willet moves even further to the Roman East, and discusses the relationship between urbanisation, demography (town and country) and agricultural processes in *Asia Minor* (Figure 1.1, region 6). On the basis of four case studies – the cities of Kyaneai, Sagalassos, Ephesos and Pergamon – the author questions whether the noticeable increase in the number of cities in *Asia Minor* between the 2nd century BC and the 3rd century AD also represents an urban demographic growth, and if so, how this growth impacted agriculture and land use in their territories. Through the use of archaeology, epigraphy and a selection of historical and comparative sources, he argues that demographic growth did take place in both town and country, but that in most cases these towns did not outgrow their agricultural potential. For bigger towns such as Ephesos and Pergamon, however, the situation might be different. In any case, an increase in agricultural productivity as a result of these processes is likely, but the Malthusian axiom seems once again ill-adapted to frame these developments. In the end, this leads the author to discard the so-called low-equilibrium trap in favour of a "gradually improving equilibrium".

Implications and future prospects

Ever since the publication of Gibbon's *Decline and Fall* in 1788, scientists have been fascinated by the end of Rome and the reason(s) why it ended (e.g. Simkhovitch 1916; Huntington 1917; Tainter 1988; 2000; 2014). As Gibbon himself has phrased so eloquently, however, the wonder is not that Rome eventually fell, but rather that it managed to last for so long (Gibbon 1776-1788). Some of the answers necessarily lie in the ways in which the Romans dealt with the population-land equilibrium. The papers in this volume hopefully testify to the potential of archaeology – if integrated within a holistic and multidisciplinary approach – as a tool for reconstructing such trajectories on a regional scale. While some arguments and conclusions necessarily remain tentative, the picture that emerges from the selected case studies is a positive one; that is, one that shows how the Romans dealt actively with demography and resources in many different ways, either by using more (marginal) lands (northern Gaul, *Asia Minor*), adapting their farming strategies (central Adriatic Italy), ramping-up and/or specializing production (Delos, Laetanian Spain), or by combining a variety of strategies (*Germania*). In all cases, the solutions point to an intelligent and maximal use of the local environment. Obviously, from a modern point of view, these solutions all had their intrinsic limits – and an inevitable expiration date – but on the face of it, they were at least successful in establishing and/or notably prolonging regional equilibria between people and natural resources. These first observations allow for some cautious optimism when it comes to assessing the level and impact of population pressure on land use and the food economy in these areas and times of the Roman world (Flohr 2019).

We are aware that the collection of papers presented here is not comprehensive. Much remains to be done – and here we formulate a clear call for many more regional archaeological datasets and case studies, especially for North Africa and the East – but we are convinced that the present volume has established a helpful framework, both conceptually and methodologically, to further tackle the fundamental link between population, natural resources and regional developments in the Roman world. The six papers in this book have shown in particular how in-depth regional studies can contribute to the understanding of the diversity in land exploitation – and in the human drivers and responses to it – that existed within the Roman Empire. At first sight, globalization theory may seem useful to investigate this 'diversity within a new shared (Roman) cultural framework' (Versluys 2014, 14). But as Pitts and Versluys have rightly pointed out, globalization is a multi-sided theory, with variations and adaptations of the concept giving rise to different interpretations of the phenomenon (Pitts and Versluys 2015, 10-13). It comes as no surprise then, that the application of globalization to the ancient economy is quite problematic. Indeed, Neville Morley has rightly argued that a key marker of modern globalization, that is the compression of space and time, did not really occur in Roman times, at least not in such a way that it caused a radical shift of economic activities from a local to a regional scale, and certainly not from a regional to a global scale (Morley 2015, 56). Production remained oriented primarily towards local markets, while imports were percentagewise often at their peak in the early phases of Roman conquest, only to be replaced later by local imitations and regional productions. On the other hand, what we can take away from globalization theory is the fact that economic changes were not always a process directed by the state. The distinct local responses pointed out in the contributions in this volume, often rooted in native traditions (cf. Ouzoulias; Van Limbergen), amply attest to this phenomenon. An increase in specialization due to changed consumption patterns – and hence changes in demand – is also clear from several cases (cf. Groot; Dodd), but these may also be interpreted as the economic consequence of cultural globalization (Morley 2015, 61). Finally, an increase in scale (cf. Oliveras et al.; Willet) and consumption surely makes the Roman world differ from preceding eras, but not necessarily in a way that this resulted in a significant reduction of journey time (time compression) in a totally interconnected world (space compression); both indispensable markers of economic globalization.

If not globalized, the Roman world certainly became much more integrated along the way. Yet we should not envision this integration process as total and unified, but rather as having led to a series of separate but interlinked market systems, organized around specific products, and to a large degree steered by the state (Tchernia 2016; Van Limbergen 2019). As Groot's chapter in particular has suggested, this distinct form of market integration may in part have resulted from an increase in (urban) demand for manufactured goods and raw materials. This proliferation of the non-agricultural sector was then precisely possible because of population growth and the larger availability of (rural) labour (e.g. Erdkamp 2015; 2020). Particularly lucrative and environmentally determined products such as wine and olive oil were another important part of such demarcated (long-distance) supply networks, and this may help to explain the remarkable recovery of Delos in Late Antiquity (Dodd), or the rise of the wine industry in *Hispania Citerior Tarraconensis* in the late 1st century BC (Oliveras et al.). Together with the central Adriatic area (Van Limbergen), the Laetanian case also provides a clear example of the apparent rise in rural settlement numbers that seems to characterize much of the Western Roman world in the Late Republic and the Early Empire (cf. Jongman 2017). Still, more so than anything else, the papers in this volume all highlight the intrinsic link between regional food production and consumption, and stress how this deep connection remained the prime driver for the dynamics between population and land. Depending on factors like the rate of urbanisation and the nature of the territory, this process could seemingly play out better in some areas (Willet) than in others (Ouzoulias), but always in distinct and sometimes even unique ways (Van Limbergen). It is precisely this diversity in human responses to the wider phenomenon of demographic growth in the Roman world that this volume wants to emphasize.

Finally, it remains difficult to determine what kind of economic growth accompanied this growth in population. Delos left aside, the archaeological data discussed in this volume clearly show signs of aggregate economic growth of local economies; that is, more (urban and rural) people meant more consumption, and hence more production. It is less clear, however, how this process impacted overall standards of living. How well-off were people in these areas? The matter of per capita growth in the Roman world remains heavily debated (e.g. Frier 2001; Jongman 2007; Erdkamp 2020), but most of our archaeological samples are insufficient in size to give definite answers (cf. De Haas et al. 2011). The chapters in this volume are, alas, no exception. Still, based on the data presented here, the potential for achieving real growth at least seems to have differed from one region to the other. Indeed, some material and epigraphic remains appear to reflect higher levels of urban and rural prosperity (e.g. *Asia Minor*, central Adriatic Italy) than others (e.g. northern Gaul, *Germania, Hispania*), and this is again a strong reminder of the different cultural backgrounds (Greek-Hellenistic in the Mediterranean, Celtic in the North) and subsequent regionality of such developments in the Roman world. But whatever the nature of these developments, they all seem to have been connected to investments in some way: investments in agricultural specialization (wine in Hispania and Delos, cattle breeding in *Germania*), in the extension and/or reorganisation of land (central Adriatic Italy, Gaul), in public and private building programs (e.g. Asia Minor), and in infrastructure such as harbours and roads. So if we were to draw up the balance sheet right now, we suggest that most of the areas discussed in this volume were able to sustain trajectories that involved more than just aggregate economic growth (cf. the 'moderate growth' models explored in De Haas et al. 2011; Launaro 2011; Poblome et al. 2011), but that is as far as the current evidence can bring us. In any case, we hope that with this volume we can at least push the long-standing debate on growth and sustainability in the Roman world into promising new directions.

Bibliography

Boserup, E. *The conditions of agricultural growth. The economics of agrarian change under population pressure.* Chicago: Aldine Publishing Co, 1965.

Bottazzi, G., Bronzoni, L. and Mutti, A. *Carta Archeologica del Comune di Paviglio. 1986-1990.* Poviglio: Centro stampa Poviglio, 1990.

Campagnoli, P. and Giorgi, E. "Assetto territorial e divisioni agrarie nel piceno meridionale. I territori di Cluana, Pausulae, Urbs Salvia e Asculum." *Rivista di Topografia Antica* 14 (2003): 35-56.

Costanza, R, Graumlich, L. and Steffen, W. *Sustainability or Collapse? An Integrated History and Future of People on Earth.* London: The MIT Press, 2007.

Curtis, R. *Coping with crisis: the resilience and vulnerability of pre-industrial settlements.* Burlington: Routledge, 2014.

De Haas, T. "The agricultural colonization of the Pomptinae paludes: surveys in the lower Pontine plain." *Bollettino di Archeologie OnLine. Volume Speciale* (2010): 1-14.

De Haas, T., Tol, G. and Attema, P. "Investing in the colonia and ager of Antium." *Facta* 5 (2011): 111-144.

Erdkamp, P. "Agriculture and the various paths to economic growth." In *Ownership and Exploitation of Land and Natural Resources in the Roman World*, edited by P. Erdkamp, K. Verboven and A. Zuiderhoek, 18-39. Oxford: Oxford University Press, 2015.

Erdkamp; P. "Economic growth in the Roman Mediterranean world: An early good-bye to Malthus?" *Explorations in Economic History* 60 (2016): 1-20.

Erdkamp, P. "Population, Technology and Economic Growth in the Roman World." In *Capital, Investment, and Innovation in the Roman World*, edited by P. Erdkamp, K. Verboven and A. Zuiderhoek, 39-66. Oxford: Oxford University Press, 2020.

Faulseit, R.K. *Beyond collapse: archaeological perspectives on resilience, revitalization and transformation in complex societies.* Carbondale: Southern Illinois University, 2015.

Fisher, C.T., Hill, J.B. and Feinman, G.M. *The Archaeology of environmental change. Socionatural legacies of degradation and resilience,* Tuscon: University of Arizona Press, 2009.

Flohr, M. "Skeletons in the cupboard? Femurs and food regimes in the Roman world." In *The Routledge Handbook of Diet and Nutrition in the Roman World*, edited by P. Erdkamp and C. Holleran, 273-280. London/New York: Routledge, 2019.

Foxhall, L. "Feeling the earth move: cultivation techniques on steep slopes in classical antiquity." In *Human Landscapes in Classical Antiquity: Environment and Culture*, edited by G. Shipley and J. Salmon, 68-97. Londen/New York: Routledge, 1996.

Frassine, M. *Palus in Agro. Aree umide, bonifiche e assetti centuriali in epoca romana.* Pisa and Rome: Fabrizio Serra Editore, 2013.

Frier, B.W. "More is worse. Some observations on the population of the Roman Empire." In Debating Roman Demography, edited by W. Scheidel, 139-160. Leiden & Boston: Brill, 2001.

Gibbon, E. *The Decline and Fall of the Roman Empire.* New York: Modern Library, 1776-1788.

Grigg, D.B. *Population Growth and Agrarian Change. An Historical Perspective*. Cambridge: Cambridge University Press, 1980.

Hanson, J.W. *An Urban Geography of the Roman World, 100 BC to AD 300*. Oxford: Oxford Unibversity Press, 2016.

Huntington, E. "Climatic change and agricultural exhaustion as elements in the fall of Rome." *The Quarterly Journal of Economics* 31.2 (1917): 173-208.

Jongman, W. "Gibbon was right: The decline and fall of the Roman economy." In *Crises and the Roman Empire*, edited by O. Hekster, G. de Kleijn and D. Slootjes, 183-199. Leiden/Boston: Brill, 2007.

Jongman, W. "The Benefits of Market Integration: Five Centuries of Prosperity in Roman Italy." In *The Economic Integration of Roman Italy. Rural Communities in a Globalizing World*, edited by T.C.A. De Haas and G.W. Tol, 15-27. Leiden/Boston: Brill, 2017.

Launaro, A. "Investing in the countryside: villas and farms, landowners and tenants." *Facta* 5 (2011): 15-30.

Lo Cascio, E. and Malanima, P. "Cycles and stability. Italian population before the demographic transition (225 B.C. – A.D. 1900)." *Rivista di Storia Economica* 21.3 (2005): 5-40.

Malthus, T.R. *An essay on the principle of population*. London: J. Johnson, 1872 (1798, 4th ed.).

Morley, N. "Demography and development in classical antiquity." In *Demography and the Graeco-Roman World. New Insights and Approaches*, edited by C. Holleran and A. Pudsey, 14-36. Cambridge: Cambridge University Press, 2011.

Morley, N. "Globalisation and the Roman Economy." In *Globalisation and the Roman World. World history, connectivity and material culture*, edited by M. Pitts and J.M. Versluys, 49-68. Cambridge: Cambridge University Press, 2015.

Paci, G. "Terre dei Pisaurensi nella valle del Cesano." *Picus* 16-17 (1996-1997): 115-148.

Pelgrom, J. "The Roman rural exceptionality thesis revisited." *MEFRA* 130.1 (2018): 69-103.

Pitts, M. and Versluys J.M. "Globalisation and the Roman World: Perspectives and Opportunities." In *Globalisation and the Roman World. World history, connectivity and material culture*, edited by M. Pitts and J.M. Versluys, 3-31. Cambridge: Cambridge University Press, 2015.

Poblome J., Malfitana, D. and Lund, J. "Investing in Roman Antiquity. Modelling moderate growth." *Facta* 5 (2011): 9-14.

Potter, T. *The Changing Landscape of South Etruria*. London: Elek, 1979.

Quilici, L. and Quilici Gigli, S. (eds.) *Interventi di Bonifica Agraria nell'Italia Romana*. Roma: L'Erma di Bretschneider, 1995.

Scheidel, W. "Progress and problems in Roman demography." In *Debating Roman Demography*, edited by W. Scheidel, 1-82. Leiden & Boston: Brill, 2001.

Simkhovitch, V. "Rome's fall reconsidered." *Political Science Quarterly* 31 (1916): 201-243.

Tainte, J.A. *The Collapse of Complex Societies*. Cambridge: Cambridge Universitey Press, 1988.

Tainter, J.A. "Problem solving: complexity, history, sustainability." *Population and Environment: A Journal of Interdisciplinary Studies* 22.1 (2000): 3-41.

Tainter, J.A. "Collapse and sustainability: Rome, the Maya, and the modern world." *Archaeological Papers of the American Anthropological Association* 24 (2014): 201-214.

Tchernia, A. *The Romans and Trade*. Oxford: Oxford University Press, 2016.

Traina, G. *Le Valli Grandi Veronesi in Età Romana*. Pisa: Giradini, 1983.

Van Limbergen, D. "The interlocked economy of the Roman Empire." *JRA* 32 (2019): 675-679.

Van Limbergen, D., Vermeulen, F., Taelman, D. and Carboni, F. "Rural settlement dynamics in the Potenza corridor between 900 BC and AD 600." In *The Potenza Valley Survey: settlement dynamics and changing material culture in a central Adriatic valley between Iron Age and Late Antiquity*, edited by F. Vermeulen et al., 112-157. Rome: Academia Belgica, 2017.

Versluys, J.M. "Understanding objects in motion. An *archaeological* dialogue on Romanization." *Archaeological Dialogues* 21/1 (2014): 1-20.

Walsh, K., Attema, P. and De Haas, T. "The Pontine Marshes (Central Italy): a case study in wetland historical ecology." *BABesch* 98 (2014): 27-46.

2

The expansion of agricultural land into marginal areas in northern Gaul

Pierre Ouzoulias

Centre National de Recherche Scientifique

Abstract: This chapter offers both a critical reappraisal of recent syntheses on ancient demography and a presentation of alternative reflections. A first issue concerns the neo-Malthusian historiographic outline, which sees demographic growth of ancient societies eventually blocked because of the archaic and non-evolving character of agriculture and the law of diminishing returns. These societies, because they were hampered by the low equilibrium trap, would have remained at a steady state determined by technology and carrying capacity. According to this economic model, the production capacity of the land is often considered as an exogenous factor and is thought to be determined mainly by its natural characteristics. On the contrary, Ester Boserup considers demographic pressures as the cause of agricultural intensification and of the exploitation of new land. In this chapter, I apply the Boserupian paradigm to analyse the expansion of agricultural land into four marginal areas of northern Gaul. In these four case studies, household farms seem responsible for an agricultural expansion by means of adjustments of the ager and improvements of the agricultural capability of the soil. In the end, I offer a reflection on the social and economic circumstances of these processes and on the capacity of the Gallo-Roman societies to 'make agricultural lands'.

Keywords: Demography, Gaul, marginal land, Malthus, Boserup

The Malthusian trap revisited

In *The Cambridge Economic History of the Greco-Roman World*, Walter Scheidel outlined a model of interconnectivity between population growth and agro-economic conditions, which – according to him – represented a key mechanism for understanding ancient societies (Scheidel 2007). Scheidel based his model on the fundamental principle of the homeostatic regulation of population size according to the available natural resources. The level of balance between these two elements was hereby determined by both the technology and the lifestyle of these societies. As such, by adopting some of the key concepts of the work of Thomas Malthus, Scheidel argued how the endogenous accommodation of demography was assured by mechanisms that both lowered fertility (preventive checks) and increased mortality (positive checks). The overall stability of this natural system could be disturbed by natural exogenous factors, such as climate change, or by human factors, such as wars and political developments.[1]

This model, as outlined by Scheidel, explicitly considers the growth of ancient populations to have been strongly constrained by the archaism and the immobility of their farming systems. This means that agriculture always remained blocked in its technical development, as such forming a bottleneck for demographic growth, because an increase in productivity would always have been limited by the 'law of diminishing returns'. This 'law' was first formulated by Anne Robert Jacques Turgot (Turgot 1766), and – under the influence of Thomas Malthus' seminal work (Malthus 1798) – further developed and theorized by David Ricardo (Ricardo 1817). Its basic principle sees additional inputs – once they surpass the optimum of production – leading to nullify the marginal yields. Applied to agriculture, the law considers inputs in labour, capital and amendments unable to obtain – over a certain threshold – benefits that are proportional to the additional efforts applied per area unit. In more prosaic terms, we could say that this is what Pliny the Elder observed when stating that "*bene colere necessarium est, optime damnosum*" ("good farming is essential, but superlatively good farming spells ruin") (Pliny, *NH*, 18.38, English translation by H. Rackham).

Likewise, David Ricardo's land rent theory considers this 'law' to determine the economic conditions for the cultivation of marginal lands (Ricardo 1817). This implies that – without major technological changes – the larger the area under cultivation, the less fertile the newly added

---

[1] This chapter is the English translation of a French article that originally appeared in the journal Gallia (Ouzoulias 2014).

plots, and thus the more costly their cultivation will become. In other words, the intensity of the cultivation of these lands is inversely proportional to their fertility. So when a community expands its territory, this expansion becomes blocked once the last marginal cultivated plots yield as much as the work they require. Beyond this threshold – on the condition that the technical conditions remain unchanged – all additional exploited plots theoretically cost more than they can ever yield. The value of the produced goods then becomes lower than the value of the invested inputs (Cordonnier 2010: 17-25).

This means that there is a proportional relationship between the price of agricultural commodities and the amount of labour required to cultivate marginal soils. Indeed, as the demand for food must be satisfied, the prices imposed on the market are those of the goods from the most demanding soils in capital and labour. Land holdings with more fertile soils benefit from this general increase in prices and so their profits increase. This means that their owners now have the opportunity to obtain higher rents from their farmers. A differential rent is then perceived by the former. This rent owes its existence to differences in soil fertility, and as such to setting the value of agricultural products based on the least favourable production conditions (Ricardo 1817).

At the macroeconomic level, Thomas Malthus argued how this sharp increase in global food prices profoundly impacted the behaviour of the population, and how the homeostatic processes described above automatically came into play to limit the growth of the population, or to reduce its size to a level more compatible with subsistence production levels under more profitable economic conditions. David Ricardo even went a step further in his analysis, as he argued how the concomitant increase in the price of subsistence, wages and rent ultimately led to the cessation of capital accumulation, thus slowing down economic growth, and in the end halting population growth. As such, these powerful constraints 'blocked' a society in its development, and brought it back to a so-called 'stationary state', from which only technological progress would allow it to escape. It is this balance between the population and its productive capacity that determined the optimum of settlement or the 'carrying capacity' (Lee 1987; 1992).

The concept of carrying capacity originates from the field of animal biology, where it is used to designate the maximum size that an organism can achieve in a given environment. Placed within the field of human sciences, this concept describes a system ruled by the interaction of the natural specifics of terroirs, the technical systems of the societies that exploit them, and the efficiency of their institutions. Without major changes in their technical and social organisation, the economic growth of these societies was limited, as gains in productivity and hence in the growth of food surpluses were always bound to be lower than the increase in the population. This situation is referred to as the 'low equilibrium trap' or the 'Malthusian trap'. For Walter Scheidel, there can be no doubt that this trap determined the working of ancient economies (Scheidel 2007: 55-56). This is because only sustained and endogenous growth – that is, self-sustaining intensive growth – would have allowed pre-industrial societies to escape from the pull of demographic growth (Myint 1966: 90-91). But this kind of growth is driven by significant technological progress; a process inherently hampered in antiquity by weak tax rates and productive investments. Also, ancient agriculture would never have been able to generate sufficient surpluses that allowed the development of an important broad artisanal sector; that is, the sector responsible for most of the necessary technological innovation within a society.

Still, Walter Scheidel admits that in certain regions and periods, a particularly high population density could have favoured a virtuous cycle of development. He believes that, in these territories, the processes described by Ester Boserup, and discussed later, would have induced the owners of the land to increase agricultural productivity. As such, he believes that the use of slaves and the expansion of villas managed following the principles described by the Latin agronomists would have effectively contributed to the improvement of cultivation techniques and, consequently, reinforced the regional disparities of economic development and settlement in the ancient world. This logic is clearly at work in the homeostatic regulation model, which establishes a strong relationship between the density of the population, the technological level, and the overall productive capacity of a society (Scheidel, 2007: 54).

Walter Scheidel, however, thinks that this could only have consisted of a minor adaptation of a system that, overall, experienced modest growth. So – in the absence of major technological and agronomic progress – the moderate increase in ancient population size was only made possible through the diffusion and subsequent adaptation of a more efficient productive system (Scheidel 2007: 85). This classic thesis is shared by many historians (e.g. Garnsey and Saller 1994: 103; Inglebert 2005: 51). Other scholars have outlined still darker scenarios. For example, Wim Jongman has argued how population growth and the limited possibilities for extending the *ager* would have resulted in an increase in land rents, a fall in labour productivity and, finally, a decline in the living standards of the working classes, and as such an increase in social inequality (Jongman 2006: 241).

Most of these models are tributary to the traditional view of classic economists; a view that considers the 'law of diminishing returns' constraining growth and maintaining the level of development at a steady state (see *supra*). Only sustained and continuous technological improvement could have ensured a permanent renewal of capital, and thus have avoided that profits in productivity were totally absorbed by the increase of the population. Interestingly, more novel theories of growth – which tend to surpass this dogma of the trending lowering of the marginal productivity of capital – still accept this Malthusian idea of a close interdependence of the growth rate of population and the growth rate of technology (Kremer 1993: 681).

This conceptual framework remained an efficient tool to understand and explain the ancient economy, as long as one held to Moses Finley's doctrine of a stagnated technology and a structural limit of agricultural productivity (Finley 1965: 29-30). Indeed, the inadequacy of technological progress was a solid explanation for the supposed lethargy of the ancient population and its inability to free itself from the 'Malthusian trap'. At the same time, it was possible to allow for the existence of moderate economic growth in antiquity, provided that it resulted from an overall increase in the factors of production, and not in per capita productivity. In other words, overall expansion was the result of more workers being mobilised, but not of their ability to produce more with the same means (Clark 2007: 22-24).[2] Antiquity was thus part of a long period in the history of the Western economy, during which per capita income only experienced minor variations, with rapid growth only occurring from the Industrial Revolution onwards (Kremer 1993).

More than half a century of historical and archaeological data make it increasingly clear that this ideological construction is based on fragile foundations. But old habits die hard. So Peter Temin has recently argued how we can no longer deny significant changes in the living conditions of ancient populations, while at the same time holding on to the thesis of an ancient economy subject to strong Malthusian constraints. He admits that there was a considerable increase in the overall volume of produced goods, and that there was a considerable improvement in the living standard of individuals. He also considers these substantial developments to be the fruit of undeniable technological progress, especially in the field of agriculture (cf. Raepsaet 1995; Bowman and Wilson 2013; Temin 2013: 196, 217, 222; Greene 2000; Wilson 2002). In his examination of several theories on economic development, he even cites the pioneering work by Paul Romer, who has stressed the possibility of endogenous growth on the basis of potentially unlimited marginal productivity, made possible through the continuous accumulation of knowledge (Romer 1986). But Temin does not consider the possibility of such self-sustaining growth occurring in antiquity. On the contrary, Temin still sees the evidence of population growth and technological progress as compatible with a model of the ancient economy based on subsistence and thus of a population subjected to Malthusian constraints. According to Temin, this apparent contradiction may be explained by a delay in the triggering of the 'Malthusian brakes' (Temin 2013: 225). This implies that, prior to the full display of these Malthusian effects, the Roman Empire was able to escape their hold for a while, and as such experience a phase of economic and demographic development. In the end, however, by progressively strengthening, these processes did halt these mechanisms of developments, and hence plunged the Roman world into a deep crisis. In other words, growth was only a transitory phenomenon before an eventual return to the Malthusian equilibrium in Late Antiquity (Temin 2013: 236-239).

This ingenious if somewhat forced construction allows Temin to reaffirm the heuristic validity of the Malthusian model while at the same time integrating the archaeological data that seemingly challenged this same model (Temin 2013: 239). Even if Temin's stubbornness for hanging on to the Malthusian model seems illogical (Erdkamp 2014), his attempt has above all the merit of unveiling the neoclassical presuppositions of his economic analysis. Indeed, many of Temin's ideas conflict with Moses Finley's theses, while at the same time he shares Finley's Malthusianism. This reveals the often artificial character of the Manichean opposition between 'modernists' and 'primitivists' (Andreau 2010: 22). So Peter Temin supposes that in Antiquity markets of land, capital and labour existed, and that these markets were in balance, and that the productivity of these three factors depended on their marginal yield. Taken by this conceptual straitjacket, Temin cannot consider the possibility of increasing returns, as they would be incompatible with the neoclassical dogma of the traditional competitive market equilibrium, on which he builds his model.

As Dominique Guellec has shown, if one accepts the possibility of increasing returns, this means accepting the idea that different territories can have different paths towards growth based on their socio-economic characteristics. This also means that – even when subjected to identical constraints and/or opportunities – these regions can have divergent developments. What this basically implies, is that 'history matters', and that the transformation of a particular socio-economic entity is closely linked with its previous states (Guellec 1992: 49). So one major feat of the theories on endogenous growth has been to put the socio-historical dimensions of development processes at the foreground. In this way, these theories align with the ideas of the institutionalist stream. It is therefore deplorable that historians like Walter Scheidel – who openly support and share this school of thought – have not gone further in their critical evaluation of the neo-Malthusian models (Étienne 2011). In fact, in the seminal Cambridge Economic History of the Greco-Roman World – co-directed by Walter Scheidel, Ian Morris and Richard Saller – Philippe Leveau is the only contributor to finally propose an evolutionary model of the Western Roman provinces that considers the incorporation of these regions into a new political entity not to have led to a standardized economic situation, but rather to an increasing territorial disparity, driven by complex interactions between traditions and multiple forms of economic development. This innovative hypothesis leads

---

[2] Richard Saller thus defends the position of Moses Finley on economic growth in antiquity. He stresses the necessity to clearly distinguish between aggregate growth and per capita growth in production (Saller 2005: 226-228). This concept of growth seems heavily inspired by the neoclassical model as outlined by Robert Solow, who conceives growth subjected to evolutions in technology and demography, both considered as exogenous variables (Solow 1960).

Leveau to challenge the validity of the Neo-Malthusian claims (Leveau 2007; Ouzoulias 2011a).

How to proceed then? It seems especially important to now focus a bit more on the status of the land in these doctrines. In fact, of the three factors of production – that is, labour, capital and land – natural resources have repeatedly been considered a relative exogenous, unreproducible and therefore rather unimportant variable within these classical theories (Temin 2013: 209). Without any significant progress in agricultural techniques, population and economic activity reach a certain balance that depends on the amount of fertile land available. The balance of this system is broken when these lands are exploited more intensively, or when lower quality soils become cultivated. In the end, what Thomas Malthus and David Ricardo did, was to intelligently translate the ancient fear that nature would never be able to satisfy the hunger of a constantly growing population. This fear is perhaps best expressed by Tertullian in an eschatological vision regarding Africa:

> *"Surely it is obvious enough, if one looks at the whole world, that it is becoming daily better cultivated and more fully peopled than anciently. All places are now accessible, all are well known, all open to commerce; most pleasant farms have obliterated all traces of what were once dreary and dangerous wastes; cultivated fields have subdued forests; flocks and herds have expelled wild beasts; sandy deserts are sown; rocks are planted; marshes are drained; and where once were hardly solitary cottages, there are now large cities. No longer are (savage) islands dreaded, nor their rocky shores feared; everywhere are houses, and inhabitants, and settled government, and civilized life. What most frequently meets our view (and occasions complaint), is our teeming population: our numbers are burdensome to the world, which can hardly supply us from its natural elements; our wants grow more and more keen, and our complaints more bitter in all mouths, whilst nature fails in affording us her usual sustenance."*
> (English translation by Peter Holmes)
>
> Tertullian, De Anima, 30, 3-4

Nevertheless, the work of Thomas Malthus is far more complex than this worried conjecture. Its most cautious commentators have emphasized that it is not always of great doctrinal unity (Platteau 1984; Cairo 1984). Moreover, as shown by Hervé Le Bras, the neo-Malthusian theoreticians – including the aforementioned Ronald Lee, have retained from his analyses only the idea of an irreducible gap between the growth of the population and the ceiling of resources (Le Bras 2003: 50). However – as explained by Le Bras on the basis of a more rigorous reading of the *Essay on the principle of population* – Thomas Malthus explicitly conceives that the demographic pressure can induce farmers to intensify their production and thus provide a source of livelihood for a new rise of the population. Le Bras specifically notes how the logic of this process of catching up is not without recalling the spirit of the mechanisms described by Ester Boserup in a work that is abusively considered as the antithesis of the *Essay on the principle of population*. Therefore, some of Ester Boserup's ideas should now be presented, as they will form the conceptual framework for the archaeological observations in the rest of this chapter.

Agricultural intensification according to Ester Boserup

Ester Boserup's 1965 work *"The Conditions of Agricultural Growth"* cannot be directly opposed to the book of Thomas Malthus, because it does not proceed of the same method of research at all. The Danish economist did not set out to write a theoretical treatise on the relationship between population and agricultural growth. She simply drew general lessons from field observations that fundamentally altered our analytical perspectives regarding the relationship between rural communities and their territories. Indeed, her experiences as a specialist in the economy of development found that the peasant societies under population pressure first increase their investment in capital and labour on the most intensively exploited land, and then, when their marginal yield decreases, shift their efforts to the least cultivated portions of the lands within their territories. This agricultural intensification is not generally the result of a radical transformation of the technical systems, but it is more surely obtained by a reduction of the fallow. Ester Boserup thus defined it as "a gradual progression towards land-use systems that allow a given area to be cultivated at shorter intervals" (Boserup 1970: 67). As a result, as both agricultural productivity and the cultivated area increase, so does the amount of work devoted to each of the cultivated hectares. Without major technical changes, the supply of labour is hereby growing faster than production. Therefore, she estimates that population growth does not lead to an increase in unemployment, but rather to an intensification of the work of the peasant family (Boserup 1970: 183).

What at first sight might seem a rather banal empirical observation conceals on closer examination a radical refutation of Neo-Malthusian theories. For the latter, land is most often considered as an almost constant exogenous element, whose quantity and fertility determine, without change, carrying capacity. Ester Boserup imposes a decisive change of perspective by considering the cultivated area as a variable closely related to both the density of the population and the socio-economic capacities of the farming communities to modify their environment. The increase of the population is then at the origin of a creative pressure which compels peasant societies to act on their environment by adapting their technical systems and by increasing their productive capacities. In a very innovative way, she therefore considers that *"the rate of [population growth], which can be borne by the own efforts of a certain rural community, is inversely proportional to the density of the population living in a given territory"* (Boserup 1970: 184). This original idea recalls the theses of Alfred Sauvy; who considers low population densities as an obstacles to development, but demographic pressure as a source of progress (Sauvy 1963: 302).

Ester Boserup showed full awareness of the innovative nature of her ideas in a seminal *Annales* article, in which she invited historians of pre-industrial societies to reject Neo-Malthusian simplifying abstractions in order to better study the complex interactions between the environment, population, and agricultural practices (Boserup 1974). At the same time, several medieval historians were involved in a debate that evoked the theses of Ester Boserup. This is not the place to present this debate in its entirety; it suffices to recall Emmanuel Le Roy Ladurie (1978), who responded to criticisms of Robert Brenner (1976) in affirming his attachment to a homeostatic Neo-Malthusian and Neo-Ricardian model, in the end quite close to the model defended by Walter Scheidel for antiquity. A few years later, Michel Morineau reproached the French historian for his vision of immobile campaigns, and opposed it to a very "Boserupian" conception of a rural history animated by "incessant change in the relationship between men and the earth" and the perpetual adaptation between resources and the population (Morineau 1984: 224).

The ability of men to alter their environment has long been considered one of the basic features of the Roman world. The drainage of the Fucine Lake under emperor Claudius is often presented as the archetype of this development work, devised at the highest level of the State and achieved through the mobilization of considerable resources and ingenious techniques. In this case, the work of 30, 000 men for 11 years resulted in the cultivation of 50 km² of new land by large landowners, who had probably financed this undertaking to a considerable extent (Leveau 1993). In the Narbonnaise region, the laying out of centuriations in very vast territories is usually considered as yet another large-scale demonstration of this so-called "Roman genius". Allegedly, this strategy provided a rigorous framework for Rome's colonization of the conquered territories, and thus favoured the extension of the cultivated space by the implementation of a more rational form of agriculture. In association with the villa system, it thus testified to a proper "Romanization of the countryside", whose beneficial effects on agricultural productivity were undeniable and formed the solid base for the demographic growth revealed through the progress in urbanisation (Le Roux 1998: 200-202). Still, without questioning the scope of these programs of rational management of land resources, nor the reality of the development work of the *ager* at the expense of wild space, Philippe Leveau believes that it would be excessive to assume that the intervention of Rome led to a "complete restructuring of the rural area" whose privileged and even exclusive instrument would have been the centuriation (Leveau 2010: 139, 148).

In the remaining part of this chapter, I would like to further develop this line of thought by presenting examples of agricultural conquest that seem to follow other modalities; modalities that question in particular the classic historiographic opposition between Romanized and indigenous forms of land development. In doing so, it will hopefully become clear that a correct analysis of the original conditions of these extensions of the cultivated soil requires questioning the validity of some of the concepts and economic categories presented in the first part of this chapter. As such, I share the methodological choices affirmed by Alan Bowman and Andrew Wilson, and their willingness to give more importance to the archaeological documentation in the debates on the ancient economy. Like them, I am also convinced that the use of highly synthetic models tend to overlook the variety and heuristic richness of local situations (Bowman and Wilson 2013: 8-10), which at times can lead to 'histories without sources'; only to preserve the coherence of theoretical constructions considered intrinsically superior to the patient proofing of minor events (Andreau 2010: 48).

In what follows, I will discuss some of the available archaeological evidence for agricultural expansion and intensification in northern Gaul, with a particular focus on four micro-regions: 1) the Haye forest; 2) the Châtillon forest; 3) the Vosges foothills; and 4) the Brie Boisée district (Figure 2.1). I focus on the various ways in which Gallo-Roman peasant communities seemingly 'created' new farmland in areas with significant soil and/or terrain constraints (Simon 1985: 245).

Case studies

The Haye forest

The Haye forest is a large woodland of ca. 120 km² that covers a limestone plateau enclosed within a broad loop of the Moselle River near the city of Toul in the department of Meurthe-et-Moselle (Figure 2.1). The area has altitudes between 350 and 450 m, and is characterised by thin calcareous or marl-limestone soils (Bajocian) with a generally low agricultural potential. In Antiquity, the plateau lay in the northern part of the territory of the *civitas* of the *Leuci*, close to the *civitas* of the *Mediomatrici* (Georges-Leroy et al. 2007; 2009; 2011; 2012; 2013; Georges-Leroy, Lafitte et al. 2013). This part of Belgian Gaul was one of the most economically developed regions of eastern Gaul, with the area lying close to the major traffic arteries of the Moselle River and the Roman road that linked Langres with Trier, passing through Toul and Metz (Kasprzyk and Nouvel 2011).

The area has been covered by a thick forest layer since at least the 12th century, and its presence has allowed for the conservation of a vast field system, recently revealed by means of airborne laser (Lidar) over a total surface of ca. 116 km² (Georges-Leroy et al. 2011; 2012, 2013b). This field system consists of walls made with recovered stones from plot borders and terraced slopes. In the southern part of the forest (ca. 5,200 ha), these agrarian structures (ca. 3,000 ha) seem spatially and chronologically connected with some 50 Gallo-Roman buildings from the second half of the 1st century AD, and with a Roman road that crosses the entire massif from north to south, probably

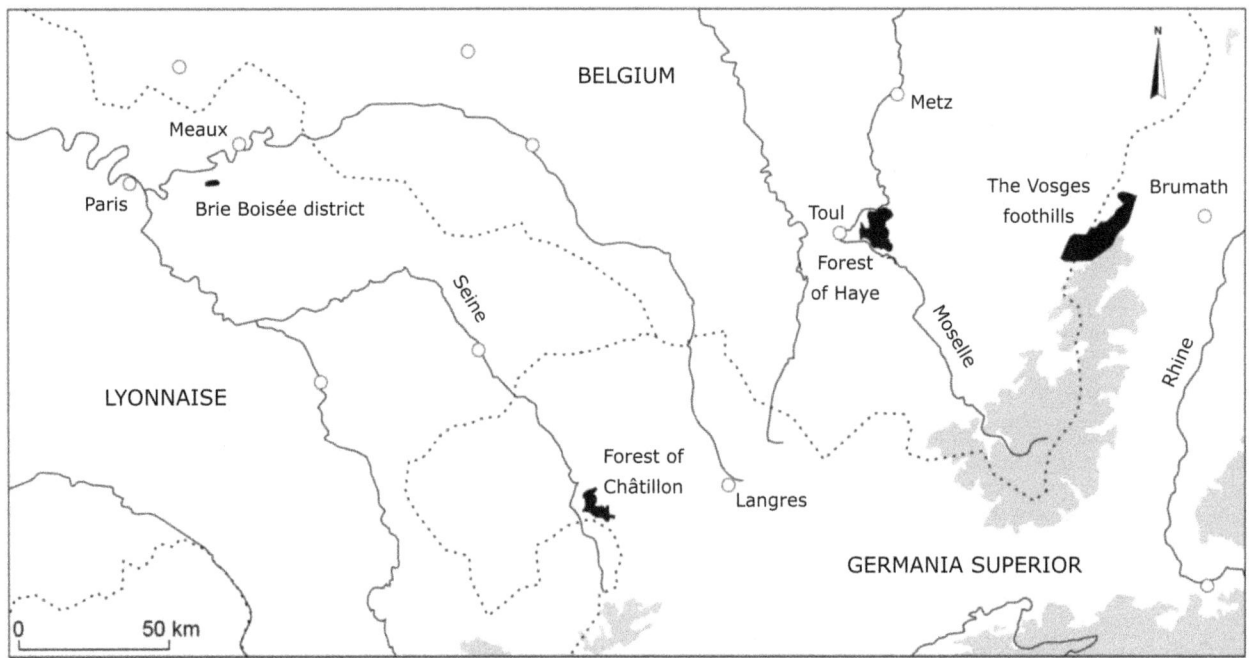

Figure 2.1: Map with the localisation of the four study areas (©P. Ouzoulias 2014, CNRS).

constructed in the first half of the 1st century AD. There is also evidence to suggest that this systematic cultivation of the Haye plateau was – at least in some places – preceded by important land clearance operations, even if we cannot rule out the possibility that the area was already loosely exploited in the Late La Tène period (3rd-1st century BC).

Two major features of the Gallo-Roman occupation of the Haye massif deserve mentioning here. First, the area was agriculturally exploited through a network of mostly small and medium-sized farms, with large farms or 'villas' – so characteristic for the territory of the *civitas* of the *Mediomatrici* – being totally absent (Ouzoulias 2011b). There might have been somewhat larger and more luxurious farms towards the edges of the plateau – as attested by the presence of painted plaster and/or architectural fragments – but real 'villas' have not been identified (Georges-Leroy et al. 2013a). I should mention, however, that the rural site density in the Haye massif is lower than the one registered in the well-surveyed loamy soils of the Lorraine plateau.[3] Second, the occupation period of these farms seems short, with most of them disappearing towards the end of the 2nd century AD, or in the first half of the 3rd century AD. A few isolated farms still seem active in the 4th, or at the latest at the beginning of the 5th century AD. The associated field system was probably abandoned within the same timeframe.

In sum, all data thus seemingly point towards an intensive and relatively short-term cultivation of the Haye plateau, despites the area's low agricultural potential. This intensification took the form of a fairly dense network of smaller farms, whose occupants put a lot of effort in building an extensive network of walled and terraced field plots. This impression of a quite intensive cultivation system is reinforced by the results of phosphorus soil content mapping in an area of 210 ha in the southern part of the massif, which indicate a strong presence of organic matter both on these farms and in their surrounding fields (Figure 2.2). This can be explained by the presence of cattle within the buildings, and thus by the use of their manure for fertilizing the cultivated plots (Georges-Leroy et al. 2009: 31).

I thus fully agree with Murielle Georges-Leroy in discarding traditional interpretations of these marginal lands being exploited for either occasional pasture or cultivation through long-duration fallow. Indeed, she is correct in stressing how these small farms in the Haye Massif testify to a phase of agricultural development long underestimated, both in terms of its geographical extension and its productive capacity. Our ignorance of this 'land exploitation system' is in part due to the practical difficulty of recognizing such systems in areas covered by thick layers of forest, but above all to the prevailing historiographical tradition, built around villa-centric interpretations of agricultural expansion and intensification under 'Romanization'. In fact, other fossil traces of field systems have now been revealed in the forests along the Bajocian coast, from Neufchâteau to Metz, and more recently also in the south of the department of Meuse (Georges-Leroy et al. 2013b: 120-121). The same goes for the woodlands of Warndt (Zender 2017). Many of these ancient field systems seem associated with smaller Gallo-Roman farmsteads, and the extension of their traces over very large areas forces us to significantly reconsider

---

[3] The 85 farms identified in the undisturbed sectors of the Haye Massif amount to a density of 1.1 sites/km², while a density of 3.38 sites/km² has been determined for the area to the south-east of Metz (Georges-Leroy et al. 2013a, 192).

The expansion of agricultural land into marginal areas in northern Gaul

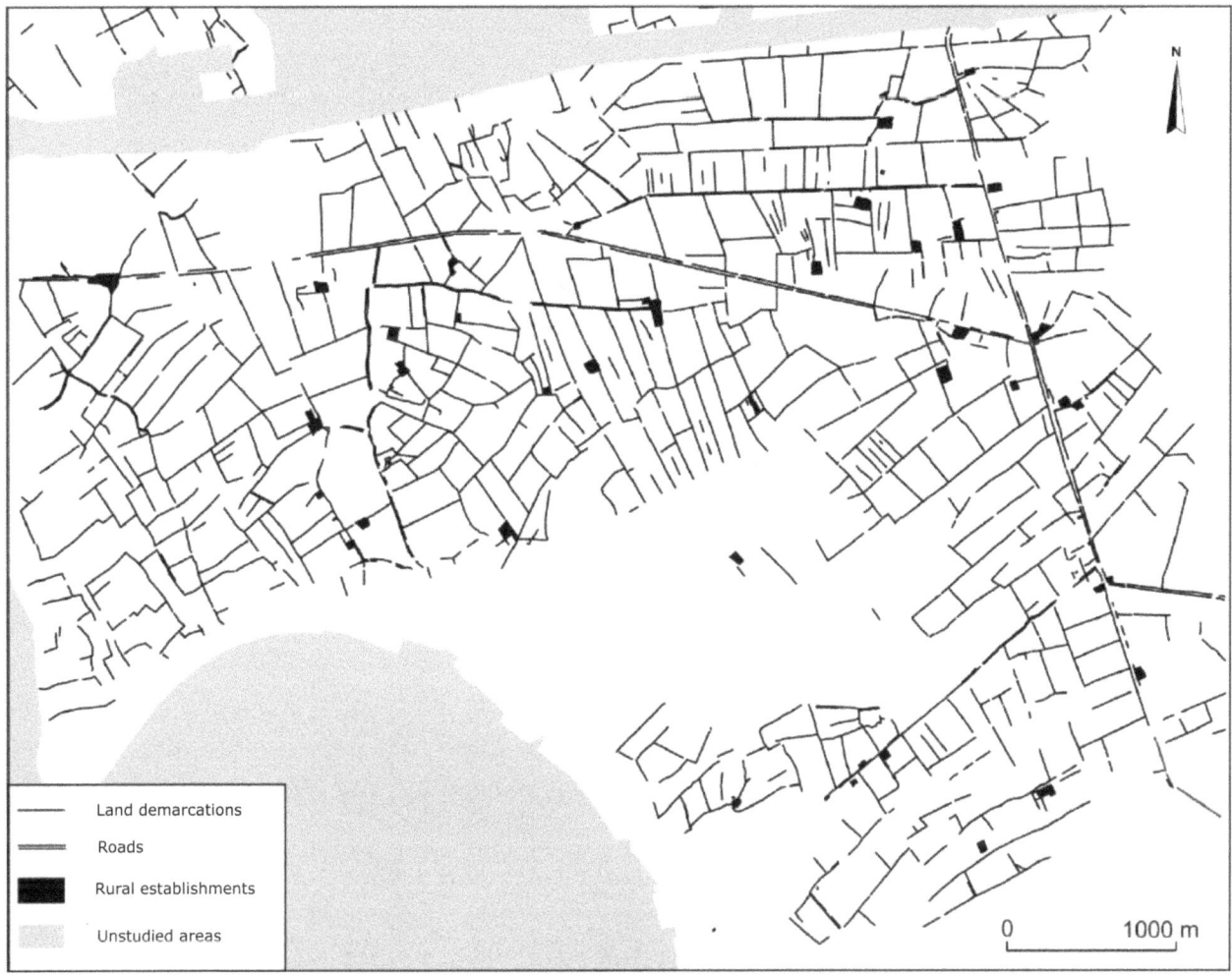

Figure 2.2: Map of the southern part of the Haye forest, with the indication of land demarcations and rural settlements (© Georges-Leroy et al. 2013a, 182, fig. 2).

our views on the agricultural boom accompanying the economic development of the *civitates* of eastern Gaul in the High Empire. We can further explore this changing view through a second case study.

The Châtillon forest

The forest of Châtillon-sur-Seine in the north of the Côte-d'Or lies on a medium-altitude plateau of about 400 m high (calcareous Bartonian substratum) immediately to the east of the Seine River valley (Figure 2.1). The plateau is covered by extremely shallow soils, especially along the edges of the slopes, where they almost disappear to uncover the underlying limestone layers. The area is located along the southwestern limit of the territory of the *civitas* of the *Lingones*, and was crossed by an important Roman road that connected the valleys of the Saone and the Seine Rivers, passing through the Lingonian settlements of Beneuvre and Vertault (*Vertillum*). Archaeological research has a long tradition in this region, but has only recently become more systematic thanks to the efforts of Dominique Goguey and Yves Pautrat, whose survey work has revealed much about the exploitation of the ancient landscape (Goguey and Benard 2006; Pautrat and Goguey 2007; Goguey and Pautrat 2009; Goguey et al. 2010).

Like in the Lorraine area, their research has revealed a series of fossil traces related to a walled field system that covers an area of about 6,500 ha (Figure 2.3). Once again, these field plots seem linked to rural habitations dating between the beginning of the 3rd century BC and the mid-3rd century AD. Many of these farms were made out of durable materials, with some even providing evidence for quality architecture and comfort facilities that are usually associated with small Roman 'villas'. The presence of these large farms – together with the attestation of two important Latinian shrines[4], and the remains of statues[5] and inscriptions[6] – thus testifies,

---

[4] The sanctuary of Essarois was founded during the La Tène D1b (120-90 BC), and abandoned towards the 4th century AD. That of Tremblois at Villiers-le-Duc was occupied from the La Tène D1a (150-120 BC) until the end of the 4th century AD (Izri and Nouvel 2011). I also refer here to the Carte Archéologique of Gaul devoted to these two sanctuaries (Provost 2009a: 315-318; 2009b: 405).

[5] The two sanctuaries have delivered several fragments of statues. Elsewhere, in the commune of Villiers-le-Duc, on the plateau, elements of a statue dedicated to Jupiter and a sculpture representing two cows have been found (Provost 2009b: 401-405).

[6] The sanctuary of Essarois has delivered nine votive inscriptions, of which three were dedicated to Appolo Vindonus (ILingons: 300-302). Three other inscriptions have been found in the commune of Villiers-le-Duc, two of them in the sanctuary of Tremblois and a funerary stele in a place called Gros Murger (ILingons: 311-313).

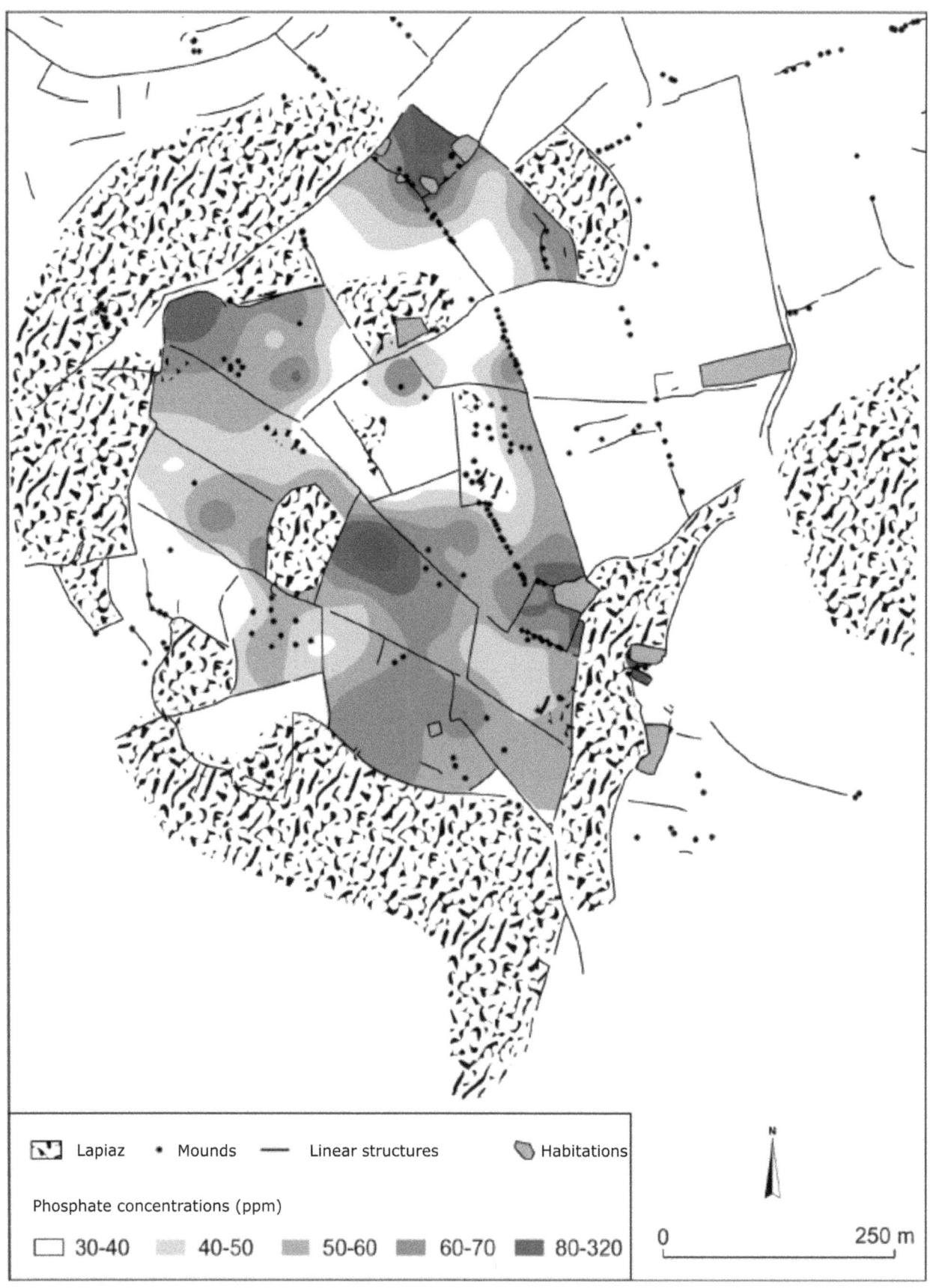

Figure 2.3: Phosphate concentrations within the Roche Chambain plot at Rochefort-sur-Brévon (© A. Giosa, Université Paris-I).

despite of the low agricultural potential of the soils, to an intensive exploitation of the area, similar to what has been observed in other territories of the *Aedui* and the *Lingones*. We can thus firmly state that the farmers who lived on this plateau were not disconnected from the wider processes of cultural, societal and agricultural change that touched the region towards the end of the Iron Age.

The expansion of this field system towards the slopes of the plateau, however, seems to have been carried out in a later stage, responding as such to different formal and economic arrangements. The field system uncovered at Roche Chambain in Rochefort-sur-Brévon is exemplary. It consists of a network of about fifty enclosures that are mostly oriented along a north-south axis, materialized by an ancient road. Based on the available chronological information, these field plots seem the result of developments within a more confined period, perhaps between the 2nd century BC and the 2nd century AD. It is also important to note that the farms registered in this area are of a more modest nature, both in size and in the quality of the materials used.

Interestingly, these field plots have delivered the remains of agricultural tools (e.g. ploughshare, plough coulter, harrow teeth) that point towards relatively intensive farming practices. Chemical soil analyses carried out by Alain Giosa support this idea, as the attested phosphate levels suggest that many plots had once benefited from fertilization and cultivation methods aimed at modifying the agricultural capacity of the soil (Figure 2.3). In addition, the construction of terraces will have enabled the containment of colluvium, and as such an increase in soil potential (Giosa 2012).

Despite the constraints imposed by the natural improvement, and despite the relatively 'coarse' nature of some rural units, it seems no longer acceptable to consider this area as only having been exploited on a seasonal basis by Gallic populations for purposes of pasture (Goguey et al. 2010: 179-180). On the contrary, it seems that the rural settlement patterns in the Châtillon forest once again testify to a process of agricultural intensification. This observation firmly contradicts the popular historiographical argument that considers traditional rural populations in Gaul forced to use more marginal lands because of the expansion of the new Roman villa system (Grenier 1985: 752). This idea has come under sustained critique through the discoveries in the forest of the Vosges foothills. I will now direct my attention to this area.

The foothills of the Vosges Mountains

The "culture of the Vosges foothills", as it has been termed by François Pétry (1977, 1978, 1979, 1997), occupies a vast region which extends, in the northern part of the Vosges Mountains, from Donon in the south to Eckartswiller in the north, in the departments Moselle and Bas-Rhin. It corresponds to the Vosges sandstone hills, whose altitude is between 300 m and 600 m. In antiquity, the border between the *civitas* of the *Mediomatrici* in Belgian Gaul and that of the *Triboci* in Germania Superior (after the reform of Domitian) ran to the west or the summits of these reliefs. According to François Pétry, these territories were covered by the large *silva Vosagus*, mentioned on the *Tabula Peutingeriana*.

More than hundred examples of field plots with small habitations have been identified in the foothills of the Vosges Mountains. The examples of Wasserwald in the commune of Haegen in particular have been intensively surveyed and excavated by François Pétry, and their organization is nowadays well-understood (Figure 2.4). In this area, plot delimitations (walls) are roughly perpendicular to a path of nearly 800 m in length that crosses the summit of the (flattened) land (Favory 2011: 398-399). The habitations within these plots are sometimes

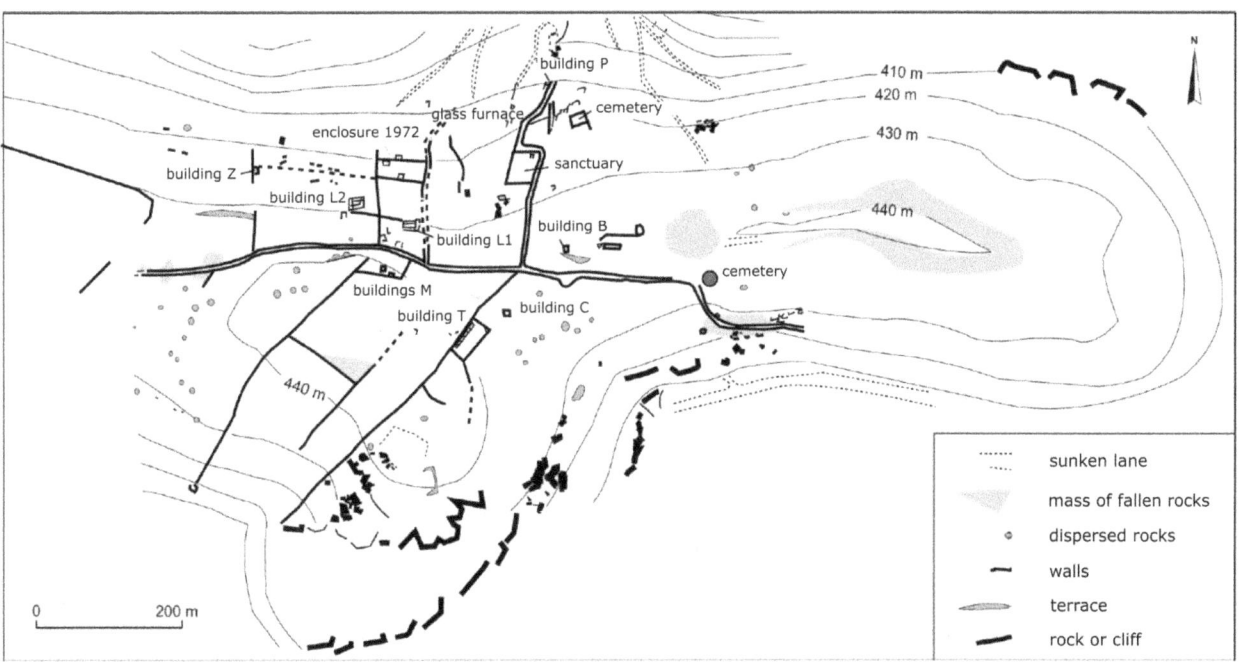

Figure 2.4: The ensemble of Wasserwald at Haegen (© N. Meyer, INRAP, after Pétry 1977, fig. 3).

surrounded by small enclosures. Most buildings were made with perishable materials, but some of them are somewhat larger constructions with several circulation spaces and rooms under a single roof (Figure 2.4: buildings L1 and L2). Both the field plots and the farms seem to have been founded in the last decades of the first century BC, and occupied until the mid-3rd century AD.

According to François Pétry, these habitations were occupied by primitive populations, living in quasi-autarky and pushed back into marginal lands because of the advent of the Romans and their civilized 'centuriation agriculture' in the Alsace plain and on the Lorraine plateau (Pétry 1978: 18; 1997: 403). The retrieval of three boundary stones in the Saverne massif – possible delimiting an (*ager*) *publicus* – led François Pétry to suppose that, at a later stage, the people living in the surroundings of the Saverne pass were incorporated into an Imperial estate for reasons of better control (by the Roman authorities) (Pétry 1981). The inscriptions on these stones, however, are very difficult to read, and there is little comparable material to confirm such a hypothesis. If there was indeed *ager publicus* in the area, the texts do not specify on which authority it depended. Alternatively, one could also consider that these small farmers in the Saverne were actually encouraged to settle in this area, perhaps by a public authority. This is a reasonable argument, as the area is quite strategic, that is, crossed by an important road that connects the Alsace plain with the Lorraine plateau, as such passing the Vosges region at its narrowest extension.

François Pétry's interpretations were firmly rooted in a historiographical tradition, which considered mountainous areas like the Vosges foothills as having served as refuges for conquered populations that had been driven out of their natural habitat by the Roman mode of production; a mode that they were unable to assimilate with (cf. Philippe Leveau for Africa and Gaul; 1977; 2003: 337). The archaeological documentation of the Vosges foothills has received renewed attention through the work of Dominique Heckenbenner, Nicolas Meyer and Antonin Nüsslein (Heckenbenner and Meyer 2004; Meyer, in press; Meyer and Nüsslein 2014), and their findings now force us to reconsider some of François Pétry's interpretations. In particular, it is now clear that the material culture of these farmers is not compatible with a view of people living in autarky. Also, even if these farmers had much more modest productive possibilities than the proprietors of more fertile lands, the coins and objects found in their dwellings clearly confirms their ability to generate surpluses for obtaining goods from the market. Finally, their funerary and votive steles – together with the repertory of statuary and the use of epigraphy – strongly hint at the presence of a culture that borrowed many of its key features from the Roman cultural sphere. In fact, François Pétry already observed that the Vosges foothills area had provided numerous examples of the so-called Anguiped horseman god iconography (Pétry, 1997: 401). The widespread distribution of this iconographic theme in Gaul and the Germanic provinces (Bauchhenss and Noelke 1981),

and especially its important role in Roman provincial art in the 2nd-3rd century AD, as well as its relation with the Imperial ideology (Van Andringa 2002: 190-191) are all convincing elements to argue that the rural communities in the Vosges foothills did not consist of poor people, captured in a traditional way of life and pushed helplessly into the margins of their territories (Pétry 1997: 403).

For a fourth and final case study, I would now like to present some features of the ancient occupation of the Brie Boisée district. Both the traces of these interventions and the environment in which they were carried out, differ substantially from those studied above, but the desire to improve the agricultural capacity of the soil seems to equally respond to an economic logic that is very similar to the one I have tried to highlight so far.

The ditch-marked land divisions of the Brie Boisée district

The Boisée district is a small area towards the western end of the Brie plateau, situated in the north-western part of the Seine-et-Marne department. The area is partly covered by forests that lie on clay soils that are hydromorphic and often difficult to plough (Roque 2003: 34). In antiquity, this territory belonged to the *civitas* of the *Parisii*, in an area that lay close to the *civitas* of the *Meldi*. The long-term occupation of this micro-region is now much better understood thanks to rescue archaeology carried out for the development of the new town of Marne-la-vallée. One particularly interesting outcome of this work is the observation that this territory was only sparsely occupied between the end of the La Tène D2 period and the mid-1st century AD, but became settled with many small farms from the second half of the 1st century AD onwards (Buchez and Daveau 1996), even if the density of these rural sites and their productive capacity remain lower compared to other regions of the Île-de-France (Ouzoulias and Van Ossel 2001).

Even more interestingly, these constructions seem contemporary with the realisation of a vast network of ditch-marked land divisions and ponds covering tens of acres (Figure 2.5) (Berga 2008). The bases of this network were laid out towards the end of the 1st century BC, or in the first half of the 1st century AD. Almost all habitations and most of these ditches seem to have been abandoned before the end of the 3rd century AD, with only some sparse occupation attested until the end of the 5th century AD. Palynology shows that this part of the plateau was recovered by forests in Late Antiquity (Boulen 2010). Many of these ditches are several hundred meters in length, with the biggest examples being 2.5 m deep and 8 m wide. As these ditches closely follow the slope lines and seem to fit into each other, they were presumably created for draining the surrounding soils (Desrayaud 2008). For the sector of Bussy-Saint-Georges alone, Johann Blanchard has estimated that some 46,000 m³ of land had to be removed for digging these ditches (Blanchard 2013). Several traces suggest that parts of this network underwent

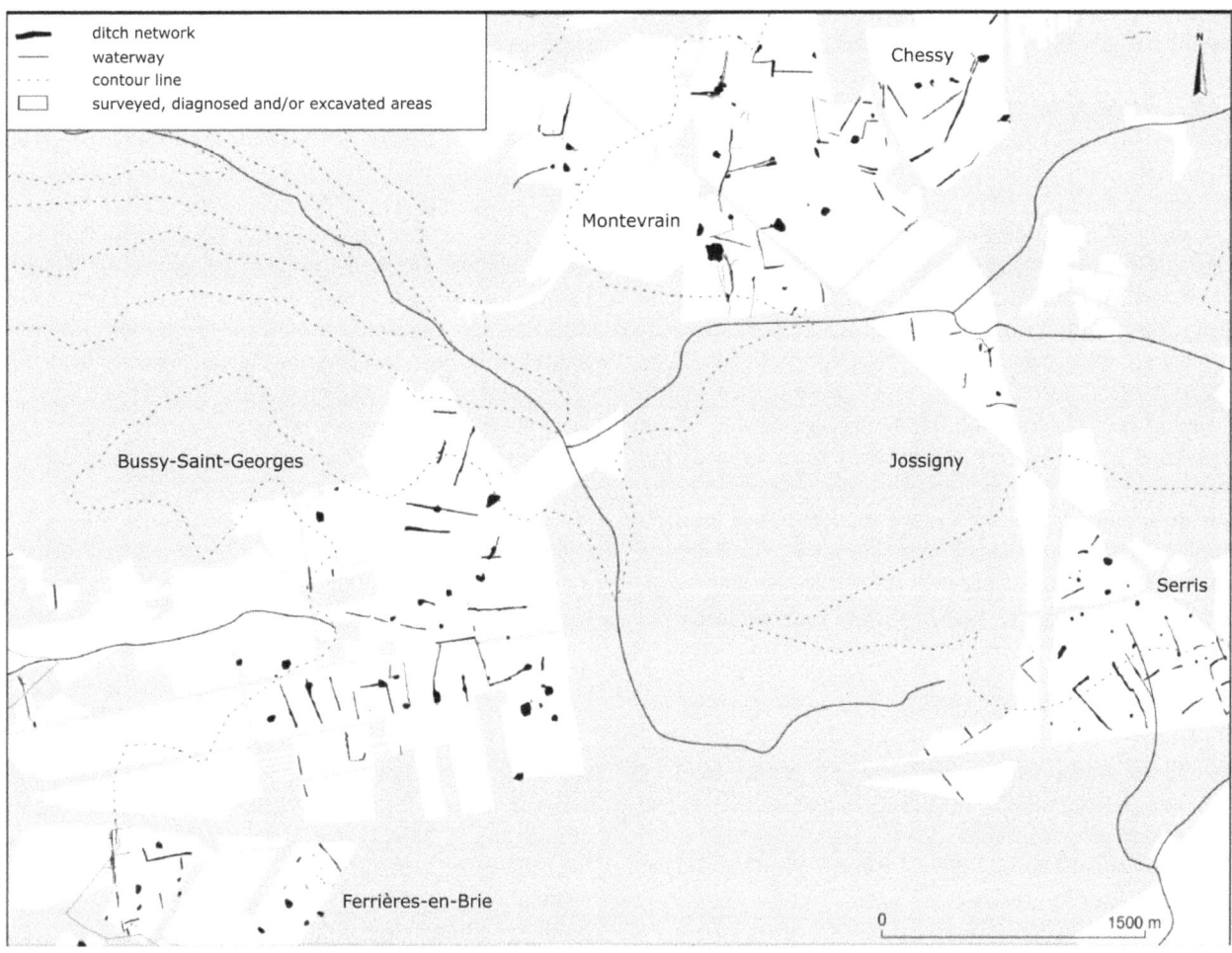

Figure 2.5: Large ditch networks in Bussy-Saint-Georges, Ferrières-en-Brie, Montévrain, Chessy, Jossigny, Serris (Sein-et-Marne) (© F. Barenghi and J. Blanchard, Inrap, after Blanchard et al. 2013, fig. 30).

restoration or adaptation works in the course of the 2nd century AD; a phase which seems to coincide with both the disappearance of old farms and the creation of new ones.

It is clear that the peasant communities made considerable efforts to improve the lands in this part of the Brie, and their work seems comparable to that carried out in the aforementioned territories. This means that, in these four regions, small farmers spent a lot of time and effort in enhancing the agricultural capacities of the soils on which they had settled. They made these investments in line with their environment and their socio-economic organisation. These observations are not without interest for the seminal questions posed in the first part of this chapter. Let us remember that – without actually exploring this line of research – Bruce Frier and Walter Scheidel already considered that archaeology could provide indirect indicators for assessing population growth (Frier 2001: 156; Scheidel 2007: 49). In the remaining part of this chapter, I would like to briefly show how the field data mentioned above provide a solid basis for a critical reappraisal of the concepts that these two authors have proposed; and that they invite us to develop alternative heuristic perspectives.

The nature of cultural intensification

The interventions that have been registered in the four regions discussed above surely were not carried out in areas that had been completely untouched by agricultural activity. Philippe Leveau has rightly pointed out how processes of agricultural intensification in antiquity were seldom carried out in totally unoccupied areas (Leveau 2003: 337). Likewise, for later periods, Ester Boserup remarked how 'newly cultivated' lands had often been exploited in the past for pasture or long-duration fallow (Boserup 1970: 10). The Châtillon forest and the Brie plateau were certainly occupied in some way before their intensive exploitation in classical antiquity; it is also doubtful that the Vosges foothills were completely unoccupied during the Second Iron Age. Indeed, the fossil traces found in the Saverne pass lie in the immediate surroundings of the Gallic *oppidum* of Fossé des Pandours; that is, the most important fortified settlement of the *Mediomatrici* in the Late La Tène period (Fichtl 2002: 319-320). This means that the lands in the surroundings of this 170 ha large agglomeration necessarily had been exploited in at least extensive ways. Unfortunately, the available archaeological data currently do not allow for capturing the nature of this exploitation. Similarly, the earliest archaeological evidence from the Haye forest only goes back to the turn of the era.

More research on these Gallo-Roman ditch networks underlying these forests is highly needed for better capturing the conditions of these areas before they became intensively cultivated, but it is already clear that the attested developments reflect major changes in the agricultural use of these soils. From an agronomic point of view, however, it is important to stress that the agricultural improvement of these lands was not achieved by adopting new cultivation techniques. Indeed, drainage ditching, stone removal, terrace culture and even manure spraying were all possible by using traditional tools and practices (Malrain 2010). We are not witnessing an 'agricultural revolution' here, nor the introduction of new and better methods that allowed for a more efficient use of the soil, but we are rather looking at a change in the cropping system by organizing the cultivated area in a different way, and by increasing the amount of work per surface unit. In other words, the agricultural expansion was achieved by increasing the volumes of the factors of production: capital and labour. This process fits perfectly with the agricultural intensification model of Ester Boserup; that is, the expansion of the cultivated area results from an increase in labour, rather than the introduction of new and better agricultural technology. It is also telling that in all these regions the family farm remained the basic unit of this intensification process.

The household as an instrument of agricultural intensification

I have stressed earlier how the agricultural units attested on these soils are not 'villas' – even if some of them can be seen as more modest versions of this type of agricultural holding – but rather farms ruled by peasant households (Ouzoulias 2010). I am convinced that this is not a coincidence, but rather an economic necessity. Family farms can mobilize their members to an extent that cannot be achieved through wage or servile labour. Indeed, in order to survive, the peasant family can push its self-sufficiency to the physical limits of all its members. Also, the amount of work devoted to each surface unit can theoretically be increased until its marginal yield becomes zero or even negative (Myint 1966: 124).

What all this means, is that the 'law of diminishing returns' does not apply to this type of farming under the same economic conditions envisioned by David Riccardo for a farming enterprise in which the farmer has to pay a rent to his landlord, as well as wages to his labourers (Riccardo 1847: 43). On the contrary, precisely because a peasant family does not have to pay for the work of its members, it can continue to cultivate the land, even when profit versus input becomes marginally low (Caanov 1990: 95). So the peasant household appreciated its degree of self-exploitation in function of an empirical assessment of its fundamental annual equilibrium. This equilibrium depends on its productive capacity, the degree of its vital needs, but also the land rent and the need to rebuild its capital.

At a more theoretical level of analysis, Karl Marx showed how the logic of David Ricardo's demonstration of the "law of diminishing returns" and the land rent is based on the principle that the market value of agricultural products cannot be less than their value of production. However, for the author of Capital – both in agriculture and industry – there are always producers who sell at prices lower than their costs of production. There is thus no theoretical impossibility to conceive the existence of individual sub-profits and negative differential rents (Tran 2003: 186). In other words, *"the general price of production is not set marginally on the least favourable production conditions"* (Tran 2003: 192).

Precisely because of its high capacity for self-exploitation, a peasant family has the ability to provide products for a value that does not account for all the work required to produce them. In such case, Karl Marx already stated that it was as if a part of the surplus labour carried out by peasants who work in the least favourable conditions is provided free or charge to society, and does not enter into the fixing of prices of production, or into the creation of value in general (Marx 1960: 185). What this means for my case study here, is that it is possible that agricultural intensification on marginal lands may not necessarily have resulted in an increase in the price of food – as is supposed by many authors who follow David Ricardo in his idea that the wages must be proportional to the productivity of marginal labour income.

What this position in fact reveals is the difficulty for historians to think of pre-industrial economies without the wages element, to appreciate the organisation of family farming, and to assess its proper role in economic development. I have repeatedly called for a reassessment of its role in the Gallo-Roman countryside, and thus of its contribution to the development of agriculture in these Northern provinces (Ouzoulias 2009; 2010). The specific socio-economic status of the family farms occupying the marginal lands surveyed in this chapter reinforces this call. We should be able to increase our understanding of their functioning by further exploring the technical and agrarian conditions under which they managed to cultivate these relatively poor soils.

In my discussion of the intensification processes described above, it is necessary to make a clear distinction between on the one hand proper farming activities – that is, e.g. the identified practices in the Haye and Châtillon forests – and on the other hand preparatory works of the *ager*, such as the construction of ditches, low delimitation walls and/or terraces. One may indeed wonder under which circumstances the latter were initially realized. The extension of the ditch network identified on the Brie plateau suggests a belonging to a wider organizational scheme outstripping the dimensions of the farm. Likewise, the field system in the Haye forest seems generally organised around major axes of great length (Favory 2011: 402). Could these arrangements reflect allotment schemes designed and implemented by the Roman authorities, or by external individuals that managed the access to these lands? This could very well be, but François Favory warns us for neglecting the capability of rural communities to autonomously structure their agrarian space, and to endow it with field systems adapted to their needs (Favory 2012: 124).

If the latter hypothesis is to be accepted, this implies that substantial efforts were made by these peasant families within a relatively short time span, precisely at a point in their existence when economic viability was rather precarious. It is important to remember that, on these poor soils, it will have taken several years to establish a level of fertility sufficient to satisfy the subsistence needs of these farmers. The question then rises how these families coped with these major constraints. Considering that these soils were initially covered by shrub or tree vegetation, slash-and-burn cultivation may have been a good way to launch a first crop cycle. But we currently do not have sufficient archaeological data to support such a hypothesis, nor to fully understand the applied cropping systems on these soils in general. What is clear, however, is that these soils were more demanding in work than those more fertile. Indeed, cultivating these marginal lands was generally more difficult and less profitable. How could we otherwise explain their abandonment in Late Antiquity?

Saltuarius and saltus

One and two makes that specific economic circumstances were required for peasant families to settle in these areas. The available written evidence from other provinces of the Roman Empire shows that rent relief is often a crucial element of incentive policies for cultivating agricultural land. This is not the place to present all the possible measures known from our sources, but it suffices to stress how in Roman Africa the *Lex Manciana* – perhaps enacted in the Flavian period (Le Glay 1968: 226) – fixed the rents of the colonists at one third of the harvest for cereals, and a quarter or a fifth of the harvest for beans. This was a general law meant to regulate sharecropping on lands of various statuses, but the share contract in itself here presented was particularly well suited for exploiting uncultivated or less productive lands. As has been shown by the Henchir-Mettich inscription, this law thus regulated the work of peasants who desired to cultivate un-allotted lands; that is, lands previously unassigned because of their low fertility.[7] The same law is also an important element within the arrangements fixed in the *Lex Hadriana* to encourage crop extension on Imperial estates (Jérôme France 2014).[8] This law is known through several inscriptions from the region of *Thugga* (Dougga) in northern part of present-day Tunisia. These texts concern several Imperial *salti*, and more importantly attest the existence within these estates of abandoned, uncultivated land, covered with wood or swamps, stressing at the same time the Imperial desire to bring them under cultivation following the rules of the Lex Manciana; perhaps for coping with an "increase in population" (Peyras 1995: 112, 120). The agricultural 'conquest' of marginal lands of the *salti* in the Bagradas basin (Medjerda) is undeniably linked with a widespread and long-lasting growth of both the urban and rural population (Lassère 1977: 651; Lepelley 1989: 19).

It is important to emphasize that these extensions of the *ager* were the work of peasant families. It is clear that the colonists from the Henchir-Mettich inscription are part of a small community, here represented by the *magister* and the *defensor*, who engraved the regulations enacted by the Imperial prosecutors. Henriette Pavis d'Escurac has already established that *coloni* from other *salti* were organised in similar ways. In addition, she quotes a text (CIL, VIII, 23022, dated in AD 164/165) by which the plebs fundi, under the conduct of his magister, consecrates a wall in the sanctuary of Cereres (Escurac-Doisy 1967: 68-70). Denis Kehoe uses the epigraphic evidence to argue that the farmers involved in the agricultural conquest of the so-called *rudes agri* in the Bagrades valley belonged to the higher peasantry (social stratification) (Kehoe 1988: 106). Finally, the peasants mentioned in the inscription from Souk-el-Khemis (*decretum de saltu Burunitano*) present themselves as weak countrymen that earn a living with their hands, while at the same time directly soliciting the intervention of the emperors in their conflicts with the *procuratores* and the *conductores* of the *saltus Burunitanus* (Charles-Picard and Rougé 1969: 218-223).

We obviously cannot directly compare the exploitation regulations of the marginal lands in the *Bagradas* basin with those of the foothills of the Vosges mountains, but it is notable that the material conditions of the Saverne farmers were far from miserable, and that they produced sufficient surplus to devote some of their income to managing their habitations and building small monuments and/or sanctuaries (see *supra*). They probably also had to form small, relatively hierarchical societies, perhaps even structured by communal institutions. Also, let us not forget François Pétry, who suggested to link the boundary stones found in the Saverne forest with the *ager publicus* of an Imperial estate, perhaps administered by the *procurator* mentioned on an inscription discovered at the foot of the massif.

A recent epigraphic discussion of Monique Dondin-Payre now allows for exploring my argument somewhat further. A few years ago, the Musée d'Archéologie nationale at Saint-Germain-en-Laye acquired a bronze wheel-bearing Jupiter with an inscribed base, found at Rontecolon in Cenves, a commune located in a now forested mid-mountain region to the south of the City of Eduens, and thus in the Roman province of *Gallia Lugdunensis*. From the inscription, it is

---

[7] "*eos agros qui subcesiva sunt*" (CIL, VIII, 25902). This inscription is dated around AD 116-117 (Chouquer and Favory 2001: 391; Charles-Picard and Rougé 1969: 212).

[8] In short, this law allows the *potestas occupandi*, that is, the possibility of a private appropriation of the ager publicus and its transmission to the heirs (Peyras 1999; Chouquer 2010: 160-161).

clear that the figure had been offered to Jupiter Optimus Maximus by a *saltuarius*, named *Criciro*, and had been produced by the *faber Sabellus*. Criciro was an employee of the *Prisciacenses*, a community present in the area from Gallo-Roman times until today (Dondin-Payre and Chew: 2010). Monique Dondin-Payre sees the Criciro of the inscription as responsible for the surveillance of the *saltus*, that is, of dry and wet wastelands, forests and pastures; all zones that without being cultivated can be exploited, have economic value and are susceptible to degradation (naturally or because of man). While she acknowledges the fact that the word '*saltus*' can have various meanings, she firmly favours the rural meaning for the term, thus designating uncultivated spaces, as opposed to '*ager*' (Vigouroux 1962: 211). She thus considers Criciro a 'ranger' who carried out a surveillance mission for the community of the Prisciacenses as their employee. This community would have exploited a territory whose name possibly survives in the *ager Prisciacensis* mentioned in texts of the 10th century (Dondin-Payre and Chew 2010: 81-85).

It is difficult to follow this too limiting interpretation, which excludes the agricultural land dimension from the word, as '*saltus*' is also undoubtedly used in texts for designating farms, estates or other entities to ensure their state or fiscal management. Indeed, in the Gromatic texts, Agennius Urbicus reminds us that, in his time, the lands in Italy were divided between those of colonies (*ager colonicus*), of municipalities (*ager municipalis*), of *castella*, of *conciliabula* and of private saltus (*saltus privatus*) (*De controv. Agror.* 20.1-3) (Campbell 2000). Even more so, the *Liber Coloniarium* mentions that in the province of *Calabria*, land that had not been distributed by the Gracchi, was assigned as '*saltus*' (*Liber Coloniarum I, Provincia Calabria*, 211, 1-8 La). Also, in the aforementioned African inscriptions – which were not discussed by Monique Dondin-Payre because she considered them "linked with the specific case of the African colonate" – it is very well possible that *saltus* appears as a term for land divisions within estates. The *Sermo procuratorum Imperatoris* mentions centuriated lands being farmed in the *saltus Blandianus* and *Udensis*, and the inscription from Souk el Khemis evokes the fate of the *coloni* of the *saltus Burunitanus*, who no longer supported the abuses by the *conductores* (see *supra*). Finally, let us not forget the brilliant career of the knight T. Flavius Macer, who was, amongst other things, an Imperial procurator of the *saltus Hipponiensis* and *Thevestinus* (Christol 1994).

It goes without saying that we cannot directly transpose the status of the Imperial African *saltus* to Gaul, but we should also not rule out the idea that Criciro was indeed the *saltuarius* of a *saltus Prisciacensium*, devised as a property constituency run by a public collective or a private *possessor*. *Prisciacenses* could then have been the generic name used by these peasants as an organizing body under which they could employ Criciro; much like how the *coloni* from the Henchir-Mettich inscription made use of the services of a *defensor*. Whatever it may be, it is clear that the landscape in question was not covered by forests in antiquity, and that Criciro was given the task to guard it.

The terms '*saltus*' and '*saltuarius*' appear in two texts recorded in the census of Monique Dondin-Payre regarding the Vosges mountains. The first text comes from a sanctuary dedicated to Hercules, located in the commune of Deneuvre, at the foot of the western slope of the southern part of the Vosges, towards the eastern end of the *civitas* of the *Leuci*. The plaque in question is highly damaged, but the text seemingly mentions a *saltus* or *saltuarius*. The second text comes from the Pfälzerwald massif, on the western foothills of the northern Vosges, towards the north-eastern edge of the *civitas* of the *Mediomatrici*. This funerary stele was reused in a late fortification of the Heidelsburg, that is, the modern town of Waldfischbach-Burgalben. The text is dedicated to the *saltuarius* T. Publicius Tertius (Finke 1927: 328), most likely a freed slave of the community that managed the saltus (Lazzaro 1963: 61). Other freedmen of this kind are known from the *civitas* of the *Mediomatrici*, like M. Publicius Secundanus, a freedman and *tabularius* of the *nautae* of the Moselle (Lazzaro 1993: 99, n° 45).

Monique Dondin-Payre assumes that this stele may originate from the Waldfischbach cemetery in the valley, but the original publisher of the text – Karl Zangemeister – noted the presence of "numerous remains of implantations" near its place of discovery, like for instance that of T. Publicius Tertius. I therefore consider it fully plausible that these stelae came from a funerary ensemble like those found in the Saverne forest, the Wasserwald or the Croix-Guillaume (Meyer, in press, 41). If this interpretation is correct, then this part of the Pfälzerwald could also have been occupied by settlements such as those found at the foothills of the Vosges in the eastern part of the *civitas* of the *Mediomatrici*. The presence of the *saltuarius* T. Publicius Tertius in Heidelsburg would then shed new light on the socio-economic conditions of the communities living on these marginal lands. While all this is not sufficient to completely revise François Pétry's theories, the few elements that I have presented in these paragraphs at least justify the exploration of alternative explanations.

The economic determinants of economic performance

Whatever the agrarian regime responsible for the agricultural intensification discussed above, it is clear that this process considerably increased the value of these lands. So, on the whole, the individuals or communities that controlled the access to these lands must have had considerable incentives to cultivate them, even if the rent of the land was zero. In a way, the value of these improvement works was gradually integrated into land capital. In his critique of David Ricardo's differential rent theory, Karl Marx presents this process: "For newly cultivated soil, the artificially created portion of fertility through capital investment is still absolutely distinct from the fertility of the soil", but "this artificial fertility appears after a certain lapse of time as productivity originating from the soil, the soil itself having been transformed, [...] and the process by which this transformation has been effected having disappeared, no longer visible". Marx thus considers it illusory to base an analysis of land rent on the "primary and original" faculties of the soil (Marx 1975: 283-284).

To return to the core issue of this chapter, it must be stressed that the productive capacity of the cultivated soil, and hence the extension of the agricultural land, have undeniably social dimensions. In other words, we cannot conceive the land and its fertility as independent of the incorporated capital (Tran 2003: 246). As a consequence, the use of the concept of carrying capacity – when insufficiently disconnected from its original and naturalistic presuppositions – can either lead to an underestimation of the ability of peasant societies to modify their environment, or to confusing what belongs to the composition of soil physics and the ability of farmers to improve its fertility. We therefore have to dismantle the logic of the classical economic model which regards land as a non-cumulative capital, and which sees nature and population as two independent variables. There is thus a need to better understand the complex relationship between the degree of intensification of peasant work, land rent, agrarian regimes, the composition of small farm capital, the value of land, and the value of exchange of agricultural surplus, as well as the forms of their marketing. I lack the space here to undertake a deep methodological examination, but I can make some observations concerning the examples presented in this chapter.

Family farms rarely operate in total self-sufficiency. Indeed, they have to acquire what they do not produce themselves, this in in order to maintain their buildings, to renew their tools and/or keep their livestock. This means that they have to dispose of their surpluses in order to meet these requirements, pay the rent of their lands and build their stocks. While I deliberately have not stressed the contribution of artisanal activities on these farms in this chapter, multi-activity was an essential characteristic of small-scale farming. Stone-, wood- or leather work often provided income in addition to agriculture. It is indeed quite plausible that the use of forest resources or stone and/or mineral extraction ensured complementary profits for the peasant families in the forests of Vosges, Haye and Châtillon, in addition to the time they necessarily had to devote to sustenance. However, as important the contribution of these crafts to the economic balance of these farms was, it did surely not diminish their dependence on external demand. Forced to dispose of their surpluses for ensuring longevity, their development was therefore linked with the existence of a market capable of consuming these surpluses. For this reason, I consider it likely that the agricultural conquest and development of these marginal lands was preceded by an increase in overall consumption; a consumption that could no longer be satisfied by the available land, and thus inherent to a concomitant increase in population. These marginal soils could thus have been used to supply this additional population, and the new surpluses by these farmers could have contributed to maintain a virtuous cycle of growth. Indeed, even if the productive capacities of these small farms were modest, the size of these newly cultivated areas certainly had a considerable multiplier effect on the overall volume of food produced (Georges-Leroy, Lafitte et al. 2013: 189).

This considerable extension of the ager thus seems connected to the rise of urbanisation, which affected both cities and smaller settlements, and as such lead to an unprecedented increase in the number of farms in lands that were already being cultivated. So what this process highlights, is the connection between the rise of urbanisation, the increase in the number of farms and the conquest of marginal land. On the other hand, the agricultural abandonment of these territories could be indicative of a decline in consumption, and thus of a reduction of the population.

Indeed, I consider the causes of both the expansion and retraction of cultivation in these areas as linked by the same economic process. In essence, we may assume that the less the sale or exchange of their products became, the more the farms on these marginal lands had to increase their production. For farmers engaged in a process of agricultural intensification, this extra activity added to an already important overwork. The risk was then that the decline in the remuneration of their products was not offset by an amplification of the self-exploitation of the peasant family, already very high. In other words, when the agricultural income became too low, it was no longer possible to renew capital to maintain the level of production (Demont and Jouve 2000: 103). We may reasonably assume that this minimum threshold of farming income was higher on marginal lands. Viewed as such, the small farms in these areas would have been much more sensitive to the changing conditions in the trade of agricultural products. In other words, they would have experienced difficulties long before the effects of economic downfall became felt on the holdings located on more fertile lands. One may wonder what happened to these small farmers who could no longer be fed by these marginal lands.

Therefore, I consider the analysis of the developments in the occupation of marginal lands as one of the means to better understand the rise and fall of populations, the cycles of demographic pressure and the reversals of economic conjuncture. But such a large-scale survey would first of all require an inventory of the areas in which such agricultural conquests developed. As we have seen, material proofs of these processes have been recognized in the *civitas* of the *Lingones*, in Upper Germany, in the *civitas* of the *Mediomatrici* and that of the *Leuci*, in Belgian Gaul, and in the civitas of the *Parisii* in the Lyonnaise region. Other traces deserve to be better studied, such as those in the *civitas* of the *Carnutes* (Vigneau 2007), again in the Lyonnaise region, and those in the *civitas* of the *Bituriges Cubi* (Laüt, 2007), in Aquitaine. Observations of similar operations have also been observed in western and northern France, and in Saarland. The forest remains the privileged environment of their study, because of its remarkable retention of the ager's footprint and thus facilitates their recognition in large areas. But the example of the Brie Boisée district, presented above, reveals that the identification of marginal lands can also be carried out according to other archaeological protocols. The recent increase in discoveries, favoured by the use Lidar suggests that these systems may have an unsuspected spatial extension. The importance of this phenomenon of the widespread occupation of marginal lands could then, if confirmed, considerably renew the historical appreciation of the demographic dynamism of Gaul during the High Empire. It would be similar to the one observed in certain parts of Italy, for which Elio Lo Cascio has stressed how the extension of cultivated fields in Antiquity was realized in soils which were not brought under cultivation before the twentieth century (Lo Cascio 2004).

To conclude, it seems that the development of agriculture in land with low productive capacity is a response to a significant increase in the demand for food and therefore both a consequence and the index of population growth. Smaller marginal-land farms probably benefited of this process because the productive potential of the more fertile soils approached its limits, the surge of which would require a qualitative evolution of cropping systems. Far from being a "Malthusian trap", population growth can thus be considered, under these conditions, as the "Boserupian motor" of agricultural intensification. It is clear that, in northern Gaul, this expansion accompanied developments closely linked to – or intensified by – the Roman conquest of the area; with regional and chronological differences that demand closer examination (Leveau 2003: 337; 347). Finally, we must resolutely abandon the idea that the supposed archaism of agriculture was an insurmountable obstacle to economic development and population growth. Soil fertility and the improvement of agricultural techniques are not determinants external to societies but, on the contrary, phenomena that contribute to their evolution.

Bibliography

Amable, B., and Guellec, D. "Les théories de la croissance endogène." *Revue d'économie politique* 102, no. 3 (1992): 313-377.

Andreau, J. *L'Économie du monde romain*, coll. Le Monde, une histoire. Paris: Ellipses, 2010.

Apicella, C., Haack M.-L., and Lerouxel, F. (eds.). *Les Affaires de Monsieur Andreau : économie et société du monde romain*, coll. Scripta Antiqua 61. Bordeaux: Ausonius, 2014.

Bauchhenss, G., and Noelke, P. *Die Iupitersäulen in den germanischen Provinzen*, coll. Beihefe der Bonner Jahrbücher 41. Cologne: Rheinland-Verlag, 1981.

Berga, A. "Le réseau hydrographique artificiel antique de Marne-la-Vallée" In *La Seine-et-Marne*, coll. CAG, 77/1 and 77/2, edited by J.-N. Griffisch, D. Magnan, D. Mordant, 133-137. Paris: Académie des inscriptions et belles-lettres, 2008.

Blanchard, J., Cammas, C., Genin, M., and Verdin, P. "Un réseau de fossés antique original au nord-ouest du plateau briard : nouvelles observations réalisées à Bussy-Saint-Georges (Marne-la-Vallée, Seine-et-Marne)." *Revue archéologique du Centre de la France* 52 (2013): 191-230.

Boserup, E. *Évolution agraire et pression démographique*, coll. Nouvelle Bibliothèque Scientifique. Paris: Flammarion, 1970.

Boserup, E. "Environnement, population et technologie dans les sociétés primitives." *Annales. Économies, Sociétés, Civilisations* 29, vol. 3 (1974): 538-552.

Boulen, M. "Synthèse des analyses polliniques pour la période romaine sur le secteur de la bordure nord-ouest du plateau briard." *Revue archéologique d'Île-de-France* 3 (2010): 133-148.

Bowman, A., and Wilson, A. "Introduction: Quantifying Roman Agriculture." In *The Roman Agricultural Economy: Organization, Investment and Production*, Oxford Studies on the Roman Economy, edited by A. Bowman and A. Wilson, 1-32. Oxford: Oxford University Press, 2013.

Brenner, R. "Agrarian Class Structure and Economic Development in Pre- Industrial Europe." *Past & Present* 70 (1976): 30-75.

Buchez, N., and Daveau, I. "La mise en place d'un réseau d'établissements au début de l'époque romaine à Marne-la-Vallée : origines et aspects des nouvelles installations." In *De la ferme indigène à la villa romaine, Actes du 2e colloque de l'association Ager, Amiens, 23-25 sept. 1993*, coll. No spécial à la Revue archéologique de Picardie 11, edited by D. Bayard and J.-L. Collart, 221-231. Amiens: Revue archéologique de Picardie, 1996.

Caânov, A. V. *L'Organisation de l'économie paysanne*, Traduction d'A. Berelowitch. Paris: Librairie du Regard, 1990.

Caire, G. "Un ou deux Malthus ?" *Revue économique* 35, vol. 4 (1984): 623-634.

Campbell, J.B. *The Writings of the Roman Land Surveyors: introduction, text, translation and commentary*. London: Society for the promotion of Roman Studies, 2000.

Charles-Picard, G. and Rougé, J. *Textes et documents relatifs à la vie économique et sociale dans l'Empire romain, 31 avant J.-C.-225 après J.-C.* Paris: Société d'Édition d'Enseignement supérieur, 1969.

Chouquer, G., and Favory, F. *L'Arpentage romain : histoire des textes, droit, techniques*. Paris: Errance, 2001.

Christol, M. "Le blé africain et Rome : remarques sur quelques documents", *Publications de l'École française de Rome* 196, vol. 1 (1994): 295-304.

Clark, G. *A Farewell to Alms: a Brief Economic History of the World*. Princeton: Princeton University Press, 2007.

Cordonnier, L. *L'Économie des Toambapiks : une fable qui n'a rien d'une fiction*. Paris: éd. Raison d'agir, 2010.

Demont, M., and Jouve P. "Évolution d'agro-écosystèmes villageois dans la région de Korhogo (Nord Côte-d'Ivoire): Boserup versus Malthus, opposition ou complémentarité?" In *Dynamiques agraires et construction sociale du territoire*, edited by P. Jouve, and M.-C. Cassé, 93-108. Montpellier: Cnearc, 2000.

Desrayaud, G. 2008 "Parcellaires fossoyés du Haut-Empire des plateaux de Brie : Jossigny/ Serris et Moissy-Cramayel (Seine-et-Marne) : approche méthodologique de l'étude des réseaux." *Revue archéologique du Centre de la France [En ligne]* 47 (2008) [URL : http://racf.revues.org/1161].

Dondin-Payre, M., and Chew, H. "Un saltuarius dévot de Jupiter Optimus maximus dans le Mâconnais." *Gallia* 67, vol. 2 (2010): 69-98.

Erdkamp, P. "How modern was the market economy of the Roman world?", *Œconomia* 4-2 (2014): 225-235

Escurac-Doisy, H. d'. "Notes sur le phénomène associatif dans le monde paysan à l'époque du Haut-Empire." *Antiquités africaines* 1, vol. 1 (1967): 59-71.

Étienne, R. "Compte rendu de The Cambridge Economic History of the Graeco-Roman World, Cambridge (2007), Table ronde, Nanterre, 13 févr. 2010." *Topoi Orient Occident* 17, vol. 1 (2011): 7-14.

Evans, J. K. "Wheat Production and its Social Consequences in the Roman World." *The Classical Quarterly* 31, vol. 2 (1981): 428-442.

Favory, F. "Les parcellaires antiques de l'est de la Gaule." In *Aspects de la romanisation dans l'est de la Gaule*, coll. Bibracte 221, edited by M. Reddé, P. Barral, F. Favory, M. Joly, J.-P. Guillaumet, J.-Y. Marc, P. Nouvel, L. Nuninger, and C. Petit, 385-416. Glux- en-Glenne: Bibracte - Centre archéologique européen, 2011.

Favory F. "Les parcellaires antiques de Gaule médiane et septentrionale ." In *Des hommes aux champs : pour une archéologie des espaces ruraux du Néolithique au Moyen Âge*, coll. Archéologie et culture, edited by V. Carpentier, and C. Marcigny, 111-130. Rennes: Presses universitaires de Rennes, 2012.

Fichtl, S. "Oppida et occupation du territoire à travers l'exemple de la cité des Médiomatriques." In *Territoires celtiques : espaces ethniques et territoires des agglomérations protohistoriques d'Europe occidentale, Actes du XXIVe colloque international de l'AFEAF, Martigues, 1-4 juin 2000*, edited by D. Garcia, and F. Verdin, 315-328. Paris : Errance, 2002.

Finke, H. "Neue Inschriften." *Bericht der römisch-germanischen Kommission* 17 (1927) : 1-107 and 198-231.

Finley, M. I. "Technical Innovation and Economic Progress in the Ancient World." *The Economic History Review* 18, vol. 1 (1965): 29-45.

France, J. "La lex Hadriana et les incitations publiques à la mise en valeur de terres dans l'Empire romain au IIe siècle p. C." In *Les Affaires de Monsieur Andreau : économie et société du monde romain*, coll. Scripta Antiqua 61, edited by C. Apicella, M.-L. Haack and F. Lerouxel, 89-96. Bordeaux: Ausonius, 2014.

Frier, B. "More is worse. Some observations on the population of the Roman Empire." In *Debating Roman Demography*, edited by W. Scheidel, 139-159. Leiden: Brill, 2001.

Galor, O. *Unified Growth Theory*. Princeton: Princeton University Press, 2011.

Garnsey P., and Saller R. P. *L'Empire romain : économie, société, culture*, coll. Textes à l'appui. Paris: Éditions La Découverte, 1994.

Georges-Leroy, M., Bock, J., Dambrine, É., and Dupouey, J.-L. "Le massif forestier, objet pertinent pour la recherche archéologique : l'exemple du massif forestier de Haye (Meurthe-et-Moselle)." *Revue géographique de l'Est [En ligne]* 49, vol. 2-3 (2009) [URL : http://rge.revues.org/1931].

Georges-Leroy, M., Bock, J., Dambrine, É., and Dupouey, J.-L. 2011 "Apport du lidar à la connaissance de l'histoire de l'occupation du sol en forêt de Haye." *Archéo Sciences. Revue d'archéométrie* 35 (2011): 117-129.

Georges-Leroy, M., Bock, J., Dambrine, É., Dupouey, J.-L., and Étienne, D. "Lidar Helps to Decipher Land-Use History in Lorrain, France." In *Understanding Landscapes, from Land Discovery to their Spatial Organization (Comprendre l'espace de peuplement, de la découverte des territoires à leur organisation spatiale), Proceedings of the XVIth World Congress of the International Union of Prehistoric and Protohistoric Sciences, Florianopolis, Brazil, 4-10 Sept. 2011*, BAR International Series 2541, edited by F. Djindjian, and S. Robert, 115-122. Oxford: BAR Publishing, 2013.

Georges-Leroy, M., Bock, J., Dambrine, É., Dupouey, J.-L., Gebhardt, A., and Laffite J.-D. "Les vestiges gallo-romains conservés dans le massif forestier de Haye (Meurthe-et-Moselle) : leur apport à l'étude de l'espace agraire." In *Des hommes aux champs : pour une archéologie des espaces ruraux du Néolithique au Moyen Âge*, coll. Archéologie et culture, edited by V. Carpentier, and C. Marcigny, 157-180. Rennes: Presses universitaires de Rennes, 2012.

Georges-Leroy, M., Heckenbenner, D., Laffite, J.-D., and Meyer, N. "Les parcellaires anciens fossilisés dans les forêts lorraines." In *La Mémoire des forêts, Actes du colloque « Forêt, archéologie et environnement », Velaine-en-Haye, 14-16 déc. 2004*, edited by J.-L. Dupouey, É. Dambrine, C. Dardignac, and M. Georges-Leroy, 121-131. Paris: ONF, INRA, DRAC, 2007.

Georges-Leroy, M., Lafitte, J.-D. and Feller, M. "Des paysages ruraux antiques contrastés dans les cités des Leuques et des Médiomatriques : effet de source ou répartitions typologique et spatiale différentes des établissements?" In *Paysages ruraux et territoires dans les cités de l'Occident romain: Gallia et Hispania*, edited by J.L. Fiches, R. Plana-Mallart and V. Revilla Calvo, 181-194. Montpellier: Presses Universitaires de la Méditerranée, 2013.

Giosa, A. *Étude géoarchéologique d'un paysage agraire : caractérisation des sols d'un parcellaire conservé sous forêt (Roche Chambain, Rochefort, Côte- d'Or)*, Mémoire de Master 2. Paris: Université Panthéon-Sorbonne-Paris-I, 2012.

Goguey, D., and Bénard, J. "Les enclos dans l'organisation de l'espace protohistorique et gallo-romain sous les forêts du Châtillonnais." In *Les Espaces clos dans l'urbanisme et dans l'architecture en Gaule romaine et dans les régions voisines : hommage à Raymond Chevallier, Actes du colloque inter- national, Faculté des lettres et des sciences humaines, Limoges, 11-12 juin 2004*, coll. Caesarodunum, edited by R Bedon, Y. Liébert, and H. Mavéraud- Tardiveau, 159-179. Limoges: PULIM, 2006.

Goguey, D., and Pautrat, Y. "Des parcellaires sous les forêts du Châtillonnais." In *La Côte-d'Or : d'Argencourt à Alise- Sainte-Reine*, coll. CAG 21/1, edited by M. Provost M., 168-182. Paris: Académie des inscriptions et belles-lettres, 2009.

Goguey, D., Pautrat, Y., Guillaumet, J.-P., Thevenot, J.-P., and Popovitch, L. "Dix ans d'archéologie forestière dans le Châtillonnais (Côte-d'Or) : enclos, habitats, parcellaires." *Revue archéologique de l'Est* 59, vol. 1 (2010): 99-209.

Greene, K. "Technological Innovation and Economic Progress in the Ancient World: M. I. Finley Re-Considered ." *The Economic History Review* 53, vol. 1 (2000): 29-59.

Grenier, A. (1934, 2nd ed.) *Manuel d'archéologie gallo-romaine -II-2- L'Archéologie du sol : navigation, occupation du sol*, coll. Grands manuels Picard. Paris: A. Picard, 1985.

Guellec, D. "Croissance endogène : les principaux mécanismes." *Économie & prévision* 106, vol. 5 (1992): 41-50.

Guellec, D., and Ralle, P. *Les Nouvelles théories de la croissance*. Paris: La Découverte, 2003.

Hecht, J. "« Traduttore traditore ? » : les traductions de l'"Essai" de Malthus en langue française." In *Malthus, hier et aujourd'hui*, edited by A. Fauve-Chamoux, 75-92. Paris: éd. du CNRS, 1984.

Heckenbenner, D., and Meyer, N. "Les habitats et les parcellaires du piémont vosgiens." In *La Moselle*, coll. CAG 57/1, edited by P. Flotté, and M. Fuchs, 177-179. Paris: Académie des inscriptions et belles- lettres, 2004.

Inglebert, H. "Le monde romain et la civilisation romaine." In *Histoire de la civilisation romaine*, coll. Nouvelle Clio, edited by H. Inglebert, 35-75. Paris: Presses universitaires de France, 2005.

Izri, S., and Nouvel, P. "Les sanctuaires du nord-est de la Gaule : bilan critique des données." In *Aspects de la Romanisation dans l'Est de la Gaule*, coll. Bibracte 21, edited by M. Reddé, P. Barral, F. Favory, M. Joly, J.-P. Guillaumet, J.-Y. Marc, P. Nouvel, L. Nuninger, and C. Petit, 507-532. Glux-en-Glenne: Bibracte - Centre archéologique européen, 2011

Jongman, W. "The Rise and Fall of the Roman Economy: Population, Rents and Entitlement." In *Ancient Economies, Modern Methodologies: Archaeology, Comparative History, Models and Institutions*, coll. Pragmateiai 12, edited by P. F. Bang, M. Ikeguchi, and H. G. Ziche, 237-254. Bari: Edipuglia, 2006.

Kasprzyk, M., Nouvel, P. 2011 "Les mutations du réseau routier de la période laténienne au début de la période impériale : apport des données archéologiques récentes." In *Aspects de la Romanisation dans l'Est de la Gaule*, coll. Bibracte 21, edited by M. Reddé, P. Barral, F. Favory, M. Joly, J.-P. Guillaumet, J.-Y. Marc, P. Nouvel, L. Nuninger, and C. Petit, 21-41. Glux-en-Glenne: Bibracte - Centre archéologique européen, 2011.

Kehoe, D. P. *The Economics of Agriculture on Roman Imperial Estates in North Africa*, coll. Hypomnemata 89. Göttingen: Vandenhoeck & Ruprecht, 1988.

Kolendo, J. (1962, 2nd ed.) *Le Colonat en Afrique sous le Haut- Empire*, coll. Centre de recherches d'histoire ancienne, 107. Paris: Les Belles Lettres, 1991.

Kremer, M. "Population Growth and Technological Change: One Million B.C. to 1990." *The Quarterly Journal of Economics* 108, vol. 3 (1993): 681-716.

Lassère, J.-M. *Ubique populus : peuplement et mouvements de population dans l'Afrique romaine de la chute de Carthage à la fin de la dynastie des Sévères (146 a. C.-235 p. C.)*, coll. Études d'Antiquités africaines. Paris: éd. du CNRS, 1977.

Laüt, L. 2007 "Caractérisation des sites antiques dans les forêts du Berry et du Bourbonnais." In *La Mémoire des forêts, Actes du colloque « Forêt, archéologie et environnement », Velaine-en-Haye, 14-16 déc. 2004*, edited by J.-L. Dupouey, É. Dambrine, C. Dardignac, and M. Georges-Leroy, 99-107. Paris: ONF, INRA, DRAC, 2007.

Lazzaro, L. *Esclaves et affranchis en Belgique et Germanies romaines d'après les sources épigraphiques*. Paris: Les Belles Lettres, 1993.

Le Bohec, Y. *Inscriptions de la cité des Lingons. Inscriptions sur pierre. Inscriptiones latinae Galliae Belgicae -I- Lingones*. Paris: Comité des travaux historiques et scientifiques, 2003.

Le Bras, H. "Malthus ou Boserup : validité et continuité historique des modèles démo-économiques." *Mathématiques et sciences humaines (Mathematics and Social Sciences)* 41, vol. 164 (2003): 45-62.

Lee, R.D. "Population Dynamics of Humans and Other Animals." *Demography* 24 (1987): 443-465.

Lee, R.D. "L'autorégulation de la population : systèmes malthusiens en environnement stochastique." In *Modèles de la démographie historique*, coll. Congrès et colloques 11, edited by A. Blum, N. Bonneuil, and D. Blanchet, 149-174. Paris: Institut national d'études démo- graphiques, Presses universitaires de France, 1992.

Le Glay, M. "Les Flaviens et l'Afrique." *Mélanges d'archéologie et d'histoire* 80, vol. 1 (1968): 201-246.

Lepelley, C. "Peuplement et richesses de l'Afrique romaine tardive." In *Hommes et richesses dans l'Empire byzantin -I- IVe-VIIe siècle*, coll. Réalités byzantines, edited by C. Morrisson, and J. Lefort, 18-30. Paris: P. Lethielleux, 1989.

Lepelley, C. 1990 : "Ubique Res publica : Tertullien, témoin méconnu de l'essor des cités africaines à l'époque sévérienne." In *L'Afrique dans l'Occident romain (Ier siècle av. J.-C.-IVe siècle ap. J.-C.), Actes du colloque de Rome, 3-5 déc. 1987, École française de Rome*, Publications de l'École française de Rome 134, no editor, 403-421. Rome: EFR; 1990.

Le Roux, P. *Le Haut-Empire romain en Occident d'Auguste aux Sévères, 31 av. J.-C.-235 apr. J.-C*, Nouvelle histoire de l'Antiquité 8. Paris: Le Seuil, 1998.

Le Roy Ladurie, E. "A Reply to Professor Brenner." *Past & Present* 79 (1978): 55-59.

Leveau, P. "L'opposition de la montagne et de la plaine dans l'historiographie de l'Afrique du Nord antique." *Annales de Géographie* 86, vol. 474 (1977): 201-205.

Leveau, P. "Mentalité économique et grands travaux hydrauliques : le drainage du lac Fucin aux origines d'un modèle." *Annales. Économies, Sociétés, Civilisations* 48, vol.1 (1993): 3-16.

Leveau, P. "Inégalités régionales et développement économique dans l'Occident romain (Gaules, Afrique et Hispanie)." In *Itinéraire de Saintes à Dougga, Mélanges offerts à Louis Maurin*, coll. Mémoires 9, edited by J.-P Bost, J.-M. Roddaz, and F. Tassaux, 327-353. Bordeaux: Ausonius, 2003.

Leveau, P. "The Western Provinces." In *The Cambridge Economic History of the Greco-Roman World*, edited by W. Scheidel, I. Morris, and R. P. Saller, 651-670. Cambridge: Cambridge University Press, 2007.

Leveau, P. "La centuriation des territoires des cités romaines d'Arles (Arelate) et d'Aix- en-Provence (Aquae Sextiae) : un retour historiographique." *Revue archéologique de Narbonnaise* 43 (2010): 129-154.

Lo Cascio, E. "Peuplement et surpeuplement : leur rapport avec les ressources naturelles." In *Espaces intégrés et ressources naturelles dans l'Empire romain, Actes du colloque de l'université de Laval- Québec, 5-8 mars 2003*, edited by M. Clavel-Lévêque, and E. Hermon, 135-152. Besançon: Presses universitaires de Franche-Comté, 2004.

Malrain, F. "L'économie agraire en Gaule septentrionale." In *Comment les Gaules devinrent romaines*, edited by P. Ouzoulias, and L. Tranoy, 59-72. Paris: La Découverte, 2010.

Malthus, T. R. *An essay on the principle of population*. London: J. Johnson, 1798.

Marx, K. *Le Capital. Livre troisième. Tome troisième*. Traduction de C. Cohen- Solal et de G. Badia, coll. Le Capital VIII. Paris: Éditions sociales, 1960.

Marx, K. *Œuvres. Économie -I- Édition établie par M. Rubel*, coll. Bibliothèque de la Pléiade 164. Paris: Gallimard, 1965.

Marx, K. *Théories sur la plus-value -2- Chapitres VIII à XVIII et annexes*. Paris: Éditions sociales, 1975.

Meyer N. "Le contexte Tres Tabernae- Tabernis et ses environs." In *Au grès du temps : les collections lapidaires celtes et gallo-romaines du musée archéologique de Saverne*, edited by F. Goubet, F. Jodry, N. Meyer, and N.Weiss, 7-43. Saverne: Société d'histoire et d'archéologie de Saverne et environs, 2015.

Meyer N., Nüsslein A. 2014 "Une partie de la campagne gallo- romaine du Haut-Empire des cites des Médiomatriques et des Triboques préservée par la forêt : les habitats et parcellaires des Vosges du Nord (Moselle et Bas-Rhin) de part et d'autre du seuil de Saverne." In *Les Parcellaires conservés sous forêt , Workshop 2 de RurLand* (coll. Programme européen « Rural Landscape in North-Eastern Roman Gaul » dirigé par M. Reddé) [URL http://hal.archives-ouvertes.fr/hal-01007619].

Morineau, M. "Malthus au village." In *Malthus, hier et aujourd'hui*, edited by A. Fauve- Chamoux, 221-231. Paris: éd. du CNRS, 1984.

Myint, H. *Les Politiques de développement, traduit de l'anglais par Teresa Marcy*, coll. Développement et civilisations. Paris: éd. Économie et humanisme, Les Éditions ouvrières, 1966.

Nicolet, C. (1977, 3rd ed.) *Rome et la conquête du monde méditerranéen (264-27 avant J.-C.) -I- Les Structures de l'Italie romaine*, coll. Nouvelle Clio, L'Histoire et ses problèmes 8. Paris: Presses Universitaires de France, 1987.

Ouzoulias, P. "Place et rôle de la petite exploitation dans la Gaule romaine : un débat en cours." *Revue archéologique* 47, vol. 1 (2009): 149-220.

Ouzoulias, P. "Les campagnes gallo-romaines : quelle place pour la villa ?" In *Comment les Gaules devinrent romaines*, edited by P. Ouzoulias, and L. Tranoy L., 189-211. Paris: éd. La Découverte, 2010.

Ouzoulias, P. "Compte rendu de "The Cambridge Economic History of the Graeco- Roman World, Cambridge (2007)", Table ronde de Nanterre, 13 févr. 2010 (chap. 24 : Ph. Leveau, 'The Western Provinces'", *Topoi Orient Occident* 17, vol. 1 (2011a): 121-134.

Ouzoulias, P. "La villa dans l'est des Gaules : un témoin de la romanisation ?" In *Aspects de la romanisation dans l'est de la Gaule*, coll. Bibracte 221, edited by M. Reddé, P. Barral, F. Favory, M. Joly, J.-P. Guillaumet, J.-Y. Marc, P. Nouvel, L. Nuninger, and C. Petit, 475-485. Glux- en-Glenne: Bibracte - Centre archéologique européen, 2011b.

Ouzoulias, P., and Van Ossel, P. "Dynamiques du peuplement et formes de l'habitat tardif : le cas de l'Île-de-France." In *Les Campagnes de la Gaule à la fin de l'Antiquité, Actes du IVe colloque de l'association AGER, Montpellier, 11-14 mars 1998*, edited by P. Ouzoulias, C. Pellecuer, C. Raynaud, P. Van Ossel, and P. Garmy, 120-135. Antibes: APDCA, 2001.

Pautrat, Y., and Goguey, D. 2007 :"État actuel des connaissances sur les sites archéologiques forestiers du Châtillonnais : l'exemple des parcellaires." In *La Mémoire des forêts, Actes du colloque « Forêt, archéologie et environnement », Velaine-en-Haye, 14-16 déc. 2004*, edited by J.-L. Dupouey, É. Dambrine, C. Dardignac, and M. Georges-Leroy, 133-146. Paris: ONF, INRA, DRAC, 2007

Pétry, F. "Structures agraires archaïques en milieu gallo-romain." *Bulletin des Antiquités luxembourgeoises* 8 (1977): 117-158.

Pétry, F. "Introduction archéologique." In *Commission régionale d'Alsace : Bas- Rhin, Canton Saverne*, Inventaire général des Monuments et des richesses artistiques de la France, no editor, 14-20. Paris: Imprimerie nationale, 1978.

Pétry, F. "Une population marginale face à la civilisation romaine dans l'est de la Gaule aux Ier et IIe siècles ." *Bulletin des Antiquités luxembourgeoises* 10 (1979): 95-142.

Pétry, F. 1981 "Ager publicus et ager privatus : note à propos d'un abornement antique de la forêt de Saverne (Bas-Rhin)" in *Frontières en Gaule*, coll. Caesarodunum XVI, no editor, 21-27. Tours: Université de Tours , 1981.

Pétry, F. "Les agglomérations des sommets vosgiens." In *Les Agglomérations secondaires de la Lorraine romaine*, Institut des sciences et techniques de l'Antiquité 161, edited by J.-L. Massy, 399-405. Paris: Les Belles Lettres, 1997.

Peyras, J. "Les grands domaines de l'Afrique mineure d'après les inscriptions." In *Du latifundium au latifondo : un héritage de Rome, une création médiévale ou moderne ?, Actes de la table ronde inter- nationale de Bordeaux, 17-19 déc. 1992*, Publications du Centre Pierre-Paris 25, no editor, 107-128. Paris: De Boccard, 1995.

Peyras, J. "La potestas occupandi dans l'Afrique romaine." *Dialogues d'histoire ancienne* 25, vol. 1 (1999): 129-157.

Platteau, J.-P. "Malthus et le sous-développement ou le problème de la cohérence d'une théorie." *Revue économique* 35, vol. 4 (1984): 635-666.

Provost, M. *La Côte-d'Or -1- Alésia (d'Argencourt à Alise-Sainte-Reine)*, coll. CAG 21/1. Paris: Académie des inscriptions et belles- lettres, 2009a.

Provost, M. *La Côte-d'Or -3- De Nuits-Saint- Georges à Voulaines-les-Templiers*, coll. CAG, 21/3. Paris: Académie des inscriptions et belles-lettres, 2009b.

Raepsaet, G. "Les prémices de la mécanisation agricole entre Seine et Rhin de l'Antiquité au XIIIe siècle." *Annales. Histoire, Sciences sociales* 50, vol. 4 (1995): 911-942.

Ricardo, D. *On the principles of political economy and taxation*. London: J.M. Dent & sons, 1817.

Romer, P. M. "Increasing Returns and Long-Run Growth." *Journal of Political Economy* 94, vol. 5 (1986): 1002-1037.

Roque, J. *Référentiel régional pédologique de l'Île-de-France à 1/250 000 : régions naturelles, pédopaysages et sols*. Paris: INRA, 2003.

Saller, R. P. "Framing the Debate Over Growth in the Ancient Economy." In *The Ancient Economy: Evidence and Models*, Social Science History, edited by J.G. Manning, and I. Morris, 223-238. Stanford: Stanford University Press, 2005.

Saumagne, C. "Les domanialités publiques et leur cadastration au premier siècle de l'Empire romain." *Journal des Savants* 1, vol. 1 (1965):. 73-116.

Sauvy, A. (1952, 3rd ed.) : *Théorie générale de la population -I- Économie et croissance*, coll. Bibliothèque de sociologie contemporaine. Paris: Presses universitaires de France, 1963.

Scheidel, W. 2001. "Progress and problems in Roman demography." In *Debating Roman Demography*, edited by W. Scheidel, 1-82. Leiden: Brill, 2001.

Scheidel, W. "Demography." In *The Cambridge Economic History of the Greco-Roman World*, edited by W. Scheidel, I. Morris, and R. P. Saller, 38-86. Cambridge: Cambridge University Press, 2007.

Simon, J.-L. *L'Homme, notre dernière chance : croissance démographique, ressources naturelles et niveau de vie*, Libre échange. Paris: Presses universitaires de France, 1985.

Solow, R. M. *Growth Theory – An Exposition*. Oxford: Oxford University Press, 1960.

Temin, P. *The Roman Market Economy*, The Princeton Economic History of the Western World. Princeton: Princeton University Press, 2013.

Tran, H. H. *: Relire « Le Capital » : Marx, critique de l'économie politique et objet de la critique de l'économie politique*, Cahiers libres. Lausanne: éd. Page Deux, 2003.

Turgot, A. *: Réflexions sur la formation et la distribution des richesses*. 1766.

Van Andringa, W. *La Religion en Gaule romaine : piété et politique (Ier-IIIe siècle apr. J.-C.)*, Les Hespérides. Paris: Errance, 2002.

Vigneau, T. 2007 "Biodiversité et archéologie : une étude interdisciplinaire en forêt de Rambouillet (Yvelines, France)." In *La Mémoire des forêts, Actes du colloque « Forêt, archéologie et environnement », Velaine-en-Haye, 14-16 déc. 2004*, edited by J.-L. Dupouey, É. Dambrine, C. Dardignac, and M. Georges-Leroy, 163-172. Paris: ONF, INRA, DRAC, 2007.

Vigouroux, C. "Le Saltus arverne, complexe économique." *Revue archéologique du Centre de la France* 1, vol. 3 (1962): 211-220.

Wilson, A. "Machines, Power and the Ancient Economy" *The Journal of Roman Studies* 92 (2002): 1-32.

Zender, S. "Römische Siedlungsplätze und alte agrarstrukturen im Warndt." In *Archäologie in der Großregion – Beiträge des internationalen Symposiums zur Archäologie in der Großregion in der Europäischen Otzenhausen vom 14. – 17. April 2016.*, edited by M. Koch, J. Bonifas, J. Wiethold and A. Zeeb-Lanz, 265-282. Nonnweiler: Europäischen Akademie Otzenhausen, 2017.

3

Farming for a growing population: developments in agriculture in the provinces of *Germania*

Maaike Groot

Freie Universität Berlin

Abstract: Population growth and urbanisation were two factors in the Roman period that caused a growing demand for food. This chapter investigates how farmers responded to the increased pressure on production, and what strategies they developed to satisfy the growing demand for food. The focus will be on animal husbandry, but some examples from arable farming will be included, since in a mixed farming system, the two components are interdependent. The study will be limited to the provinces of *Germania Inferior* and *Germania Superior* in the first three centuries AD. The limiting factors on agricultural production were the carrying capacity of the land and the amount of agricultural labour. The demand of agricultural products not only concerns food, but also raw animal materials for crafts and industry. Moreover, there was the need to pay taxes, whether in kind or in money. This paper discusses the possible ways in which farmers can increase production (intensification, extensification and specialisation) and will then look at the evidence for this in the provinces of *Germania*. Both provinces experienced a transformation of the agricultural economy from mainly subsistent to surplus-producing. All three main ways to increase production are found in farming. Finally, the role of animals in providing raw materials for craft and industry is discussed.

Keywords: animal husbandry, arable farming, *Germania Inferior*, *Germania Superior*

Introduction

During the Roman period, the population of north-western Europe grew significantly. This put pressure on agricultural production. There are two main theories that tackle the effect of population growth on food production: Malthus's doom scenario (Malthus 1798), devised before the Industrial Revolution, in which population growth would inevitably lead to famine and disease, and Boserup's more modern scenario (Boserup 1965), based on her eyewitness accounts in development countries. According to Boserup's theory, a growing population would lead to agricultural intensification. Boserup's theory better explains the developments observed in agriculture in the north-western provinces in the first two centuries AD: farmers changed the way they farmed in response to the growing demand for food.

The aim of this paper is to investigate developments in animal husbandry that were responses to the population growth and urbanisation occurring in the Roman period. I will focus on two of the north-western provinces: *Germania Inferior* and *Germania Superior*. The large size of this research area and the differences within and between the two provinces will allow me to showcase different developments in agriculture. The main research questions of this paper are how agrarian communities adapted to an increased pressure on production, and what strategies they developed to satisfy the growing demand for food. I will discuss some of the changes in animal husbandry and arable farming that can be regarded as such responses. It is not my aim to provide an overview of agriculture in the region, since these have been published elsewhere (E.g. Deschler-Erb 2017; Groot 2016a; Groot and Deschler-Erb 2015; Groot and Deschler-Erb 2017; Kreuz 2005; Lauwerier 1988; Peters 1998; Pigière 2015; Pigière 2017; Pigière and Lepot 2013). While animal husbandry forms the main focus, I will also include some examples from arable farming, as they were complementary and interdependent systems in the Roman period. I will focus on the first three centuries AD, since this is the period with the most significant population growth in the research area. As my previous research areas are the central Netherlands and northern Switzerland, the paper is biased towards these areas.

Within the province of *Germania Inferior*, there are significant differences in landscape: from the coastal dunes, peat, and river delta in the central Netherlands to the sandy and loessic soils of the south-eastern Netherlands, to the river plain in the part of the province that is now in modern Germany. Frequent flooding was a risk in the western part of the province; this in combination with the dominance of peat and heavy clay made this part of the province less suitable for arable farming but suitable for livestock raising. The river plain of the Rhine continues into the province of *Germania Superior*, but this province also includes the hilly areas of the Vosges and Jura. Agriculture was mainly concentrated at altitudes below 500 m, on the river plain, in valleys and on the Swiss plateau, where soils were fertile. The river Rhine formed the border of the two provinces, although in the early 2nd century in the part of *Germania Superior* that is now northern Switzerland, the border was moved to the north. The Rhine formed a convenient transport route for goods.

The transformation of the agricultural economy

Before the start of the Roman occupation, the provinces of *Germania Inferior* and *Germania Superior* were mostly characterised by a subsistence economy based on mixed farming. Urbanisation was limited to a handful of larger conglomerations of settlement in the later province of *Germania Superior*, such as Basel-Gasfabrik and Basel-Münsterhügel, and oppida such as Altenburg-Rheinau. In these cases, some surplus production may have occurred in the surrounding countryside to feed the proto-urban centres. In what would become *Germania Inferior*, only small settlements existed.

At the beginning of the Roman occupation, urbanisation became more widespread and the population started to increase. The major towns in the two provinces discussed in this paper varied in population from circa 5,000 to over 20,000 people.[1] Population growth not only occurred as a result of the growing towns, but also occurred in the countryside. This has been well documented for the central Netherlands, where rural settlement density increased during the Roman period (Vossen 2003; Van Dinter et al. 2014). The rural population of the *civitas Batavorum* has been estimated at 20-40,000 people in the Early Roman period and over 50,000 people in the Middle Roman period (Vossen 2003; Willems 1984: 234-237).[2] Besides the growing urban and rural population,

there was the large military population, stationed along the border of the Empire. Buringh et al. consider the military the most important driver of the local economy in the north-western provinces (Buringh et al. 2012: 23). This is not surprising, since military supply in the north-western provinces mostly relied on local production (Thomas and Stallibrass 2008: 9). While luxury and exotic products were transported over large distances, bulk goods such as meat and cereals were acquired from as close as possible, to keep cost and time of transport low. Just for the western part of *Germania Inferior* (the Lower Rhine delta from the coast to the fort of Vechten), the military population has been estimated at a minimum of 4550 for the period AD 40-69 and 7700 for the period AD 70-140 (soldiers and inhabitants of the military vici) (Van Dinter et al. 2014). For *Germania Inferior* as a whole, the military population consisted of a maximum of 42,000 soldiers around AD 16-17 and decreased to 20,000 from the early 2nd century onwards (Alföldy 1968: 137-143, 149-152, 160-162; Polak 2009). Overall, the Lower Rhine area was the most densely settled area of north-western Europe around AD 150 (Buringh et al. 2012: 18).

However, it is not just population growth that is the issue. If population growth was limited to the rural population, then the finite amount of available land may have caused problems, but the number of agricultural workers would have increased, so that labour shortage would not have been a problem. However, in the Roman period, it was not just the rural population that increased, but there was also a large increase in the urban population. In this case, the area of land available for agriculture was still finite, but the number of agricultural workers remained stable while the pressure to produce more food and raw materials grew. So in fact there were two limitations on production: carrying capacity and the amount of agricultural labour.

Furthermore, the pressure on agricultural production did not only concern food. Agricultural communities produced raw materials for several industries and crafts located in the towns, such as tanning, textile production, horn working, bone working and marrow and grease extraction. It is important to also take into account the production of raw materials for such industries. Urbanisation brought with it an increasing separation between production and consumption, both in terms of food and the processing of raw materials, such as wool and hides. Before the Roman period, rural people produced their own textile and leather, but during the Roman period, these crafts were transformed into industries, located in the towns (e.g. Groot 2008: 70-73).

Taxation forced farmers to produce an agricultural surplus that could be sold at the market to raise money to pay taxes, or taxes could be paid in kind by the surplus itself. In this respect, taxation stimulated agrarian production, trade and monetisation (Hopkins 1980). Imported material culture and coins of all denominations are found in rural settlements, so it is certain that market transactions

---

[1] Nijmegen: 5000 civilians in late 1st century AD. Mentioned in Willems (1990: 71) and Willems and Van Enckevort (2009: 74), but the estimate is from Brunsting (1937); Augst: 10,000 to 15,000 around 200 AD. Bossart et al. (2006). Bossart et al. base their calculations on the actual area of Augst that was built on and a settlement density of 135-250 people per hectare (estimate by Vercauteren, based on 18th-century French towns). They also used the inhabited area to calculate the number of households (one household estimated to be on average 0.02 hectare, with 5 to 8 people per household). This paper also gives figures for other towns, although these are probably too high: Xanten: 11-14.600, Köln: 14-19.200, Mainz: 21-27.600, Worms: 9750-13.000.

[2] Willems used two methods to calculate the rural population of the *civitas Batavorum*. The first is based on historical sources on the number of Batavian soldiers serving in the Roman army in the pre-Flavian period, leading to a minimum population required to supply this number. The second is based on the reconstructed number of settlements for the Early Roman period (900), the average number of houses per settlement (3-4), and the number of people per household (5-8). Vossen evaluates these calculations and concludes that the recruitment method is less suitable, as neighbouring tribes may have contributed soldiers to Batavian troops as well. He calculates the rural population for the Middle Roman period for two microregions (using the same assumptions as Willems) and extrapolates this to the entire *civitas*.

existed.[3] Using money not only facilitated transactions between farmers and consumers, it also allowed more complex and indirect transactions. It was no longer necessary to exchange one product for another directly; a farmer could now sell agrarian produce for money, and buy another product from a third person with that money.

To conclude, the pressure on agricultural production of food and raw materials increased significantly during the Roman period as a result of population growth, urbanisation and the growth of industry. This, in combination with taxation, the transformation of household crafts to urban industries and an increasingly monetarised society caused a transformation of the agricultural economy.

Possible responses to increased demand for food

When faced with an increased demand for food, farmers in pre-industrial agricultural economies could respond in different ways to increase production (Groot and Lentjes 2013; Groot 2016a: 18). The first way is by increasing the yield per unit through intensification. For livestock, animal size can be increased through selective breeding, improved nutrition or by postponing slaughter to the optimum slaughter age. This will not only increase meat yield but also the power of draught cattle (allowing them to plough heavier soil or pull heavier loads). Yield of arable fields can be increased by applying manure and weeding and irrigating crops. The second way is by increasing the number of units through expansion or extensification. This can be achieved by increasing herd size or cultivating a larger acreage of arable land. Increasing the acreage under cultivation would imply an increased need for agricultural labourers, while a larger herd can be achieved with little extra labour if animals were kept extensively. Finally, specialisation in agricultural products leads to a more efficient production. Specialisation in pre-industrial societies was generally relative, as farmers still produced most of their own food, and is therefore hard to detect. By moving crafts to town and increasing their scale to an industrial level, production became more efficient. Finally, there are also external factors enabling agricultural production, such as an improved infrastructure, which would facilitate the transport of goods to the market.

Evidence from the provinces of *Germania*

Intensification

An increase in the size of livestock is found in many regions of the Roman Empire (Albarella et al. 2008; Colominas et al. 2014; Dobney et al. 1996: 31-33; Groot 2008: 74, 91-93; 2009: 368, 384-385; 2016a; Lauwerier 1988: 166-167; Lepetz 1996; MacKinnon 2010; Peters 1998; Schibler and Schlumbaum 2007; Schlumbaum et al. 2003; Teichert 1984; Valenzuela et al. 2013). While this increase also occurs in other species of livestock (Albarella et al. 2008; Breuer et al. 2001; Groot 2016a: 122, 140; Johnstone 2008: 135; Junkelmann 1990: 39), it is best studied for cattle. Two questions can be asked about this size increase: first, why were larger animals desirable, and second, how was the size increase achieved?

Larger cattle would have been desirable for two reasons. First, they would give a higher meat yield per animal. Second, larger draught cattle would be able to plough heavier soils and pull heavier loads. If a desire for larger ploughing cattle was the main reason behind the size increase, this would reflect intensification of arable farming.

A size increase can be achieved in three ways[4]: 1) by improved nutrition. Livestock stabled over winter may have received higher-quality food. In the Late Iron Age Netherlands, farmhouses already had stable sections, and this did not change during the Roman period. It is therefore unlikely that stabling was a cause behind the size increase in this region. Although there is some archaeological evidence for fodder, this is too scarce to establish whether it changed over time and whether it reflects improved nutrition (Lange 1990: 118-122; Kooistra 2009a: 442, 447). Pucher claims that the effect of improved nutrition on size has been overestimated in Roman studies (Pucher 2013: 29); 2) by selective breeding of existing stock. A deliberate breeding strategy for larger cattle may have been in place, by selecting the larger animals in the herd and deselecting smaller animals for breeding. Alternatively, the discontinuation of a selection for smaller animals[5] may have resulted in a size increase; 3) by importing larger animals and breeding these with the local population. Large cattle could have been imported in a deliberate attempt to improve local populations, or simply as a means of transport, pulling wagons. Of course, they would have to have been cows or bulls and not oxen to have an effect on local populations. Interbreeding with imported stock and selective breeding of local stock may have gone hand in hand.

Figure 3.1 shows the size increase in cattle during the Roman period in two regions in *Germania*: the *civitas Batavorum* and the northern Swiss lowland. The Roman data are from rural settlements (Netherlands) and villas (Switzerland). In general, cattle size increased in both regions. However, the most significant increase occurred earlier in the Swiss lowland (1st and 1st/2nd century versus 2nd/3rd century in the *civitas Batavorum*).[6] Furthermore,

---

[3] For the Lower Rhine area: Aarts (2014); Buringh et al. (2012).

[4] It is assumed here that the observed size increase does not reflect a shift in sex distribution, which in theory could cause a larger average size due to the sexual dimorphism in cattle. However, in the rural *civitas Batavorum*, there is no change in the sex ratio of cattle during the Roman period. Groot (2016a: 141). Furthermore, the fact that size increases are visible in all three anatomical planes make it unlikely that the cause is a shift in sex ratio. Thomas (2005: 79).

[5] See Roymans' hypothesis on the cultural significance of cattle in the Late Iron Age (1999). If cattle functioned as a standard of value, quantity would have been more important than size.

[6] Robeerst (2005) identified an increase in withers height in the *civitas Batavorum* occurring as early as the first decades of the 1st century AD,

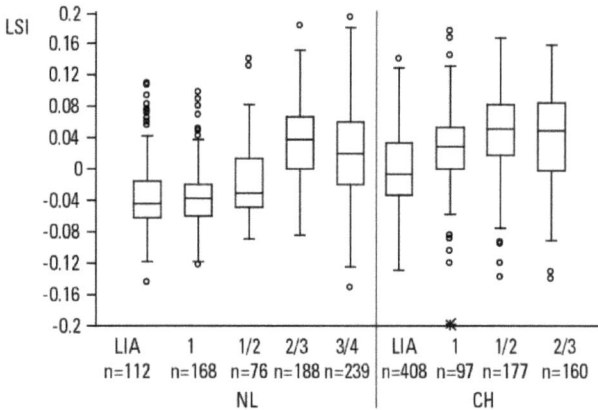

Figure 3.1: Log size index (LSI) for width, length and depth measurements from cattle from rural sites in the *civitas Batavorum* and villas in the northern Swiss lowland. The standard is a 13-year-old Hinterwälder cow (Reference collection IPNA, Basel, inventory number BS 2431). See Groot/Deschler-Erb (2016) for more details on the methodology. Sites included are listed in Groot/Deschler-Erb (2016), except the Late Iron Age sites (Netherlands: Geldermalsen-Hondsgemet, Odijk-Singel West/Schoudermantel (Zeiler 2007) and Deest (Whittaker 2002); Switzerland: Basel-Münsterhügel). Sites from the western part of the Netherlands are not included here.

cattle in the Swiss lowland were already larger before the Roman occupation than those in the *civitas Batavorum*. In fact, a size increase has been observed in the Late Iron Age, which has been attributed to Roman influence and imported cattle (Breuer et al. 1999: 217-219). For the Roman town of Augst (*Augusta Raurica*), where a size increase in cattle was also observed, this is assumed to be due to a combination of better conditions, selective breeding of local cattle and interbreeding with imported cattle, as both the maximum and minimum values increase (Breuer et al. 1999: 220-221). A changing prevalence of congenital (non-metric) traits in cattle in the *civitas Batavorum* supports the introduction of new cattle with a different genetic composition from the local population (Groot 2016a: 141-142). Import of larger cattle and subsequent interbreeding with local stock seems the best explanation for the size increase. The same scenario has been demonstrated for Roman Cologne (Berke 1997).

By providing fodder to livestock, the output of the animal could be improved. However, it is hard to say whether this is a factor in the size increase of cattle, as little is known about fodder for both the Roman and previous periods. In Wijk bij Duurstede-De Horden, fodder consisted of a combination of hay, uncleaned cereals and weeds (Lange 1990: 118-122). In Geldermalsen-Hondsgemet, water plants and grass may have been fed to livestock in summer; however, they could also be the remains of manure (Kooistra 2009a: 442, 447).

Intensification can also be observed in arable farming. Together with colleagues, I investigated the production of an agrarian surplus in two settlements in the *civitas Batavorum* (Groot et al. 2009). There was no change in the cereal spectrum during the Roman period, but the wild flora show changes in both settlements. While for the settlement of Wijk bij Duurstede-De Horden, an increase in grassland and meadow plants was observed (reflecting an increasing emphasis on animal husbandry), for Tiel-Passewaaij, the opposite was found: an increase in arable, farmyard and garden plants (reflecting an increasing emphasis on arable farming). Despite this difference in development, both settlements show an increase in storage capacity as measured by the number and size of granaries. This apparent increase in the overall amount of cereals produced is not in proportion to changes in the local population, so it seems that a surplus of cereals was produced in both settlements. Since large granaries are not found in all settlements, it is possible that settlements where they do occur acted as collection sites for surrounding settlements, and that the production and transport of the agrarian surplus of cereals and perhaps also livestock were organised from these sites (Vos 2009: 256-257).

The $2^{nd}/3^{rd}$-century Roman villas in the German regions of Hessen and Mainfranken specialised in growing naked wheat, spelt wheat and barley (Kreuz 2005: 129, 135). Since there was no recent local tradition for the first two crops, these can be seen as new introductions. Naked wheat requires intensive care compared to other cereals and both naked wheat and spelt wheat require good feeding (Kreuz 2005: 134). This implies accompanying changes in farming methods along with the new cereal crops. This probably included crop rotation and alternation between summer and winter crops (Kreuz 2005: 161).

Indirect evidence for intensification of arable farming is found in the increase in slaughter ages of cattle in the *civitas Batavorum* during the later 1st and 2nd century AD (Groot 2016a: 129, 216). As the importance of traction and manure grew, cattle were kept alive for longer. The size increase in cattle – discussed above – may reflect a wish for larger, stronger plough animals.

More indirect evidence for agrarian intensification can be found through the study of paleopathology in livestock. A recent study found an increase in the prevalence of pathology in three rural settlements in the Lower Rhine area from the Late Iron Age to the Roman period and in two of those settlements, from the Early to Middle Roman periods (Figure 3.2; Table 3.1) (Groot 2016b). This can be explained by intensification of both animal husbandry and arable farming. Stocking animals at higher densities and higher slaughter ages of cattle due to an increased emphasis on ploughing had a negative effect on animal health.

Extensification

Extensification in farming means increasing the number of units, either of arable land or of animals. In the late 1st

---

but this was in the urban settlement of Nijmegen. The similarity in size to the cattle from Tongeren led her to suggest that animals may have been imported from this region to Nijmegen.

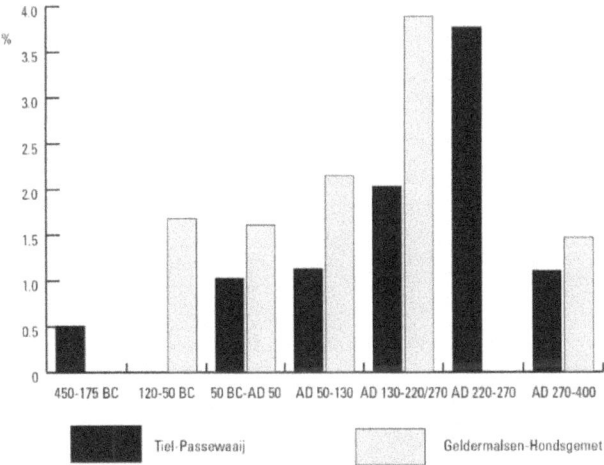

Figure 3.2: Percentage of bones with pathological changes (out of the total number of identified bone fragments) through time for the sites of Tiel-Passewaaij and Geldermalsen-Hondsgemet.

Table 3.1: Percentages of pathology for all animals in the Late Iron Age and Roman period for three rural sites in the civitas Batavorum, based on the number of identified fragments of animal bone.

Site	Late Iron Age	Roman	Factor of increase
Gendermalsen-Hondsgemet	1.78%	2.87%	1.6
Tiel-Passewaaij	0.51%	1.33%	2.6
Odijk	0.29%	0.94%	3.2

and 2nd century AD, systems of ditches were laid out in the land surrounding rural settlements in the central part of the Netherlands (Heeren 2009: 238-239, 248-250; Groot and Kooistra 2009: 3.2.2); Van Londen 2006: 183-188, 220-221); Vos 2009: 105, 115-116, 257-258). While there are several possible explanations for this, including water management, measuring land for taxation and marking land ownership, one explanation fits in with the other developments in agriculture: draining land and extending the area that was cultivated for crops onto previously marginal land. In the settlement of Tiel-Passewaaij, during the 2nd century AD long ditches were dug in the margin of the flood basins. Some of these were located on an older stream ridge in the flood basin, which was slightly higher than the surrounding land, with sandier soil. This may reflect an expansion of arable farming, as after providing drainage, the soil may have been suitable for growing crops (Groot and Kooistra 2009: 3.2.2).

The large areas of flood basin in the Lower Rhine area, especially the *civitas Batavorum*, were unsuitable for growing crops but very suitable for the extensive keeping of livestock. Already before the Roman period, most of the natural woodland had disappeared from the flood basins, and was replaced by reed marshes and wet grassland.[7] Wet grassland is maintained by grazing or mowing.

Indications for wet grassland are found in the settlements of Geldermalsen-Hondsgemet and Kesteren-De Woerd (Kooistra 1996: 49; Kooistra and Van Haaster 2001: 326). This suggests that livestock indeed grazed in the flood basins and not just on the stream ridges. Further evidence for the grazing of livestock in the flood basins is found in mineralised manure from Kesteren-De Woerd. The pig or cattle who produced the manure fed on wet, fertile grassland.[8] The absence of young lambs and young horses in certain periods in the settlement Tiel-Passewaaijse Hogeweg suggests that these animals were kept in the flood basins during the season of birth.[9] This implies extensive management of livestock (Groot 2008: 84-89). An advantage of such management is that it requires little labour, as there is no need for collecting fodder. During the Roman period, an increase in grassland, and presumably, an accompanying expansion of animal husbandry, was observed from pollen data from Wijk bij Duurstede-De Horden (Lange 1990: 146). Groot et al. quantified the herd size needed to feed the rural population of two settlements in the *civitas Batavorum*, and concluded that the stabling capacity exceeded the minimum herd size required in all periods. The extra space in the stables could be used to house cattle sold as a surplus (Groot et al. 2009: 242, table 1, 250).

Specialisation

In small farmsteads in the Roman period, specialisation is limited in extent, because farmers had to produce food for themselves in the first place, and any specialised surplus production came second. Furthermore, it made sense to avoid risk by growing different kinds of crops and keeping different kinds of animals. Even large villas are likely to have produced most of their own food, but due to the scale of production, specialisation should be more pronounced.

Because specialisation is relative, it is difficult to detect archaeologically. Indicators are one dominant species (crop or animal), slaughter ages with a focus on a particular age category, or a major change in the proportions of crops or animals or slaughter ages.

Evidence for specialisation in wool production is found in several sites. In Tiel-Passewaaij, the proportion of sheep increases in the second half of the 1st century AD (Figure 3.3). The exploitation of sheep also shows a change at this time, from a focus on meat (animals slaughtered young) to a focus on wool (animals slaughtered as adults) (Groot 2008: 70-73). High proportions of sheep are found in several other 1st-century settlements in this region, so it seems that wool production was not limited to Tiel-Passewaaij. In the

---

[7] As demonstrated by palynological research (Groot and Kooistra 2009: 3.1.2; Kooistra 2009b; Van Haaster 2004).

[8] Again demonstrated by palynological research (Kooistra and Van Haaster 2001: 318-327).

[9] Where any natural fatalities occurred, leaving no trace in the settlement (Groot and Kooistra 2009: 3.3.2). Lambs were also absent in the settlement of Geldermalsen-Hondsgemet (Groot 2009: 385). Foals were also absent in the settlement of Houten-Tiellandt (Laarman 1996: 356).

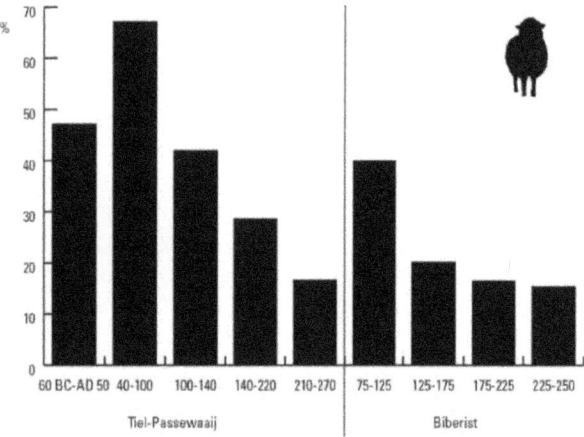

Figure 3.3: Proportion of sheep/goat (out of the total number of identified fragments for cattle, sheep/goat and pig) for the rural site of Tiel-Passewaaij (NL) and the villa of Biberist (CH).

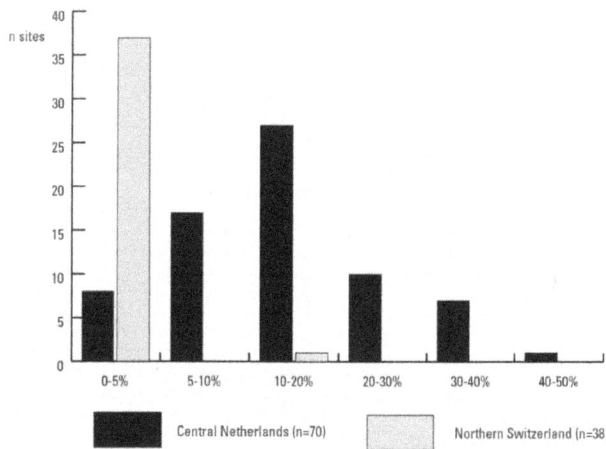

Figure 3.4: Number of assemblages per category of proportion of horse fragments (out of the total number of bone fragments for cattle, sheep/goat, pig and horse) for rural sites in the central Netherlands and northern Switzerland.

Late Iron Age, sheep were already an important part of the local agrarian economy. By shifting exploitation strategy, producing a surplus of wool was easily accomplished. The proportion of sheep declines after c. AD 100 (Figure 3.3), suggesting that the demand for wool had ended. Perhaps new supply lines for wool had been established, or perhaps the demand for wool had come from the army, and was much reduced when the 10th legion left Nijmegen in AD 102/104. Wool production has also been suggested for the villa of Biberist (Deschler-Erb 2006a: 658-659). Sheep/goat are the dominant species in the period AD 75-125, and decline after that (Figure 3.3). Slaughter ages show a development from mainly young adult and old adult to mainly infantile/juvenile from AD 75-125 to AD 175-225. The proportion of sheep killed as old adults increases from ca 35% to ca 70% in the period AD 125-175 (estimated from graph), and decreases in the period AD 175-225 to ca 30% (Deschler-Erb 2006a: 659, fig. 30/40). This can be explained as a change in exploitation for wool and milk to meat in the period AD 175-225. Since a building was found (Werkstatt B) that was probably used for washing wool and/or for fulling cloth, exploitation for wool is the most likely. It has been suggested that the wool was produced for the site of Windisch (*Vindonissa*) (Deschler-Erb and Akeret 2011: 30). In Erps-Kwerps in the *civitas Tungrorum*, 85% of sheep and goat were killed at ages older than 2 years, and 25% or more at ages older than 4 years (Pigière 2015: 172). Since sheep usually outnumber goats in the research area (Groot and Deschler-Erb 2015), this suggests that this herd was managed mainly for wool production.

An increasing proportion of horse among the domestic animals in many rural sites in the *civitas Batavorum* has been interpreted as a specialisation in horse breeding (Groot 2008: 77-91; Groot 2016a: 131-132; Hessing 2001: 162; Laarman 1996: 377; Roymans 1996: 82). The proportions of horse in the Middle Roman period are especially striking when compared to proportions in rural sites in another region, that of northern Switzerland (Figure 3.4), where they are nearly always below 5% and in the majority, even less than 1%.[10] Although horse meat was consumed by rural people in the *civitas Batavorum* (Groot 2008: 78-81; Groot 2016a: 132) this does not seem to have been the primary product of horses. During the Roman period, the proportion of young horses killed in rural sites increases, but remains much lower than that for pig, an animal only kept for meat. The young horses could reflect the culling of unsuitable animals. There was certainly a market for horses in the region, as the army stationed here demanded a regular supply (Groot 2008: 77, 90). A rough but conservative estimate for *Germania Inferior* consists of 1000 horses required annually to replace those no longer suitable for service. By keeping most of the herd extensively in the flood basin, a surplus of horses could be produced with little extra labour (Groot 2008: 84-87). While a specialisation in breeding horses is typical for this region, it is more pronounced in some settlements than in others. Furthermore, within a settlement, some households seem to have specialised in breeding horses, while others did not. In three rural sites, a similar pattern was found when species proportions were analysed at household level. All households showed high proportions of cattle, but some also had a relatively high proportion of horse, while others had a relatively high proportion of sheep (Groot 2011; 2012). The households with higher proportions of horse also had more finds of military gear and horse gear, and some of the houses showed construction elements deriving from Roman military construction. This suggests a relationship between veterans from the Roman army and horse breeding. Perhaps they could use their connections in the army to sell the horses.

Villa: specialisation and increase in scale

One development that can be seen as specialisation is the spread of the villa: an agrarian business run at a large scale, aimed at production for the urban market. While products

---
[10] Data from Groot and Deschler-Erb 2015.

varied – dependent on local environmental circumstances – there is usually a large degree of specialisation. On fertile soils, such as the loess zone between Tongeren and Cologne, villas focused on cereals. Most villas seem to have specialised in growing one or two types of cereals, with spelt wheat as the main crop (Bakels 2009: 167; Habermehl 2013). Other crops, such as pulses, vegetables and fruit, were also grown (Bakels 2009: 173-174), but it was the cereal crop that was specifically grown for the urban and military market. Possibilities for transport and access to markets influenced the location of villas, with a higher density in the area surrounding towns and forts (Bakels 2009: 194). Not only the proximity to towns, but also the proximity to main roads affected settlement choice in villa regions (Jeneson 2013: 204-218). The area of land per villa varied per region, but reflects their large scale.[11] While most villas specialised in growing cereals, some villas, for instance in the Condroz region in modern Belgium, seems to have specialised in animal husbandry (Habermehl 2013).

Archaeological and archaeobotanical research in the villa of Voerendaal in the South-eastern Netherlands provided evidence for large-scale cultivation of spelt wheat in the 2nd and 3rd centuries. This evidence consists of a large *horreum*, a threshing floor next to a building where presumably the unthreshed harvest was stored and finds of cereals (mainly spelt wheat) and chaff (Kooistra 1996: 132-133, fig. 24, 158-163, 171). Large storage buildings are also found in other villas (Habermehl 2013). In Voerendaal, only the ears of cercals were harvested, so that the straw remained in the fields, where livestock would feed on it, manuring the fields at the same time (Kooistra 1996: 171). The evidence for slaughter ages of cattle was scarce at Voerendaal, but at other villas, a dominance of older animals suggests that their main role was to support arable farming. In the villa of Biberist, Switzerland, this was the case until the last phase of the villa, where a shift towards meat production seems to have occurred (Deschler-Erb 2006a: 657). In the villa of Triengen, Switzerland, too, adult cattle dominated, so that it was concluded that arable farming formed the core business of this villa (Stopp 1997: 394, 399, 401-402). A majority of adult cattle is also found at the 3rd-century villa of Worb-Sunnhalde (Büttiker-Schumacher 1998: 98).

The villa of Hoogeloon, the Netherlands, provides evidence for the organisation of food production and transport to the market, as well as for the scale of production. The villa is remarkable due to its rather isolated position in the middle of a large cover-sand area. This area was not suitable for large-scale arable farming (Kooistra and Groot 2015). The absence of a large granary and different cereal ratios from those found in the closest towns, further suggest that arable farming was only carried out in terms of subsistence. Furthermore, the villa is not well situated in terms of access to the urban markets. The nearest towns are 60-70 km away, and there are no navigable rivers to reach them, so the only way of transport is over land.

While cereals were clearly not an agrarian product produced in Hoogeloon as a surplus, the archaeological structures, landscape and zooarchaeological data offer indications for another product. Two structures found next to the villa are a large enclosure with six compartments (37x34 m) and a wood-lined watering place. Together, these form indications for the large-scale handling of livestock. Palynological data indicate that there were wet and dry grasslands and heathlands in the region, suitable for raising livestock. It seems likely that livestock was only kept in the enclosure for short periods at a time, since otherwise they would require feeding. This fits with a scenario in which livestock was collected from settlements in the surrounding area, sorted in the enclosure, perhaps selecting suitable animals, and sent on to the town. Hoogeloon seems to have functioned as a central collecting point for livestock from the rural settlements in the coversand area. By organising the collection and transport of livestock, the villa owner was able to generate wealth.

A study of the ceramics from Hoogeloon revealed that trade routes were oriented more to the south (Tongeren and Tienen) than to the north (Nijmegen). The proportions of bones from cattle, sheep/goat and pig found in Hoogeloon are also consistent with those found in the towns in the *civitas Tungrorum* (and much less so with those found in the town of Nijmegen), which means that Hoogeloon could have supplied these towns with meat animals, or at least been one of the sources supplying the towns. Cattle were the main meat provider in the towns. Slaughter ages of cattle reveal a mixture of cattle of prime beef age and older animals. This suggests that the animals were surplus animals (young males and retired draught or breeding animals) from subsistence, mixed-farming settlements. Similar results (a small-scale surplus produced by mixed farming rather than specialised meat production) are found in York, Winchester (mainly adults) and in the provinces of *Germania*, *Raetia* and *Noricum* (O'Connor 2000: 50-51; Maltby 1994: 90; Peters 1998: 69).

Several villas in Switzerland produced smoked meat. For the villa of Neftenbach, production of smoked meat of red deer and cattle was suggested on the basis of the overrepresentation of the elements containing little or no meat and a near absence of shoulder blades (Deschler-Erb and Schröder Fartash 1999: 260-261). Building Q of the villa of Biberist has been interpreted as a slaughterhouse and production site for smoked meat. Fat from the early-3rd-century oven was chemically analysed and must have come from cattle or sheep/goat (Spangenberg 2006). Among the largest assemblage of animal bones associated with the building, pig is dominant in number (Figure 3.5), followed by an unusually high amount of red deer (Deschler-Erb 2006a: 652-653, 933-935, tables 30/2-4). Among the pig bones, the humerus is overrepresented. This bone is removed when pig shoulders are prepared for

---

[11] Bakels mentions two examples. Villas in the German Rhineland west of Cologne had access to ca 50 ha each, while villas in the loess zone in the southern Netherlands had access to ca 200 ha each (Bakels 2009: 192).

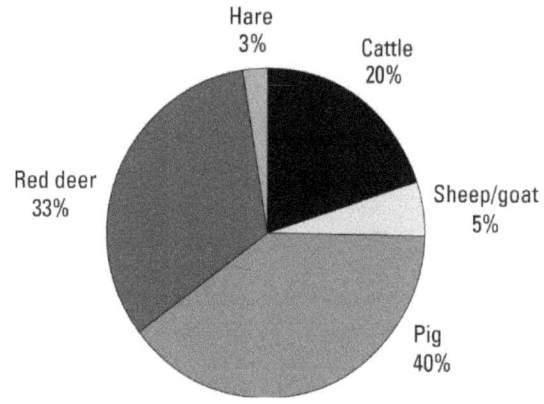

Figure 3.5: Species proportions for an animal bone assemblage associated with Building Q in the villa of Biberist (based on the total number of identified fragments, n=134). Data from Deschler-Erb (2006, 933-935, tables 30/2-4).

smoking, while the shoulder blade remained in the meat portion (Deschler-Erb 2006a: 652). The size of the oven suggests surplus production, and not just smoking for own use (Deschler-Erb 2006a: 653). In this phase of Biberist, slaughter ages of cattle show an increase in younger animals, reflecting a larger focus on meat (Deschler-Erb 2006a: 657). In the villa of Dietikon, two 3rd-century buildings with smoking ovens indicate the production of smoked meat. Unfortunately, no concentrations of animal bones were associated directly with these buildings. Animals were perhaps butchered somewhere else or the refuse was cleared away (Fischer and Ebnöther 1995: 258). Among the animal bones from the villa, long bones from pigs are underrepresented, so perhaps these parts were removed as smoked meat (Fischer and Ebnöther 1995: 262-263). Remains from smoked meat have been found in the town of Augst. The shoulder blades, which are underrepresented in the villas, remain in the meat and are therefore found at the site of consumption (Deschler-Erb 1991: 146-148).

Changes in crops and animals

In the Roman period, several new crops and animal species were introduced to the north-western provinces. Chicken was already present in Hallstatt sites in Switzerland and southern Germany (Schibler et al. 1999), but only reached the Netherlands in the Roman period. Chicken is found in most urban and military assemblages throughout the two provinces, although usually in low percentages; in rural sites, it occurs much less commonly in the western part of *Germania Inferior* (Groot and Deschler-Erb 2015; 2017). Other newcomers are herbs and orchard fruit and nuts (Bakels 2009: 159-167; Kooistra 2009b; 2012). It was not just the new plant species themselves that were introduced, but also the methods to grow and propagate them, in orchards and vegetable gardens (Bakels and Jacomet 2003: 555; Van der Veen et al. 2008: 32-33). While chicken and herbs were important for diet and culture, they did not play a role in terms of calorific contribution to the diet. An exception is the grape, which was also introduced to north-

western Europe in the Roman period, and grown at a much larger scale, for instance in the Moesel region and along the river Oise (Bakels 2009: 175-180). However, within the provinces of *Germania*, wine growing seems to be limited. So although the Roman period is characterised by several new introductions of animal and crop species, this did not result in increased food production in the north-western provinces.

Industry

In the Roman period, several crafts based on agricultural raw materials moved to the towns and were transformed into industries. The countryside remained responsible for supplying the raw materials. The separation of production and processing of raw materials led to a more efficient production of finished goods due to the larger scale and the specialisation of workers. This transformation of agrarian products into goods of higher value by craftsmen depending on local farmers for their food fits into Hopkins' 'taxes and trade' model (1980). While some of the goods were consumed locally, others could have been exported outside the region, leading to increased long-distance trade.

Although limited, there is some evidence for textile production in towns in the north-western provinces. For instance, in Switzerland, bone spindlewhorls have been found in towns, but are noticeably absent in several villas (Deschler-Erb 2012b: 125). Evidence for fulling operations is found in Oberwinterthur, where two water basins were connected to a water supply and drains. A fulling comb was found in one of the basins, providing a clue to the activities carried out here. In Kaiseraugst, a large water basin with a drain is also interpreted as a fulling basin. Three wooden vats from the same building may have been used for collecting urine used in the fulling process (Deschler-Erb 2012b: 126).

Cattle provided most of the meat consumed in towns. They reached the town alive and were butchered there by professional butchers. Most of the evidence for urban butcheries consists of animal remains. Evidence of a different nature was found in Augst: a total of five stone butchery blocks with a narrow channel along the edges and in some cases a basin to catch the blood (Deschler-Erb 2012d: 144). A Roman butchery site in Gaul gives us an idea about the scale of consumption: at least 2400 cattle were represented on this one site (` 2005). For late-1st-century urban and military Nijmegen, circa 2,500 cattle would be required per year to provide the population with 10% or circa 125 grams of meat in their daily diet (Groot 2016a: 224).

Meat could be sold fresh or processed or preserved first. Zooarchaeological evidence for smoked meat is usually found in the place of consumption, for example in Nijmegen, Augst and Xanten, and consists of ribs and shoulder blades with knife traces from removing the meat (Lauwerier 1988: 61; Deschler-Erb 1991: 146-148; Berke

1995a). Sausage making can leave zooarchaeological evidence in the form of fragmented skulls and split long bones (Deschler-Erb 2014d: 149), but such bone waste can also result from butchery or consumption of fresh meat. Smoking chambers or ovens used for smoking meat or sausages provides better evidence, and is particularly strong and well researched for Augst (Deschler-Erb 2012d: 137-142, 149). In a bone concentration from Nijmegen-Weurtseweg, fragments from the head are overrepresented; this could reflect initial butchery of cattle and processing of the head, perhaps for sausages or another kind of prepared meat (Groot 2016a: 183-185). The *canabae* in Nijmegen have also yielded evidence for processed meat production, in the form of large quantities of skull fragments and mandibles (Lauwerier 1988: 62-64). A different kind of meat preparation is known from Dangstetten and Bad Wimpfen: grilling cattle muzzles leaves marks of burning on the front part of the upper and lower mandibles (Uerpmann 1977: 266; Filgis 1988; Kokabi and Frey 1988). These specialised industries catered for large groups of people and are therefore a typically Roman phenomenon.

Urban cattle butchery provided a number of side-products that were the raw materials for other industries, namely hides, horns, bones, marrow and grease. Evidence for urban tanning has been found in the town of Augst, where a large number of sheep and goat horncores and footbones was interpreted as tanning waste (Schmid 1972: 45-46). In Oberwinterthur, three wooden vats were part of a tannery active in the 1st century AD. Organic remains found in the vats contained resin, animal hair, leather and skin remains (Deschler-Erb 2012c: 131). Leatherworking workshops were active in Windisch and Oberwinterthur, demonstrated by leather offcuts and typical tools. In Oberwinterthur, the leatherworking workshop was located next to a tannery, suggesting close cooperation between the two industries (Deschler-Erb 2012c: 135-136).

Evidence for bone working was found in the same insula in Augst as that for tanning, suggesting these industries also worked closely together. The evidence consists of primary waste (sawn-off ends of long bones) and half-products (facetted bone sticks and split shafts of long bones). The production seems to have focused on hair pins (Deschler-Erb 2012a: 117). A concentration of cattle horn cores was found in a cellar in Insula 31 in Augst. The horn cores were intact, with cut marks at the base suggesting removal of the horn. This is waste from horn working (Schmid 1968; 1972: 47-48). Concentrations of horn cores were also found in several pits in Tongeren (Vanderhoeven and Ervynck 2007: 163-166; Ervynck 2011). One of the pits was lined in clay and contained large stones: it was suggested that this pit was used for soaking hides, with the stones used to weigh the hides down. The horn cores could then be evidence for a horn working workshop and a tannery, working side by side. The scale of the operation is shown by the minimum number of cattle represented here, which is 550.

The last industry that will be mentioned here that is based on animal raw materials is the production of marrow, bone grease and glue. Marrow can be removed by long bones that are split open, while extracting bone grease and glue requires further fragmentation (Vanderhoeven and Ervynck 2007: 161-162). Bone glue is made by boiling chopped-up bones in a pot with water. Waste from glue making is typically highly fragmented, and consists of fragments of the diaphysis and joints and smaller carpal and tarsal bones (Deschler-Erb 2006b). There was a relation between bone working and glue making: the waste from bone working was used to make glue, while glue was used in making composite bone artefacts (Deschler-Erb 2012e: 159). In Switzerland, evidence for glue production is found only in towns: Augst, Oberwinterthur, Zurzach, Pfyn and Schleitheim (Deschler-Erb 2012e: 160). Glue production is also suggested for a mid-2nd century bone concentration from Cologne (Berke 1989). For a concentration of heavily fragmented bones from a large pit in Tongeren, it is left open whether the industry responsible produced marrow, marrow oil, grease or glue (Vanderhoeven and Ervynck 2007: 168-171). The pit contained long bone fragments, nearly exclusively from cattle. The minimum number of animals was reconstructed to be 160. A bone concentration from Xanten consisted only of specific long bones of cattle (humerus, radius, femur, tibia) (Berke 1995b). The fragments include parts of the joints, which Berke believes were unsuitable for glue. The selection of elements, in combination with a high degree of butchery, with bones split lengthwise, leads Berke to the conclusion that this is waste from marrow extraction. Animal bones from Nijmegen-Maasplein show a selection of the upper limb bones of cattle. This could represent butchery (removing the meat from the bones rich in meat), extraction of marrow or grease or glue making (Groot 2016a: 183-185).

A waste deposit from Cologne consists almost entirely of cattle bones (Berke 1996). Careful consideration of the skeletal elements present (mostly mandibles, shoulder blades and long bones) and absent, fragmentation and butchery marks led to the conclusion that this is a mixture of waste from smoked meat, marrow extraction and bone working.

As mentioned above, there was a strong relation between butchery and other industries relying on animal raw materials, with the exception of textile production. The meat was consumed, while the other parts of cattle were used in the tanning, horn working, bone working and glue production industries. The rural settlements in the hinterland of towns had to supply the animals and in the case of textile production, the wool. The large scale that is attested in some cases gives us some idea of the necessary increase in production; of course, this would also depend on the density and size of the towns.

Infrastructure and long-distance transport

So far, the developments discussed have been directly related to the production and processing of agrarian products. But there were other, external factors affecting food supply, which were not under the direct influence of farmers or individual craftsmen or entrepreneurs. One of these, improved infrastructure, was certainly a factor in the Roman period. The construction of roads and widespread use of ships for river transport and wagons for transport overland made it easier to transport crops and other agrarian products to the urban markets. Furthermore, it increased the area from which a town could draw its food.

Long-distance transport took advantage of growing crops in more suitable regions and then transporting them to regions that were less suitable for growing this crop. This allowed growing under optimal conditions, which would lead to higher yields and made it possible to feed more people. Alternatively, it allowed urban people to eat foods that were not locally available. An example of this is the transport of emmer wheat from the loess zone between Tongeren and Cologne to the central Netherlands, which is proven by the presence of non-local arable weeds among the crop (Kooistra 2009b: 222; Pals and Hakbijl 1992).

Conclusion

Developments in agriculture in the two provinces of *Germania* correspond to Boserup's theory that a population increase, with a growing demand for food, will lead to agricultural intensification. Taxation would have provided another stimulus to increased production and trade (Hopkins 1980). The rural and urban population used every opportunity to increase production and to make it more efficient: intensification, extensification, specialisation and developing crafts into industries. Evidence is found for all these possible responses open to farmers to increase production. There are several indications for intensification in the research area, such as an increase in the size of livestock. This could be either because of a wish for a higher meat yield or because of a wish for a higher labour output, with the first suggesting intensification of animal husbandry and the second intensification of arable farming. The size increase is likely to have been achieved by the import of larger cattle and subsequent interbreeding with local animals. Better nutrition in the form of fodder may have played a role, but at the moment there is not enough evidence on fodder. Intensification is also seen in the production of a surplus of cereals, both in villas and small farmsteads. The (re)introduction of naked wheat and spelt wheat in villas in Hessen and Mainfranken went hand in hand with changes in farming practices, such as crop rotation and the alternation of winter and summer crops. Finally, indirect evidence for intensification is found in higher slaughter ages of cattle and an increase in the prevalence of pathology of livestock.

Evidence for extensification is found in the draining of marginal land to grow crops, the extensive keeping of livestock, an increase in grassland – reflecting more grazing animals – and an increase in herd size, with a herd larger than required for subsistence. Specialisation in production is found for wool in the central Netherlands, a rural site in the *civitas Tungrorum* and a villa in northern Switzerland. Horse breeding was a specialisation specific for the central Netherlands, and related to the large presence of veterans in this region. Villas often specialised in one agrarian product or task (whether cereals, smoked meat or collecting cattle from the surrounding area) and also display an increase in scale compared to the Iron Age. While several new crops and species were introduced by the Romans, this had little impact in terms of bulk food.

Several kinds of processing of agrarian raw materials that had previously been carried out as household crafts moved to towns in the Roman period and took on the nature of industries. This led to a more efficient production due to specialised workers and a larger scale of production. The provinces of *Germania* show evidence for tanning, bone working, horn working and the production of marrow, bone grease and glue. It was professional cattle butchery that provided raw materials for these industries as well as food for urban people. The craftsmen involved in these industries consumed food produced by local farmers, who in turn bought some of the craft products on markets, thus increasing the number of economic transactions (Hopkins 1980). The improvement of infrastructure in the Roman period allowed for an easier transport of agrarian products, so that food could reach the market quicker and from more distant places. Furthermore, the long-term transport of food allowed the growing of food in its optimum place, which improved production.

Although this paper talks of 'pressure' on farmers to produce more food, this was not necessarily a bad thing. The urban markets also brought new opportunities and new products, and by selling produce for money, farmers were able to tap into this new wealth. The end of household crafts meant more time for other activities or for producing a surplus of food. A downside was that it went hand in hand with a dependency on the market for essential products such as textile and leather. In that sense, farmers were fully integrated into the Roman economy, for better and for worse.

While we find evidence for farmers responding to the growing demand for food, this is not found everywhere. This does not mean that the agrarian transformation did not occur throughout the research area; whether we find any evidence depends on the available data. Small-scale production of a surplus next to subsistence farming is notoriously difficult to identify. Farmers seem to have responded in different ways. How they changed their farming practices would have depended on a range of factors, including local environmental conditions, infrastructure, the local political situation, proximity to a market, local market demand, the available land and farm labour – arable farming is more labour-intensive than animal husbandry –, the available technology and knowledge and finally the life history and network of the farmer. By responding in ways appropriate to their situation, farmers managed to supply the Roman towns with all the food and raw materials that were needed.

Furthermore, the transformation in the agrarian economy also brought them new opportunities and wealth.

In the Lower Rhine area, the economy seems to have been disrupted around the turn of the 2nd/3rd centuries. Several factors are mentioned as possible causes: Chaucian raids in the later 2nd century, a pest epidemic, a rise in water level in part of this region (Vos 2009: 259-260), a change in recruitment (Van Driel-Murray 2003: 213-215) and soil exhaustion (Groenman-Van Waateringe 1983). However, in terms of the evidence for animal husbandry from the Lower Rhine area, there are no signs for any drastic developments around this time (Groot 2016a). Population decline in this area may have occurred, but is much more noticeable in the Late Roman period. The traditional view is that barbarian attacks and the destruction of *castella* brought an end to a stable situation in the second half of the 3rd century, with widespread abandonment of settlements and cities in the hinterland of the limes.[12] However, this belief has recently been challenged (Heeren 2016). There seems to be little evidence for discontinuity of the Lower Rhine *limes*, and although discontinuity is proven for the Obergermanische-Raetische *limes*, this was caused by violence in only two cases, and there was no lasting abandonment. Late-3rd-century coins in the hinterland of the Obergermanische-Raetische *limes* suggest that rural people were still connected to the monetary economy. Overall, rural people in the two provinces of *Germania* are likely to have contributed and taken advantage of the Roman economy at least until AD 300.

Bibliography

Aarts, J. "A frog's eye view of the Roman market: the Batavian case." In *A History of Market Performance. From Ancient Babylonia to the modern world*, edited by R.J. Van der Spek, J.L. van Zanden, and B. van Leeuwen, 394-409. London: Routledge, 2014.

Albarella, U., Johnstone, C., and Vickers, K. "The development of animal husbandry from the Late Iron Age to the end of the Roman period: a case study from South-East Britain." *Journal of Archaeological Science* 35 (2008): 1828-1848.

Alföldy, G. *Die Hilfstruppen der römischen Provinz Germania Inferior*. Düsseldorf: Rheinland Verlag, 1968.

Bakels, C.C. *The Western European Loess Belt. Agrarian History, 5300 BC–AD 1000*. Dordrecht: Springer, 2009.

Bakels, C., and Jacomet, S. "Access to luxury foods in Central Europe during the Roman period: the archaeobotanical evidence." *World Archaeology* 34, vol. 3 (2003): 542-557.

Berke, H. „Funde aus einer römischen Leimsiederei in Köln." *Kölner Jahrbuch für Vor- und Frühgeschichte* 22 (1989): 879-892.

Berke, H. "Knochenreste aus einer römischen Räucherei in der Colonia Ulpia Traiana bei Xanten am Niederrhein." *Xantener Berichte* 6 (1995a): 343-369.

Berke, H. "Reste einer spezialisierten Schlachterei in der CUT, Insula 37." *Xantener Berichte* 6 (1995b): 301-306.

Berke, H. "Die Tierknochenfunde aus den Ausgrabungen an der Jahnstrasse in Köln." *Kölner Jahrbuch* 29 (1996): 579-604.

Berke, H. "Zur Entwicklung der Rinderhaltung und Rinderzucht vom 1.-13. Jhrh. in Köln." *Anthropozoologica* 25-26 (1997): 405-412.

Boserup, E. *The conditions of agricultural growth. The economics of agrarian change under population pressure*. Chicago: Aldine Publishing Co, 1965.

Bossart, J., Koch, P., Lawrence, A., Straumann, S., Winet, I., and Schwarz, P.-A. "Zur Einwohnerzahl von Augusta Raurica." *Jahresberichte aus Augst und Kaiseraugst* 27 (2006): 67-108.

Breuer, G., Rehazek, A. and Stopp, B. "Grössenveränderungen des Hausrindes. Osteometrische Untersuchungen grosser Fundserien aus der Nordschweiz von der Spätlatènezeit bis ins Frühmittelalter am Beispiel von Basel, Augst (Augusta Raurica) und Schleitheim-Brüel." *Jahresberichte aus Augst und Kaiseraugst* 20 (1999): 207-228.

Breuer, G., Rehazek, A., and Stopp, B. "Veränderung der Körpergrösse von Haustieren aus Fundstellen der Nordschweiz von der Spätlatènezeit bis ins Frühmittelalter." *Jahresberichte aus Augst und Kaiseraugst* 22 (2001): 161-178.

Brunsting, H. *Het grafveld onder Hees bij Nijmegen. Een bijdrage tot de kennis van Ulpia Noviomagus*, unpublished PhD dissertation University of Amsterdam. Amsterdam, 1937.

Buringh, E., Luiten van Zanden, J. and Bosker, M. *Soldiers and booze: The rise and decline of a Roman market economy in north-western Europe*, Center for Global Economic History Working Paper 32. Utrecht: no publisher, 2012.

Büttiker-Schumacher, E. "Tierknochen." In *Worb-Sunnhalde. Ein römischer Gutshof im 3. Jahrhundert*, edited by M. Ramstein, 91-103. Bern: Berner Lehrmittel- und Medienverlag, 1998.

Colominas, L., Schlumbaum, A., and Sana, M. "The impact of the Roman Empire on animal husbandry practices: study of the changes in cattle morphology in the north-east of the Iberian Peninsula through osteometric and ancient DNA analyses." *Archaeological and Anthropological Sciences* 6 (2014): 1-16.

Deschler-Erb, S. "Das Tierknochenmaterial der Kanalverfüllung nördlich der Frauenthermen: Küchenabfälle einer Taberne des 2. Viertels des 3.

---

[12] See Heeren 2016 for an overview of this theory.

Jahrhunderts n. Chr." *Jahresberichte aus Augst und Kaiseraugst* 12 (1991): 143-151.

Deschler-Erb, S. "Die Tierknochen." In *Die römische Villa von Biberist-Spitalhof/SO (Grabungen 1982, 1983, 1986-1989). Untersuchungen zum Wirtschaftsteil und Überlegungen zum Umland*, Ausgrabungen und Forschungen 4, Band 2, edited by C. Schucany, 635-665. Remshalden: Greiner, 2006a.

Deschler-Erb, S. "Leimsiederei- und Räuchereiwarenabfälle des 3. Jahrhunderts aus dem Bereich zwischen Frauenthermen und Theater von Augusta Raurica." *Jahresberichte aus Augst und Kaiseraugst* 27 (2006b): 323-346.

Deschler-Erb, S. "Animal husbandry in Roman Switzerland: state of research and new perspectives." *European Journal of Archaeology* 20, vol. 3 (2017): 416-430.

Deschler-Erb, S. "Bein- und Hornverarbeitung." In *Das römerzeitliche Handwerk in der Schweiz. Bestandsaufnahme und erste Synthesen*, Monographies instrumentum 40, edited by H. Amrein, E. Carlevaro, E. Deschler-Erb, S. Deschler-Erb, A. Duvauchelle, and L. Pernet, 113-121. Montagnac: M. Mergoil, 2012a.

Deschler-Erb, S. "Textilverarbeitung." In *Das römerzeitliche Handwerk in der Schweiz. Bestandsaufnahme und erste Synthesen*, Monographies instrumentum 40, edited by H. Amrein, E. Carlevaro, E. Deschler-Erb, S. Deschler-Erb, A. Duvauchelle, and L. Pernet, 121-127. Montagnac: M. Mergoil, 2012b.

Deschler-Erb, S. "Herstellung und Verarbeitung von Leder und Pelzen." In *Das römerzeitliche Handwerk in der Schweiz. Bestandsaufnahme und erste Synthesen*, Monographies instrumentum 40, edited by H. Amrein, E. Carlevaro, E. Deschler-Erb, S. Deschler-Erb, A. Duvauchelle, and L. Pernet, 127-137. Montagnac: M. Mergoil, 2012c.

Deschler-Erb, S. "Nahrungsmittelproduktion." In *Das römerzeitliche Handwerk in der Schweiz. Bestandsaufnahme und erste Synthesen*, Monographies instrumentum 40, edited by H. Amrein, E. Carlevaro, E. Deschler-Erb, S. Deschler-Erb, A. Duvauchelle, and L. Pernet, 137-157. Montagnac: M. Mergoil, 2012d.

Deschler-Erb, S. "Leimsiederei." In *Das römerzeitliche Handwerk in der Schweiz. Bestandsaufnahme und erste Synthesen*, Monographies instrumentum 40, edited by H. Amrein, E. Carlevaro, E. Deschler-Erb, S. Deschler-Erb, A. Duvauchelle, and L. Pernet, 158-161. Montagnac: M. Mergoil, 2012e.

Deschler-Erb, S., and Akeret, Ö. "Archäobiologische Forschungen zum römischen Legionslager von Vindonissa und seinem Umland: Status quo und Potenzial." *Jahresberichte der Gesellschaft Pro Vindonissa 2010* (2011): 13-36.

Deschler-Erb, S., and Schröder Fartash, S. "Tierknochen." In *Der römische Gutshof in Neftenbach*, Monographien der Kantonsarchäologie Zürich 31/1 und 2, edited by J. Rychener, 260-264. Zürich: Zürich and Egg, 1999.

Dobney, K.M., Jaques, S.D., and Irving, B.G. *Of butchers and breeds: report on vertebrate remains from various sites in the city of Lincoln*, Lincoln Archaeological Studies 5. Lincoln: City of Lincoln Council, 1996.

Ervynck, A. "Everything but the leather. The search for tanneries in Flemish archaeology." In *Leather tanneries: the archaeological evidence*, edited by R. Thomson, and Q. Mould, 103-115. London: Archetype Books, 2011.

Filgis, M.N. "Neue Funde...: Bad Wimpfen, Kreis Heilbronn." *Archäologie in Deutschland* 3; vol. Juli-Sept. (1988): 44-45.

Fischer, M. and Ebnöther, C. "Tierknochen." In *Der römische Gutshof in Dietikon*, Monographien der Kantonsarchäologie Zürich 25, edited by C. Ebnöther, 254-163. Zürich: Zürich and Egg, 1995.

Groot, M. *Animals in ritual and economy in a Roman frontier community. Excavations in Tiel-Passewaaij*, Amsterdam Archaeological Studies 12. Amsterdam: Amsterdam University Press, 2008.

Groot, M. "Dierlijk bot en speciale deposities met dierlijk bot." In *Opgravingen in Geldermalsen-Hondsgemet. Een inheemse nederzetting uit de Late IJzertijd en Romeinse tijd*, Zuid-Nederlandse Archeologische Rapporten 35, edited by J. van Renswoude, and J. Van Kerckhove, 355-409. Amsterdam: Archeologisch Centrum van de Vrije Universiteit and Hendrik Brunsting Stichting, 2009.

Groot, M. "Household specialisation in horse breeding: the role of returning veterans in the Batavian river area." In *Fines imperii - imperium sine fine? Römische Okkupations- und Grenzpolitik im frühen Prinzipat*, Osnabrücker Forschungen zu Altertum und Antikerezeption 14, edited by R. Wiegels, G.A. Lehmann, and G. Moosbauer, 203-218. Rahden/Westf: Leidorf, 2011.

Groot, M. "Animal bones as a tool for investigating social and economic change: horse-breeding veterans in the *civitas Batavorum*." In *More than Just Numbers? The role of science in Roman archaeology*, Journal of Roman Archaeology Supplementary Series 91, edited by I. Schrufer-Kolb, 71-92. Portsmouth, Rhode Island: Journal of Roman Archaeology, 2012.

Groot, M. *Livestock for sale: animal husbandry in a Roman frontier zone. The case study of the civitas Batavorum*, Amsterdam Archaeological Studies 24. Amsterdam: Amsterdam University Press, 2016a.

Groot, M. "Paleopathologie: gezondheid en welzijn van dieren in het verleden." *Argos* 54 (2016b): 125-131.

Groot, M., and Deschler-Erb, S. "Market strategies in the Roman provinces: Different animal husbandry systems explored by a comparative regional approach." *Journal of Archaeological Science: Reports* 4 (2015): 447-460.

Groot, M. and Deschler-Erb, S. "Carnem et circenses – consumption of animals and their products in Roman urban and military sites in two regions in the Northwestern provinces." *Environmental Archaeoloy* 22, vol. 1 (2017): 96-112.

Groot, M., Heeren, S., Kooistra, L. and Vos, W. "Surplus production for the market? The agrarian economy in the non-villa landscapes of Germania Inferior." *Journal of Roman Archaeology* 22 (2009): 231-252.

Groot, M., and Kooistra, L.I. "Land use and the agrarian economy in the Roman Dutch River Area." *Internet Archaeology* 27 (2009) (*http://intarch.ac.uk/journal/issue27/5/1.html*).

Groot, M., and Lentjes, D. "Studying subsistence and surplus production." In *Barely surviving or more than enough? The environmental archaeology of subsistence, specialisation and surplus food production*, edited by M. Groot, D. Lentjes, and J. Zeiler, 7-27. Leiden: Sidestone Press, 2013.

Habermehl, D. *Settling in a Changing World. Villa Development in the Northern Provinces of the Roman Empire*, Amsterdam Archaeological Studies 19. Amsterdam: Amsterdam University Press, 2013.

Heeren, S.. *Romanisering van rurale gemeenschappen in de civitas Batavorum. De casus Tiel-Passewaaij*, unpublished PhD dissertation Vrije Universiteit Amsterdam. Amersfoort: no publisher, 2009.

Hessing, W.A.M. "Paardenfokkers in het grensgebied. De Bataafse nederzetting op De Woerd bij Kesteren." In *Opgespoord verleden. Archeologie in de Betuweroute*, edited by A. Carmiggelt, 142-172. Abcoude: Uniepers, 2001.

Hopkins, K. "Taxes and Trade in the Roman Empire." *Journal of Roman Studies* 70 (1980): 101-125.

Jeneson, K. *Exploring the Roman villa world between Tongres and Cologne*, unpublished PhD dissertation Vrije Universiteit Amsterdam. Amsterdam: no publisher, 2013.

Johnstone, C. "Commodities or logistics? The role of equids in Roman supply networks." In *Feeding the Roman Army: the Archaeology of Production and Supply in the North-West Roman provinces*, edited by S. Stallibrass, and R. Thomas, 128-145. Oxford: Oxbow Books, 2008.

Junkelmann, M. *Die Reiter Roms. Teil I: Reise, Jagd, Triumph und Circusrennen*. Mainz am Rhein: Zabern, 1990.

Kokabi, M., and Frey, S. "Neue Funde...: Bad Wimpfen, Kreis Heilbronn." *Archäologie in Deutschland* 3, vol. Juli-Sept. (1988): 43-44.

Kooistra, L.I. *Borderland farming. Possibilities and limitations of farming in the Roman period and Early Middle Ages between the Rhine and Meuse*. Assen and Amersfoort: Van Gorcum, 1996.

Kooistra, L.I. "Botanische materialen." In *Opgravingen in Geldermalsen-Hondsgemet. Een inheemse nederzetting uit de Late IJzertijd en Romeinse tijd*, Zuid-Nederlandse Archeologische Rapporten 35, edited by. J. van Renswoude, and J. Van Kerckhove, 411-457. Amsterdam: Archeologische Centrum van de Vrije Universiteit and de Hendrik Brunsting Stichting, 2009a.

Kooistra, L.I. "The provenance of cereals for the Roman army in the Rhine delta, based on archaeobotanical evidence." *Beihefte der Bonner Jahrbücher* 58, vol. 1 (2009b): 219-237.

Kooistra, L.I. "Die pflanzlichen Grundnahrungsmittel der Rheinarmee vor und nach der Gründung der Germania inferior." In *Verzweigungen, Eine Würdigung für A. J. Kalis und J. Meurers-Balke*, edited by A. Stobbe, and U. Techtmeier, 171-187. Bonn: Habelt R., 2012.

Kooistra, L.I., and Groot, M. "The agricultural basis of the Hoogeloon villa and the wider region." In *The Roman Villa of Hoogeloon and the Archaeology of the Periphery*, Amsterdam Archaeological Studies 22, edited by N. Roymans, T. Derks, and H. Hiddink, 141-162. Amsterdam: Amsterdam University Press, 2015.

Kooistra, L.I., and Van Haaster, H. "Archeobotanie." In *Kesteren-De Woerd. Bewoningssporen uit de IJzertijd en Romeinse tijd*, Rapportage Archeologische Monumentenzorg 82, edited by M.M. Sier, and C.W. Koot, 293-359. Amersfoort: ROB, 2001.

Kreuz, A. "Landwirtschaft im Umbruch? Archäobotanische Untersuchungen zu den Jahrhunderten um Christi Geburt in Hessen und Mainfranken." *Bericht der Römisch-Germanischen Kommission* 85 (2005): 97-292.

Laarman, F.J. "The zoological remains." In *Borderland farming. Possibilities and limitations of farming in the Roman period and Early Middle Ages between the Rhine and Meuse*, edited by L.I. Kooistra, 343-357. Assen and Amersfoort: Van Gorcum, 1996.

Lange, A.G. *De Horden near Wijk bij Duurstede. Plant remains from a native settlement at the Roman frontier: a numerical approach*, Nederlandse Oudheden 13. Amersfoort: ROB, 1990.

Lauwerier, R.C.G.M. *Animals in Roman times in the Dutch Eastern River Area*, Nederlandse Oudheden 12. Amersfoort: ROB, 1988.

Lepetz, S. *L'animal dans la société Gallo-romaine de la France du Nord*. Amiens: Revue Archéologie de Picardie, 1996.

MacKinnon, M. "Cattle 'breed' variation and improvement in Roman Italy: connecting the zooarchaeological and ancient textual evidence." *World Archaeology* 42, vol. 1 (2010): 55-73.

Maltby, M. "The meat supply in Roman Dorchester and Winchester." In *Urban-rural connections: perspectives from environmental archaeology*, edited by A.R. Hall, and H.K. Kenward, 85-102. Oxford: Oxbow Books, 1994.

Malthus, T.R. *An essay on the principle of population.* London: J. Johnson, 1798.

O'Connor, T.P. "Bones as evidence of meat production and distribution in York." In *Feeding a city: York. The provision of food from Roman times to the beginning of the twentieth century*, edited by E. White, 43-60. Devon: Prospect Books, 2000.

Oueslati, T. "Les ossements animaux, l'archéozoologie et les professions de l'álimentation dans le Nord de la Gaule romaine: le cas de la boucherie bovine." *Revue du Nord - Archéologie de la Picardie et du Nord de la France* 87, vol. 363 (2005): 175-183.

Pals, J.P. and Hakbijl, T. "Weed and insect infestation of a grain cargo in a ship at the Roman fort of Laurium in Woerden (Province of Zuid-Holland)." *Review of Palaeobotany and Palynology* 73 (1992): 287-300.

Peters, J. *Römische Tierhaltung und Tierzucht. Eine Synthese aus archäozoologischer Untersuchung und schriftlich-bildlicher Überlieferung*, Passauer Universitätsschriften zur Archäologie 5. Rahden and Westf: Leidorf, 1998.

Pigière, F. "Arable farming and animal husbandry in the core area of the civitas Tungrorum." In *The Roman Villa of Hoogeloon and the Archaeology of the Periphery*, Amsterdam Archaeological Studies 22, edited by N. Roymans, T. Derks, and H. Hiddink, 163-175. Amsterdam: Amsterdam University Press, 2015.

Pigière, F. "The evolution of cattle husbandry practices between the Iron Age and the Roman period in Gallia Belgica and Western Germania Inferior." *European Journal of Archaeology* 20, vol. 3 (2017): 472-493.

Pigière, F., and Lepot, A. "Food production and exchanges in the Roman civitas Tungrorum." In *Barely surviving or more than enough? The environmental archaeology of subsistence, specialisation and surplus food production* M. Groot, D.M. Lentjes, and J.T. Zeiler, 225-246. Leiden: Sidestone Press, 2013.

Polak, M. "The Roman military presence in the Rhine delta in the period c. AD 40-140." In *Limes XX, XXth International Congress of Roman Frontiers studies, Leon, (España), Septiembre 2006* (Anejos de Gladius 13-2, edited by A. Morillo, N. Hanel, and E. Martín, 945-953. Madrid: Ediciones Polifemo, 2009.

Pucher, E. "Milchkühe versus Arbeitsochsen: osteologische Unterscheidungsmerkmale zwischen alpin-donauländischen und italischen Rindern zur Römischen Kaiserzeit." *Beiträge zur Archäozoologie und Prähistorischen Anthropologie* 9 (2013): 9-36.

Robeerst, J.M.M. "Interaction and exchange in Food Production in the Nijmegen Frontier Area during the Early Roman Period." In *TRAC 2004: Proceedings of the Fourteenth Annual Theoretical Roman Archaeology Conference, University of Durham 26-27 March* 2004, edited by J. Bruhn, B. Croxford, and D. Grigoropoulos, 79-96. Oxford: Oxbow Books, 2005.

Roymans, N. "The sword or the plough. Regional dynamics in the romanisation of Belgic Gaul and the Rhineland area." In *From the sword to the plough. Three studies on the earliest romanisation of Northern Gaul*, Amsterdam Archaeological Studies 1, edited by N. Roymans, 9-127. Amsterdam: Amsterdam University Press, 1996.

Roymans, N. "Man, cattle and the supernatural in the Northwest European plain." In *Settlement and landscape. Proceedings of a conference in Århus, Denmark, May 4-7 1998*, edited by C. Fabech, and J. Ringtved, 291-300. Højberg: Aarhus University Press, 1999.

Schibler, J., Stopp, B., and Studer, J. "Haustierhaltung und Jagd." In *Die Schweiz vom Paläolithikum bis zum frühen Mittelalter – SPM IV: Eisenzeit*, edited by F. Müller, G. Kaenel, and G. Lüscher, 116-136. Basel: Archäologie Schweiz, 1999.

Schibler, J., and Schlumbaum, A. "Geschichte und wirtschaftliche Bedeutung des Hausrindes (*Bos taurus* L.) in der Schweiz von der Jungsteinzeit bis ins frühe Mittelalter." *Schweizer Archiv für Tierheilkunde* 149, vol. 1 (2007): 23-29.

Schlumbaum, A., Stopp, B., Breuer, G., Rehazek, A., Turgay, M., Blatter, R., and Schibler, J. "Combining archaeozoology and molecular genetics: the reason behind the changes in cattle size between 150BC and 700AD in Switzerland." *Antiquity* 77, vol. 298 (2003) (URL: *http://antiquity.ac.uk/projgall/schlumbaum/index.html*)

Schmid, E. "Beindrechsler, Hornschnitzer und Leimsieder im römischen Augst." In *Provincialia. Festschrift für Rudolf Laur-Belart*, edited by E. Schmid, L. Berger, and P. Bürgin, 185-197. Basel: Schwabe, 1968.

Schmid, E. *Atlas of animal bones. For prehistorians, archaeologists and quaternary geologists.* Amsterdam: Elsevier, 1972.

Spangenberg, J. "Verkohlte Masse aus dem Konservierungsofen in Gebäude Q." In *Die römische Villa von Biberist-Spitalhof/SO (Grabungen 1982, 1983, 1986-1989). Untersuchungen zum Wirtschaftsteil und Überlegungen zum Umland*, Ausgrabungen und Forschungen 4, Band 2, edited by C. Schucany, 675-677. Remshalden: Bernhard A. Greiner, 2006.

Stopp, B. "Die Tierknochen." In *Triengen, Murhubel. Ein römischer Gutshof im Suretal*, Archäologische

Schriften Luzern 7, edited by H. Fetz, and C. Meyer-Freuler, 387-413. Luzern: Kantonaler Lehrmittelverlag Luzern, 1997.

Teichert, M. "Size variation in cattle from *Germania* Romana and *Germania* Libera." In *Animals and archaeology: 4. Husbandry in Europe*, BAR International Series 227, edited by C. Grigson, and J. Clutton-Brock, 93-104. Oxford: BAR Publishing, 1984.

Thomas, R. "Zooarchaeology, improvement and the British agricultural revolution." *International Journal of Historical Archaeology* 9, vol. 2 (2005): 71-88.

Thomas, R., and Stallibrass, S. "For starters: producing and supplying food to the army in the Roman north-west provinces." In *Feeding the Roman army: the archaeology of production and supply in North-West Europe*, edited by S. Stallibrass, and R. Thomas, 1-17. Oxford: Oxbow Books, 2008.

Uerpmann, H.-P. "Schlachterei-Technik und Fleischversorgung im römischen Militärlager von Dangstetten (Landkreis Waldshut)." In *Festschrift Elisabeth Schmid*, edited by L. Berger, G. Bienz, J. Ewald, and M. Joos, 261-272. Basel: Geographisch - Ethnologische Gesellschaft, 1977.

Valenzuela, A., Alcover, J.A., and Cau, M.A. "Tracing changes in animal husbandry in Mallorca (Balearic Islands, Western Mediterranean) from the Iron Age to the Roman period." In *Barely surviving or more than enough? The environmental archaeology of subsistence, specialisation and surplus food production*, edited by M. Groot, D. Lentjes and J. Zeiler, 201-223. Leiden: Sidestone Press, 2013.

Vanderhoeven, A. and Ervynck, A. "Not in my back yard? The industry of secondary animal products within the Roman civitas capital of Tongeren, Belgium." In *Roman finds: context and theory*, edited by R. Hingley, and S. Willis, 156-175. Oxford: Oxbow Books, 2007.

Van der Veen, M., Livarda, A. and Hill, A. 2008. New plant foods in Roman Britain – dispersal and social access. *Environmental Archaeology* 13 (1): 11-36.

Van Dinter, M., Kooistra, L.I., Dütting, M.K., van Rijn, P., and Cavallo, C. "Could the local population of the Lower Rhine delta supply the Roman army? Part 2: Modelling the carrying capacity using archaeological, palaeo-ecological and geomorphological data." *Journal of Archaeology in the Low Countries* 5, vol. 1 (2014): 5-50.

Van Haaster, H. *Botanisch onderzoek aan enkele grondsporen bij de Romeinse wachttoren aan de Zandweg op de VINEX locatie Leidsche Rijn (LR31)*. Zaandam: Biaxiaal 182, 2004.

Van Londen, H. *Midden-Delfland: the Roman native landscape. Past and present*, unpublished PhD dissertation University of Amsterdam. Amsterdam, 2006.

Vos, W.K. *Bataafs platteland. Het Romeinse nederzettingslandschap in het Nederlandse Kromme-Rijngebied*, unpublished PhD dissertation University of Amsterdam. Amsterdam, 2009.

Vossen, I. "The possibilities and limitations of demographic calculations in the Batavian area." In *Kontinuität und Diskontinuität. Germania inferior am Beginn und am Ende der römischen Herrschaft*, edited by T. Grünewald, and S. Seibel, 414-435. Berlin and New York: De Gruyter, 2003.

Whittaker, W.E. *Zooarchaeological analysis of the Roman frontier economy in the Eastern Netherlands*, unpublished PhD dissertation University of Iowa. Iowa City, 2002.

Willems, W.J.H. "Romans and Batavians: a regional study in the Dutch eastern river area II." *Berichten van de ROB* 34 (1984): 39-331.

Willems, W.J.H. *Romeins Nijmegen. Vier eeuwen stad en centrum aan de Waal*. Utrecht: Matrijs, 1990.

Willems, W.J.H., and Van Enckevort, H. *Ulpia Noviomagus. Roman Nijmegen: the Batavian capital at the imperial frontier*, Journal of Roman Archaeology Supplementary Series 73. Portsmouth, Rhode Island: Journal of Roman Archaeology, 2009.

Zeiler, J.T. "Archeozoölogie." In *Oudheden uit Odijk. Bewoningssporen uit de Late IJzertijd, Romeinse tijd en Merovingische tijd aan de Singel West/Schoudermantel*, Zuid-Nederlandse Archeologische Rapporten 30, edited by E. Verhelst and M. Schurmans, 159-180. Amsterdam: Archeologisch Centrum van de Vrije Universiteit and Hendrik Brunstig Stichting, 2007.

4

Viticulture and demography in the Laetanian region (*Hispania Citerior Tarraconensis*), 1st c. BC – 3rd c. AD

Antoni Martín i Oliveras*, Víctor Revilla Calvo*, César Carreras Monfort**
and José Remesal Rodríguez*

University of Barcelona* / Autonomous University of Barcelona**

Abstract: The study of the economy and the archaeology of ancient viticulture in Roman times generally have multiple fields of knowledge and expertise with enormous possibilities for research. Most studies dedicated to the development of viticulture in antiquity combine archaeological information with written sources to determine the absolute chronology of a site, a socio-economic phenomenon, or to pinpoint a wine production centre and/or pottery activity in time and place. Intensive viticulture in Roman times in the north-eastern part of the Iberian peninsula – and specifically in the Laetanian region covering the coastal territory between the Tordera and Llobregat rivers as far as the start of the Garraf massif, and inland to the Catalan pre-coastal mountains – was a powerful phenomenon with huge economic implications which represented a cultural revolution for this region in all areas and at all levels. This chapter starts from the working hypothesis that 1) regional variability is one of the key points in understanding the changing patterns of rural settlement in any historical period; 2) the development and expansion of the wine phenomenon in the Laetanian region between the 1st century BC and the 3rd century AD was an important catalyst for the specific interaction between intra-regional and extra-regional economic networks; 3) the level of dependence of the rural population in a given area on both the regional and foreign market – the latter in this case Western Europe and the Italian peninsula with Rome – are matters that respond to a series of socio-economic patterns and behaviours which are likely to be studied and modelled economically & econometrically. This chapter aims to analyse the evolution of this complex economic system involving the production, trade and consumption of Laetanian wine between the 1st century BC and the 3rd century AD against the background of local demographic developments.

Keywords: Laetanian wine, viticulture, *Hispania Citerior Tarraconensis*, Palaeodemography.

Introduction

The Roman economy has been defined as an agrarian regime where wheat, the basic staple crop in the Mediterranean region, was mainly cultivated combined with livestock farming and cash-crops such as wine and olive oil. Possibilities for economic growth during the Roman period depended upon changes in agrarian productivity, but subject to constraints imposed by the agro-ecological environment, fluctuation in population that could affect labour availability, new techniques of cultivation and processing and the adoption of technological advances. The combination of these factors explains how advantages arose through a growing process that generated an increase of production surplus, which was traded in overseas markets.

The intensive viticulture practised during the Roman period in the ancient Laetanian region placed in the centre of Catalan coastal depression was a widespread phenomenon with huge economic implications that represented a cultural revolution for this territory in all areas and at all orders. Outstanding questions and key structural features reveal this revolution such as the land use increase, the crop regimes, the landowner tenancy, the population fluctuations, the balance between production and consumption, the investment needs, the implantation of the villa system and its evolution as a cash-crop market-oriented surplus production and the widespread of wine pressing and pottery facilities. This phenomenon appeared between the first century BC until the third century AD and can only be explained by an intensification process and the application of an *agency-oriented wine-growing specialization production* system related to a profit mentality that exploited a comparative advantage arose in productivity, a pulled force of product demand and the accessibility of larger distributed markets resulting from the reduction of production and transaction costs (Van Minnen 1998, 205-220).

This research built on a previous work in which the theoretical and epistemological framework of study was established, based on of the different variables, factors and endogenous and exogenous agents involved in every stage of the production, distribution, trade and consumption of wine in the Roman period between the 1st century BC and 3rd century AD (Martin i Oliveras 2015b). A further study must be focused on a geospatial, and geo-economic analysis, which supposes the identification of the settlement patterns, the organization of the rural habitat, the forms of production and management related to the crops capacities to obtain optimal yields for generating surpluses in a context of a growing population.

Demographic analysis and its behaviour over the time is one of the important variables to consider for making models related with the internal organization of population's evolution, the needs and availability of labour and for the identification of possible settlement patterns related to the resources management, urban development and agrarian exploitation of the territory (Isoardi 2012, 37-60).

This chapter attempts to make a first approach to this important variable which is demography, and its role in the configuration of this ancient winegrowing socioeconomic system, analysing the possibilities of implementation with a scantily available dataset where estimations will play a decisive role.[1]

Demography: an ignored variable in *Hispania*

The relations between forms of production and demography have occupied a very limited space in the studies dedicated to the Roman economy in *Hispania*. Nor have relations between agrarian systems and population structure evolution at the local or regional level.[2] More specifically, this perspective is ignored in the study of certain productive processes and associated forms of organization, such as viticulture, whose impact on the economic and social structure of certain territories seems especially important, by the quantity of resources, investments, technology and labour involved. In this sense, it has not gone beyond supposing a necessary relationship based on the use of a certain amount of labour in the various phases of a productive process whose components, from wine itself to the *amphorae* containing it, are so complex (Revilla 2015; Martin i Oliveras 2015a; 2015b; Martin i Oliveras, Martin-Arroyo and Revilla 2017). Other approaches to the agriculture-demography relationship are equally limited. On one hand, some studies have tried to evaluate generically the distribution and density of the rural habitat of some specific territories as a procedure to reconstruct the structure of the property and the land tenancy. This implies an implicit relationship between the forms of exploitation and the evolution of demography, but the terms of this relationship are not really defined (Miró 1988; Prevosti 1991). On the other hand, attempts have been made to reconstruct the dynamics of enrichment and social promotion related to certain activities affecting a limited percentage of the population. But, until now, this analysis has been necessarily limited, since it was based on the study of individual cases, whose representativeness is difficult to determine (Rodà et al. 2005). Despite the interest of these cases, their status as partial biographies limits its heuristic value (Revilla 2015, 7).

This study intends to evaluate the relations that could be generated between the development of viticulture and a set of complementary activities and the evolution of the demography in certain territories between the end of the Republic and the Early Roman Empire. In particular, it seeks to explore the impact of certain organizational forms, articulated by rational strategies, large investments in technology, resources and economic intensification processes, which involve labour needs and a precise organization of the global structure of the settlement. This impact can be assessed from factors such as habitat distribution and density, the calculation of labour needs also in agriculture and handicrafts, both fixed and variable, and its influence on living standards. These factors may be related to others more difficult to evaluate, such as the possible population growth, population mobility or the possibilities of enrichment and social promotion of certain groups related to the wine's economy such as freedmen.

The geographical sample area of study is the central coast of Catalonia. This territory presents particular characteristics that respond to certain historical conditions: A process of socioeconomic transformation from the mid-second century BC, which has a particular impact on the structure of rural settlement, a series of urban foundations in the first quarter of the first century BC until the last decades of the century that replaced former indigenous settlements; a development of new forms of economy throughout the same century, a phenomenon what appears to be a strong acceleration during the Augustan period. This last epoch, in which profound political-institutional reforms and socio-economic changes take place, provides a frame of reference to identify and contextualize the demographic dynamics. As complementary objectives, we intend to approximate the problem of the population employed in agriculture or other activities related or not to viticulture and the position of the city, understood as the centre of residence of proprietary elite that is simultaneously the leading group concerning the territory. Ultimately and therefore, it is intended to deepen in the knowledge of the Roman economy's nature.

The *baseline hypothesis* is there is a relationship between the processes of agrarian intensification and specialization, which are evident in the field of viticulture, and the structure and dynamics of settlement and population. These processes seem to promote the occupation of new

---

[1] This research is integrated in the University of Barcelona-CEIPAC-EPNet project: "Relaciones interprovinciales en el Imperio Romano. Producción y comercio de alimentos hispanos (*provinciae Baetica* et *Tarraconensis*)" (HAR2105-66771-P) and "*Production and Distribution of Food during the Roman Empire: Economic and Political Dynamics*" (ERC-2013-ADG-340828) see http://www.ceipac.ub.edu and http://www.roman-ep.net/wb/

[2] For a general overview on Roman Economy in *Hispania* see Blázquez 1978; Montenegro and Blázquez 1982. A recent synthesis but with important theoretical and methodological limitations in Lowe 2009.

territories. The extension of new productive forms requires a minimal amount of labour, necessary for the functioning of the agricultural and artisanal productive function with complex technology and the extension of the crops. In turn, this economy can boost population growth and generate processes of increasing socio-economic inequality and the promotion of some social groups as a result of an uneven distribution of incomes and wealth generated. Part of this population growth can be evidenced by the intensification and unequal density of rural habitat and urban population growth, though it is hardly perceptible through the data provided by the archaeological record. The unequal extent of new economic forms of production along with other factors, not only the administrative ones, can help us to understand the different evolution of some cities and secondary settlements, in particular their function concerning a territory and its demography, and to deepen into the global analysis of the urbanisation process of the region.

Demographic analysis: issues, methods and questions

Traditional demographic analysis in ancient agrarian societies has three main aspects to take into account: first, the *dynamics of population*, as regards fluctuation over the time; second, the *structure of population* as regards the composition of communities (gender, age, fertility, mortality), normally obtained from written census, besides demographic proxies on social stratification. Finally, the *distribution in space and time*, as regards its settlement patterns typologies and the rural and urban population distribution, distinguishing between resident and floating population especially into stationary cash-crops as vine-growing during the *vindemia* period. Behind these fluctuations in population over a long-term period, many historical and economic events may be hidden: wars, migrations, starvation, diseases and epidemics as well. Indeed, each historical or economic phenomenon leaves its mark on the shape of human society. So, demographic approaches can be variable regarding its nature and the type of data.

Demographic and palaeodemographic studies of ancient agrarian societies have been increased during the last years, trying to recognize the structure and the internal organization of population's evolution and to identify possible settlement patterns related to the resources management, urban development and exploitation of the territory. Nevertheless, some arrangements have to be made in order to apply historical demography methods to some periods of Graeco-Roman civilization. It supposes that demographic studies for these chronologies may be focussed in other geo-economics and statistical parameters as the *carrying capacity* of the environment and the maximum level of human populations supported.[3] For antiquity, historians rely mainly upon Graeco-Roman writing sources and try to obtain the maximum information from casual indications as distribution of food and money, the movements of individuals on the ground or by sea and so on. At a regional scale, some archaeological studies attempt to build theoretical population growth models, sometimes inferred by ethnographical approaches from past traditional societies or still extant. Population figures seem often to be the result of a *global estimate* and linked to the rural or urban framework variations in the inhabited area. In this sense, it is indeed common to use the density population rates for macro and meso-spatial calculations. Archaeologists have also keen to estimate food production from the extent of cultivation, with markers as likely crop-yield, daily food requirements and so on, to calculate the minimum territorial needs of self-sufficiency. For micro-spatial calculations, there are more or less sophisticated formulas to relate minimum dwelling surface to a minimum number of occupants. So, it is common to attribute to an archaeological site an overall population figure linked to its surface area, sometimes with regard to settlements of similar size where the number of inhabitants is already known by applying dwelling density formulas or archaeo-ethnographical spatial analyses. Whatever approach used, it seems that methodology based on rural settlements calculation capacity, both for population supported and for crop yields, appears to be the most promising for providing a reliable and continuous image of the past's population throughout the whole chronological period studied. Therefore, it is essential to obtain a continuous and broader demographic panorama of these societies, spatially and chronological. Such a global population's view would seem to be more easily achieved quantitatively rather than qualitatively as regards population's structure (Isoardi 2012, 39-42). This kind of archaeo-demographic studies allow us to revise basic concepts such as the relationship between the urban and rural world, as regards the urbanism development related to the implementation of an intensive agrarian production system and the different administrative status, sizes and ranges of ancient Roman cities:

Quantitative and qualitative analyses of territories have been favoured by a better knowledge of the urban perimeters and the settlement patterns of rural distribution; either from the contribution of urban archaeology, field-surveys, or cadastral studies.

Therefore, the most important interrelations that can be to take into account to make our analysis is the total number of population, its distribution -urban or rural- and its internal

Table 4.1: Sizes and ranges of ancient Roman cities according to Carreras (2014) from Morley (1996).

Range	Name ex.	Extension	Inhabitants
1st	Rome	+200 ha	1-1.5 m
2nd	Ostia	+100 ha	30.000
3rd	Mediolanum	133-83 ha	25-5.000
4th	25 cities	83-16 ha	25-5.000
5th	400 cities	16-3 ha	5-1.000

---
[3] Specially and due the lack of information, in the analysis of demographic issues during prehistoric and protohistoric periods.

configuration trends, such as -gender, age, social status and so on- for every chronological period in the area object of study. This serves to calculate, on one hand, the needs of foodstuff in terms of maintenance and self-consumption and, on the other hand, the labour availability necessary to make the different activities that allow the wine intensive system of production, distribution and trade works.

Likewise, some scholars attempted to convert the results obtained from archaeological excavations and field-surveys into demographic data. Thus, some analyses achieve to develop ranges of estimated inhabitants by settlement typologies and their sizes (Table 4.1), despite there are important methodological problems. For instance, such samples and the data obtained are partials, due to the fact that not the entire territory has been excavated or prospected and not all domestic or habitational spaces have been preserved (Carreras 1996, 59-82; Carreras 2014, 53-82).

Others studies try to quantify the food supplies necessary for cover the basic diet of the resident population in urban and rural settlements as regards the main crops and other derivate products -wheat, vegetables, wine and olive oil, as well as animal husbandry; to transform them in estimate units of land necessaries for produce it. Roman agronomists such as Cato, Varro and Columella also inform us about some environmental aspects, labour and facilities to consider for manage an agricultural holding. Issues, all of them, that can help us about the calculation of this minimal unit of land necessary for its self-sufficient maintenance (Cato *Agr.* 1-11; Varro *R.R* 1.4-1.11; Col. *R.R.* 3.3.8-9):

1 worker = 7 *iugera* (*iug*) vineyard

1 worker = 51 *modii* (*m*) wheat / 1 year

1 *iug.* x 4 *m* seed x 3 *m* wheat = 12 *m* / 1 *iug.*

51: 12 = 4.25 *iug.*

6.138 *iug.* for wheat and reposition of seed

6.138 x 3 (triennial rotation system) = 18.414 *iug.*

18.414 + 7 = 25.414 *iug.* / 1 worker ([4])

The internal quote of wine consumption has been also an important parameter to take into account, so that the balance between intra-regional and extra-regional consumption can determine the performance of the intensive productive wine's economy in our study area and the possibility to obtain surpluses for trade and benefits for increasing social position of some intervenient agents

Table 4.2: Typology of rural Roman settlements and estimation of inhabitants according to Carreras (2014) from Perkins (1999).

Types	Extension	Population
Big vicus	800 m²	80 p
Big villa	500 m²	50 p
Little vicus	400 m²	40 p
Little villa	300 m²	30 p
Big farm	100 m²	10 p
Little farm	60 m²	5 p

such us: *vilici, conductores, mercatores, negotiatores, argentari, naviculari, institores,* etc.

Endogenous and exogenous factors as economic success and wealth increment can stimulate population's rates increases, otherwise poor harvests, wars, diseases, plagues, etc.; can provoke social conflicts and economic crisis that supposes population's rates decreases.

Economic models and demographic studies: a necessary joint perspective

The study of the wine's economy can be approached through diverse perspectives that include, in one form or another, the demographic variable. Most of the recent attempts on quantification in the Roman economy were made in a wide-scale of Roman Empire and have adopted a *top-down* rather than a *bottom-up* approach (Bowman and Wilson 2013, 8; Scheidel and Friesen 2009). They often pay too little attention to territorial or chronological variation in terms of change in the forms of production, to its diachronic evolution over the time and neither make no attempt to identify and aggregate individual, local or regional production, as well as consumption and income distribution. On one hand, they do not take into account the minimum level of subsistence and, on the other hand, they do not calculate the maximum surplus capacity in an intensive and specialized winegrowing production situation as the case of study we have.

Here we try to show from a necessary joint perspective, different economic models and demographic analyses systems that could be adopted for explaining this evolution in our different scenarios of the Laetanian region.[5]

The Roman villa system

The Roman villa was defined as an autarkical agrarian production system where its main aim was to be self-sufficient in a context of a closed economy, meaning from a theoretical point of view, that no imports are brought in and no exports are sent out. Therefore, the main goal is providing consumers with everything they need from

---

[4] Martín-Arroyo (2016) p. 105-124. The *iugerum* or *iugera* is an ancient Roman unit area for land measures. It corresponds to 71 meters long and 35.5 wide, equivalent to 0.623 acre or 0.2518 hectares. The *modius* is an ancient Roman unit of Dry measures, equivalent to 8.73 litres.

[5] See Martin i Oliveras, Martin-Arroyo, Revilla 2016, 210, and the territorial scope section.

their agro-ecosystem borders.[6] It supposes that every agrarian economic unit should produce the required supplies for maintaining its own inhabitants and the population of a determinate territorial scope. Most villas were food-production centres made up from cultivated fields, meadows and forests. Watermills, cowsheds, grain dryers, wine cellars and kilns were other typical farm facilities. Artisan activities as pottery production and blacksmith works related to farming processes have been also documented. Villas produced wool, leather and tallow in addition to food. Hunting, fowling and fishing were also common activities and sources of food protein as well.

Slave-based villas existed in large number in Italy and in other provinces, especially after the wars of conquest, but free peasants and tenant farmers working for villas were the common workforce (Table 4.3). The Roman institution of slavery in the empire also provided other options and incentives. Many slaves were rewarded for their good services and there were also opportunities to earn money and buy their own freedom. A promising young slave might attend lessons of specialized studies. Thus, the *domini* could bring up secretaries, accountants, administrators, and tutors for their proper use or for renting them out. By the end of the second century AD, up of 80 % of the Roman population was composed of old citizens, freedmen (emancipated slaves) or by their descendants and at the end of the Empire, most slaves worked in domestic service rather than as labourers on the agrarian properties (Dyson 2003, 43; Johnston 2004, 7-18; Bowman and Wilson 2013, 20-22).

The landscape of north-west *Hispania Citerior* shows a diversity of situations. More fields were used for pasture than for crops because of the need for cattle, sheep and forage. Local people also managed the forest intensely for wood and wild products. Villages and hamlets were denser in the countryside during the Iberian period (V-III centuries BC) and tribal areas probably were divided into borough parcels. Each one usually had a settlement at its centre, and sometimes they were located closer to a road, stream or waterway. The Romans enhanced this agrarian system without dramatically altering it. An appropriate water source was the primary site-location factor for a villa. Cisterns for collecting rainwater and deep wells were often built and dug to ensure enough drinking and cleaning water for the family, labour and livestock. Clean water was also essential for watermills and eventually baths, the Roman indicator of a fully civilized life. The villas were also related to a broader Roman economy through a system of primary, secondary and tertiary roads. These were sometimes built or maintained by villa owners or tenants, especially if crossed their lands.

---

[6] An agroecosystem is the basic unit of study in agro-ecology, and is somewhat arbitrarily defined as a spatially and functionally coherent unit of agricultural activity, and includes the living and non-living components involved in that unit as well as their interactions (Gliessman, Engles & Krieger 1998, 3-14).

Table 4.3 Hypothetical distribution of free & slave population of the Roman Empire (in millions & %) from Scheidel (2007), p. 6. Italy: Scheidel 2004b (free population), 2005a (slave population). Egypt: Scheidel 2001: 246-7 (total population); Alexandria (guess; cf. Scheidel 2004a): 350,000 free + 150,000 slaves; other cities (above): 890,000 free + 110,000 slaves; villages (above): 4,220,000 free + 280,000 slaves. Other provinces: Scheidel 2007 (total imperial population, provincial breakdown, and urbanisation rates); low estimate: slaves are 10% of urban population (~ Egypt) and 6% of rural population (~ Egypt); high estimate: slaves are 20% of urban population (~ Italy/Egypt mean) and 12% of rural population (~ Italy/Egypt mean).

Provinces	Urban			
	Free	%	Slave	%
Italy	1.3m	68.42	0.6m	31.58
Egypt	1.25m	87.50	0.25m	12.50
Others	4-5m	90.91-83.33	0.4-1m	9.09-16.67
Estimate	6.5-7.5m	83.33-79.78	1.3-1.9m	16.67-20.22
Aggregate	7.8-9.4m (12.93-13.84%)			
Provinces	Rural			
	Free	%	Slave	%
Italy	3.5m	85.36	0.6m	14.64
Egypt	4.2m	93.33	0.3m	6.67
Others	42-45m	94.39-90.09	2.5-5.5m	5.61-9.91
Estimate	49-52m	93.33-88.89	3.5-6.5m	6.67-11.11
Aggregate	52.5-58.5m (87.07-86.16%)			
Total	60.3-67.9m (100%)			

The establishment of viticulture in the Laetanian region is related to the thorough transformations brought about by the Roman conquest. Especially interesting in this respect is the existence of early wine production in the territory close to the indigenous *oppida* of the central Catalan coast, which survived until the mid-1st century BC. This phenomenon is already confirmed in the final third of the 2nd century BC, in connection with a global transformation of the settlement patterns and the production structures probably associated with an incipient italic population migration (Miret, Sanmartí & Santacana 1991, 47-53; Revilla 2004b, 175-202; Revilla 2010b, 139-159).

Roman viticulture spread rapidly close to the change of era and throughout the first half of the 1st century AD, covering new territories or exploiting more intensively those spaces that were already occupied. The first evidence is placing in the north coastal area situated between the *Baetulo flumen* (Besós River) and the *Arnum flumen* (Tordera River), where the villa system had been strongly established since the Augustan period, organized around two *municipia*, *Baetulo* and *Iluro*, and the small *oppidum* of *Blanda* or *Blandae*. These rivers and other minor streams connected the coastal settlements with the inland territory, ensuring access to other agricultural spaces and their resources. It also affected the plain area situated between the *Rubricatum* (Llobregat River) and *Baetulo* (Besós River), where the *deductio* of the colony of *Barcino* supposes an important territorial reorganization. During the 1st century

AD, increasing economic interests of important *gentes* from the colony would consolidate their presence in this whole area. This explains the socioeconomic development of *Barcino* over the 1st and 2nd centuries AD (Rodà *et al.* 2005, 47-57; Olesti 2006, 175-200; 2009, 141-158; Olesti and Carreras 2012, 309-333; 2013, 147-189). This specific distribution responds to different ways of exploiting the territory, characterized by a particular architecture based on the Roman *villa* concept, defined by spatial planning, differentiating between, one sector called *pars urbana* with all the services necessaries for domestic life set aside to the private spaces for the owner, and another sector used for agricultural production, either for processing, *pars rustica*, or for storing, *pars fructuaria*. In general, the technology for wine-making, including several presses along with one or more tanks for collecting the must, is found in buildings close to the residential sector. But they could also be a little further away. Some *villae* had the agricultural and artisan sectors set apart from the residential area. In some places, however, there is evidence of the simultaneous storage of a cereal production, either for personal use or for trading (Revilla 2011-2012, 87). Some of these establishments had also a pottery workshop where *amphorae* and other ceramic products were made. So far over ninety pottery workshops have been identified in the north-east of *Citerior*'s province and almost fifty of them are located in the Laetanian region.[7] These features suggest that these were places given over to intensive and specialized work processes forming part of a production structure organized from elsewhere, possibly a nearby villa. Indeed, some buildings were occupied only seasonally, during certain phases of the agricultural cycle (Burch et al. 2005; Revilla 2010a, 36-37).

The first Roman villas documented in Laetanian region were settlements with former occupation that reused stone, water sources, raw materials and transportation networks, and functioned as colonial centres. Those new agrarian centres replaced previous Iberian rural economic system in an already intensely farmed landscape. These refinements took the form of large investments and technological improvements enhancing the economic structure which included the transport of raw materials and the transformed goods to distant and large markets. The primary early-Roman modifications were technical and technological improvements, which often meant an intensification of agrarian production and its orientation towards a proto-market economy. One of the major technological innovations was a bigger plough, which could break up the heavier soil. This new plough cut deeper into the soil, and the peasant could regulate its depth. It was usually pulled by two to eight oxen. Other techno-functional innovations on cultivation, tools and machinery for transforming processes were applied in different intensive cash-crops like wheat, wine-growing and olive oil production activities. The results of these innovations were longer fields suitable for large estates, great processing facilities, population growth and wealth increase with the large-scale surplus of food produced and trade.

The Boserup model of population growth and agricultural intensification

Archaeological conceptions of agricultural productive intensification during the Roman period related with the abandonment of subsistence model, the rise of an intensive agrarian production, the spread of large-scale trade, population growth and the development of social complexity in an open economy context, are still a not resolved debate and there are different explanations for it (Morrison 1994, 111).

Boserup's model defends that intensification of production refers to an increase in the productive output per unit of land or labour (Boserup 1965, 43-44). This increase may be achieved in a different number of ways. The variables that held constants are: *land*, in reference to capacity of food production on a given area by agriculture, hunting, fishing or gathering; *labour*, about workforce needed for increasing efficiency in yields, craft production and so on; and *capital*, necessary for investments and maintenance of transforming facilities and technology (tools and machinery). Alternate temporal parameters as compulsory fallow for grain crops or growing and mature periods for viticulture, are also important factors to take into account in the economic strategies adopted for obtaining an optimal amount of productive diversity or specialized intensive production. Boserup's model also defines population pressure as a pushing force driving the intensification of agricultural production and, at the same time, as a resilience factor for balancing the change inland, in order to guarantee its sustainability. On the contrary, the Malthusian proposal defends an agroecosystem collapse (Malthus 1872; Rubin 1972, 36). Thus, while Malthus saw land and particularly the availability of arable soils as limiting factors in terms of frequency of cropping for the increasing of production that eventually would be outstripped by a decontrolled growing population; Boserup turned the duple production-population pair around the land and labour use along an adaptable extensive-intensive continuum. Other variables as the technical or technological advantage in cultivation systems, the diversity of crops as productive strategies, the importance of mobility and the access to resources, can contribute to sustaining a growing density of population by the intensification of production (Morrison 1994, 116).

The notion of *population pressure* is also associated with agroecosystem *carrying capacity*. The calculation of carrying capacity requires the specification of two main concepts related to environmental potential and agro-economic pattern (Dewar 1984, 601-602). Then, the agricultural productive intensity is not only a simple consequence of human-land ratios. Decisions taken by producers to intensify or extensify crops to face contingent conditions or negative factors such as bad harvests by adverse climatology, lack availability of labour by wars,

---

[7] For *figlinae* see Tremoleda 2007-2008, 116, fig.2; for *torcularia* see Revilla 2011-2012, 88, fig.2.

plagues or diseases, increase of land rents by increments of tax and prices, fluctuations of internal or external demand, increases or decreases on costs or sales prices and so on, respond rather economic strategies than changes in population size. It does not mean that demographic factors are not important, but they may be affected by other proximate factors and constitute only one aspect of human productive organization. Human populations possess not only size but also structure, so that population size and growth rates will be determined by age-specific fertility and mortality rates. In age-structured populations, the distributions of various age groups and the nature of the domestic cycle impinge directly on the organization of labour. Other factors as the social condition either as a free citizen, freedman or slave, the administrative situation and relationships between owner and tenant, master and servant, patron and client, will be determinant for an overall analysis of the population dynamics and the social relationships. Thus, from a socioeconomic point of view, these dynamics and relationships must be considered an aspect of the organization of labour and consumption. Often intensification or specialization practices were applied to only a part of the total productive system, specifically to the part obtaining high revenue, ideological or ritual significance (e.g. wine as prestige or consumption good), or were used in large-scale social or political loans as the *annona* system for supply grain, meat and olive oil to the Roman army and the inhabitants of Rome.[8]

The agency-based wine-growing specialisation production model (Figure 4.2)

The agency-oriented model adopted for explaining the spread of vineyard cultivation and wine-growing specialization system in the Laetanian region during the Roman period (1st century BC to 3rd century AD), is an applied version of the Heckscher-Olhin-Samuelson economic model that combines the theoretical insights of the Adam Smith´s principle of absolute advantage that refers to the ability of a party or territory to produce a greater or most efficient quantity of a good, product, or service than competitors, using the same amount of resources; and the 'Ricardian' principle of comparative advantage that considered what goods and services a party or territory should produce, and suggested what they should specialise by allocating their scarce resources to produce goods and services for which they have a comparative cost advantage.[9] Thus, there are two types of cost advantage, absolute and comparative. Absolute advantage means being more productive or cost-efficient than another territory whereas comparative advantage relates to how much productive or cost-efficient one is than another. All of this in a context of a 'Boserupian' pushing force of increasing population densities that triggered more intensive land use and a 'Smithian' market pulling force exerted by an overseas demand for wine which induced to the reallocation of land and labour towards vine-growing specialization and more capital investment in larger wine-making production centres. According to it, vine-growing specialization would have developed in those territories with more favourable factor endowment to meet the increasing demand for wine. Thus, vines tended to spread where land and labour endowments were more suitable. However, there is a set of driving forces to be considered. Some of them can be regarded as naturally given, such as agro-climatic conditions. Others depended on human agency, like migrations that changed population densities or the readiness of many landless peasants to invest their labour force in own tenancy. This could be an important factor for explaining the evolution in the productive forms and in the land use changes.

This agency-oriented model of land-use change combines what environmental historians and geographers call first-nature and second nature variables. While agro-climatic endowments are first-nature factors, the dynamic interaction between the other variables becomes a set of second-nature drivers including time-distances to the nearest seaport (network analysis), that is take as a proxy for the market-pulling force (Badia-Miró and Tello 2014, 203-226; De Soto 2010). What eventually really matters is the combination between them. Except for the demand, all variables had to move along a specific range of values, higher than a minimum but not exceeding an upper level, so as to fit with the rest in a suitable economic factor endowment. Local suitability to vine-growing was the outcome of a myriad of decisions taken by a lot of people interacting in a given set of challenges and opportunities which in turn they transformed. This agency-driven impulse set in motion self-reinforcing processes of vineyard planting works, and some technical advantages could be established within them. Planting vines in bush/head training system using the *alveus* /*goblet* technique without trellis infrastructure combined with the spur pruning system supposes and important technical advantage witch favours both grape's productivity and harvesting efficiency.[10] In this sense, Columella (*De Arboribus* IV, 1-2), indicates that these efficient planting

---

[8] Remesal 1990, 355-367; 2004, 163-182; Pons Pujol 2004, 1663. Wine was not included in the *annona* until the reign of Aurelian (270-275 AD), Conison 2012, 18.

[9] The Heckscher-Olhin-Samuelson economic model features that the best effective combination of relative factor endowments such as a favourable agro-ecological conditions, suitable factors of production (land, labour and capital), high technical expertise in the production processes, availability for applying technological advances and a good transportation network, can determine that a comparative advantage arise. Territories have comparatives advantages in those goods for which the required factors of production are relatively abundant or suitable. This is because the profitability of goods is determined by input costs. Goods that require conditions and inputs that are locally abundant will be cheaper to produce than those goods that require inputs that are locally scarce. Technical expertise and technological advances can contribute also that these comparative advantage improves being even determinant (Heckscher 1919, 497-512; Ohlin 1967; Samuelson 1948, 163-184; 1953-1954, 1-20). For absolute advantage see: Smith 1776. For comparative advantage see: Ricardo 1817.

[10] The head-trained and spur-pruned training system was one the most ancient vines driving techniques employed. Traditionally it was used in the coastal vineyards as well as in the foothills and interior valleys because it was inexpensive and very easy to manage due it not needs trellis infrastructure. An additional advantage to this system was that it could be cross-cultivated for weed control, a very important water

and pruning systems was already used by Carthaginians and was adopted in *Hispania* by native peasants and italic colonists.

André Tchernia hypothetically proposes that during the Roman period the Laetanian vineyards planted and pruned with these techniques was maintained at a height of between one Roman foot and a half (44.4 cm) and three Roman feet (88.8 cm). This fact facilitates the recollection and increased the labour productivity during the harvesting, improving its performance and reducing the production costs greatly. According to the same author, this fact, combined with strong investments in processing and storage facilities and in applying technology (tools and machinery), would explain in large part the great production capacity and the strong competitiveness of the Laetanian wines (Tchernia 1986, 127; Martin i Oliveras 2015b, 73).

This vine-growing specialization stops when one or several key variables exceed certain threshold values. For example, reaching population densities higher than the ones capable to be sustained by a still mainly agrarian economy, or exhausting the marginal lands available for planting vineyards. The result is a dilemma. Or local economies have to start a structural change for returning to an autarkical self-sufficient system in resilience, what suppose that demographic surpluses have to emigrate towards other places; or either they have to expand towards new economic activities. It should be noted that this agency-oriented model was previously applied by economic historians in a cross-sectional analysis in the same territory and in a similar context of wine-growing specialization in the mid-nineteenth century, basing on the theoretical comparative advantages, factor endowment and the impact of trade openness to international markets (Tello et al. 2008, 1-42; Tello and Badía-Miró 2011, 1-30; Badía-Miró and Tello 2013, 1-31; 2014, 203-226).

The taxes-and-trade model

Regional, inter-regional and extra-regional trade was a common feature of the Roman world. A mixture within state control and a free-market approach ensured that consumer goods produced in one territory could be exported far and wide. Foodstuffs as cereals and agrarian processed goods as wine and olive oil were exported in huge quantities. The Roman Empire included regions which were completely different from one another. Notwithstanding, all of these regions were linked through a network of trade routes (nautical and terrestrial), that supposes an important mobility of goods and people, but integrated into a larger common tax structure. All the inhabitants of these regions pay taxes directly or indirectly to the Roman state. Taxes could be raised in money, earned through regional, inter-regional or extra-regional trade, or in kind, through agrarian production surpluses obtained by the exploitation of the land. Public or private investments of capital are also an important factor to take into account. If the state does not spend so much money on infrastructure, the region has to finance this deficit through the revenues obtained by the internal and the external trade of goods produced and sold.

Hopkins' *taxes-and-trade model* is a good point of departure for further research on production and distribution economic dynamics. Insofar as it takes into account the fact that the ancient economy was in constant flux, it could be useful for analysing long-term changes and enough versatile to accommodate local circumstances. Substantial growth in the scale of an economy is impossible without quantitative and qualitative changes in means of production. Economic success often is matched by a growth in population. Hopkins thinks that mainly of regional trade were raw materials and foodstuffs in buck and interregional and international trade were processed food products, consumer and luxury goods. He also calculates, from a macroeconomic point of view, that the Roman state needs and amount of capital about 825 millions of sesterces every year in the first century AD. This money was raised mainly in the form of direct and indirect taxes and that these represented 10 % of the value of all economic activities in the Early Roman Empire and they do not seem to have increased over the time (Duncan-Jones 1994, 57-59; Hopkins 1980, 101-125; Kehoe 2013, 37).

The Roman villa system understood as agricultural economic phenomenon with tendency to intensification and specialization in the production and commercialization of larger cash-crop surpluses, reflects a profit mentality that exploited the accessibility to a larger distributive markets resulting by the lowering of production and transportation costs combined with the development of Empire-wide political and economic institutions where taxation plays a central role. This taxation affects as regards the vine-growing and wine-making production and trade, all the stages of the productive function as land tenure, yields, sales, revenues and so on. The internal relationships and the tensions generated by the competition for the control of the wealth and the resources by the state, civic community and private individuals will be a key issue to understand its evolution and the changes over the time. The fact is that there was extensive Roman commerce and trade accompanied by a highly sophisticated law body to regulate it.

The provinces were to carry on the heavyweight of administering the Empire. Roman state imposed a considerable number of indirect taxes, in particular customs and some taxes on trade –*portoria*-. There were also crop taxes as 1/10 -*decima vectigalis*- (Cic. *Verr.* IV-103), for grain and 1/5 -*quinta vectigalis*- for wine and olive oil; sales tax as 1/100 -*Centesima rerum venalium*-; property tax; emergency tax; and so on, but the burden of taxation was distributed unevenly across the economy. Other issues related to land tenancy levied the crop yields

---

conservation tool in non-irrigated vineyards (Martin i Oliveras 2015b, 71-74, fig. 16-17).

and the incomes as leases and the share-cropping practices where the yield is proportionally split between landowner and tenant. These rents could oscillate around 50 % of the yields, depending on the region and the product produced (Van Minnen 1998, 205-220; Cicero "*In Verrem*", IV.103: "*agros populo Romano ex parte décima... vectigalis fuisse*").

The territorial scope

The Laetanian region is an ill-defined area in historical terms organized around some urban centres as, *Blanda* or *Blandae*, the *municipia* of *Iluro* and *Baetulo* and the *colonia* of *Barcino* on the coast and, among others, the *municipium* of *Egara*, the secondary settlement of *Arrago* and the thermal station of *Aquae Calidae* inland. The extension and limits of these cities' territories have not been precisely defined except for the *ager Barcinonensis*, the constitution and legal status of which must have had an effect on the urban centres that were there before (Palet 1997; Palet et al. 2009, 106-123; Palet et al. 2011, 113-129; Palet et al. 2012, 341-352) (Figure 4.3).

Laetanian territory also comprised the extensive plain situated between *Baetulo* River (Besós) and the mouth of *Rubricatum* River (Llobregat), located on the southwest side of Montjuïc promontory. The first foothills of the Garraf Massif would have risen from this point. Away from the coast, the colony's *ager* would have included the lower course of the Llobregat River as far as *Ad fines* (Martorell) and the lower course of Besós River to where it joined Ripoll River and the Congost-Mogent basin, spreading across the great Vallès plain as far as the Catalan Pre-Coastal Range.

Attending the special features of the *Laeetana regio*, as regards its particular geomorphological configuration, agro-economic characteristics and demographical evolution over time, we distinguish four specific areas of study:

- Study Area 1: Barcelona Hinterland Plain-*Ager Barcinonensis* (Figure 4.4a; 4.4b).
- Study Area 2: Central Coast-*Territoria* of *Baetulo*, *Iluro* and *Blandae*.
- Study Area 3: Lower Llobregat-*Rubricatum* estuary.
- Study Area 4: Vallesian Plain-*Territoria* of *Arraona*, *Egara* and *Aquae Calidae*.

In the interest of the efficiency and therefore not duplicate unnecessary work, we give in this chapter an approach to these concepts focusing in a sample area: Laetanian central coast between Besós and Tordera Rivers, as a first demographic analysis application example. Further studies on each one of this study areas will be developed in future specific papers profiting all the quantitative and qualitative data provided by the administrative official inventories of archaeological sites and the scholar's studies, whether from doctoral theses or from synthesis works related to landscape evolution and wine-growing economy in this territory during the Roman period or by the historical economic studies analysing the spread of modern wine-growing specialization in Catalonia between the 17th-19th centuries (IPAC-"*Inventari del Patrimoni Arqueològic i Paleontològic de Catalunya*"-Generalitat de Catalunya; Prevosti 1981a; 1981b; Riera 1994; Olesti 1995; Palet 1997; Ruestes 2002; Flórez 2011; Oller 2015; Tello et al. 2008, 1-42; Tello and Badía-Miró 2011, 1-30; Badía-Miró and Tello 2013, 1-31; 2014, 203-226).

Viticulture, settlement patterns, technology and economic growth in the Laetanian region

All studies dedicated to the implantation and evolution of viticulture in the north-eastern area of *Hispania Citerior*, nowadays Catalonia, have assumed the existence of an important export phenomenon and an intense commercial relationship among these territories and several regions and cities of the Roman Empire (Miró 1988; Revilla 1995; 2004a; Martin i Oliveras, Martin-Arroyo and Revilla 2016). These markets are mainly located in the western provinces and Italy. All the scholars' situated the period between August and late Julio-Claudian and Flavian dynasty as the main stage of wine's trade contained in *amphorae*, with the highest peak in the first third of the first century AD.[11]

The identification of these trade routes and the presence of a considerable number of *Tarraconensis amphorae* in different places have been interpreted as an important demand and consumption of wines from these territories, Laetanian region included, proposing its evolution towards specialized viticulture oriented almost exclusively to overseas trade and consequently organized through market mechanisms. This view is also supported by other arguments such as the establishment of a very dense rural implantation in the region, which includes various types of settlements in agricultural production and the widespread presence of vine-growing facilities for the production and storage of wine and its derivates. All related with the presence of large landowners and investors from provincial elites and also from Italy and Rome itself with strong economic capacity and a diverse legal status and social position (Miró 1988; Revilla 1995; 2004a; 2015).

The progress on archaeological field research as shown itself to be an essential resource for defining the geography of vineyards, since the technological evidence relating to the production and storage of wine or the manufacture of *amphorae* containers can, in many cases, be located and dated with quite an accuracy. Archaeology's contribution has been also essential as regards increasing our knowledge of the rural habitat and of how the territory was occupied and exploited.[12]

---

[11] For a first approximation, see Tchernia 1986; (1986); for a general distribution areas and markets identification, see Miró 1988; for a forward study propose, see Carreras 2009.
[12] Bibliography on rural settlement is difficult to summarise. However, see Revilla, Gonzalez and Prevosti 2008-2011.

In the evolution of this wine's economy we have pointed out some specific phases, characterized by three main processes of *implantation, intensification* and *specialization* of production and its broader trade to diverse destinations:

The first *phase of inception* was developed with a progressive substitution of the precedent native agrarian system by the implantation of the "villa system" as self-sufficient production units, initially oriented versus crop diversification and afterwards to an incipient wine-making specialization to fulfil their own needs and those of the immediate territory including nearest urban settlements. The main goal of local wine producers at that time was to imitate Italian exported quality wines at a better price to satisfy the demands of a growing population. The *Citerior*'s wine trade contained in *amphorae* was initially distributed at local or regional markets by coastal shipping. At the end of this phase the distribution becomes inland in the province through the Ebro River, and interprovincial to the *Gallia Narbonensis* and the *Gallia Comata* through the so-called "*Isthme gaulois*".[13] The principal *amphorae* forms produced during this period were imitations from Greco-italic, Dressel 1 A, B, C and Lamboglia 2 shapes, until the developing of the first local form called *Tarraconense 1/Laietana 1*. The main external market of this new amphora type seems to be Gaul, with a smaller but significant presence in northern military border camps and settlements (Comas 1985, 15, Fig. 1; Nolla and Solías 1984-1995, 138; Nolla 1987, 218; López Mullor and Martín Menendez 2008a, 689-724).

Later during the second *phase of expansion*, *Citerior*'s wine was also exported inland of the European continent by river waterways and Atlantic seafaring route, associated with the diffusion of a new container: the Pascual 1 amphora, mostly manufactured in coastal Laetanian pottery workshops.[14] However, the spread of vineyards geared commercialization towards in overseas markets did not come about until the second half of the 1st century BC, specifically in the final third of that century. This is confirmed by the foundation chronologies of many pottery workshops and numerous villa-type settlements and other rural centres, equipped with facilities for pressing and storing wine production. This incipient viticulture can be found also in certain territories of the central Catalan coastline (Figure 4.1) (Revilla 1995, 122-125; 2008c, 198, 202).

The situation apparently changes during the third *phase of reorientation*, so that in the early Julio-Claudian period, Italy and Rome itself, become the preferential market of *Citerior*'s wine, while the Gaul and Northern provinces seem to be transformed in a secondary market in clear recession. This hypothesis is based on the existence of direct seafaring routes from the *Citerior* coast to the Italic ports, attested by many wrecks scattered throughout the whole western Mediterranean arc with cargoes of *amphorae* Dressel 2-4 (mainly 3-2), from *Tarraconensis*, often mixed with *dolia* working as tank-ships. Most of these shipwrecks are dated between 15 and 40/50 AD. Meanwhile, the presence of the Pascual 1 amphora in Gaul was reduced remarkably almost disappearing. For its part, the Dressel 2-4 *Tarraconense*, mainly manufactured since the change of era by the Catalonian workshops of the central coast and the pre-littoral neighbourhood, becomes the *Citerior*'s most common wine amphora form. This new recipient has a better weight-capacity ratio, which makes it more competitive and suitable as a container and as a nautical transport element. In addition, its manufacturing seems to show a greater standardization from the first decades of the first century AD (Corsi-Scialliano and Liou 1985, 170; Miró 1988; Lopez Mullor and Martín Menendez 2008b, 33-94). This fact could be related to the transformations in the producing forms of artisan work in areas such as the *Baetulo* and *Iluro* territories, the *Barcino* plain and the lower Llobregat River where also large pottery centres with a complex internal organization were attested. Moreover, the large number of amphora stamps documented in this area, many of them associated on the same container, could indicate the existence of coordinated work processes (Revilla 2007, 1183-1192; Berni and Revilla 2008, 95-111). These specialized places, which seem to work on a large-scale pottery production, are not directly linked to *villae* as other periods, maybe due to the concentration of a multiplicity of artisan units that operates jointly for supplying this wider demand of *amphorae* and other ceramic objects (Miró 1988; Carreras and Guitart 2009; Carreras, Lopez Mullor and Guitart 2013). The reorganization of artisanal activity could be related to an intensification of agricultural production in general and to winemaking activities in particular (Revilla 2004b, 175-202; 2015, 1-17). This fact is also associated with the evolution of the socio-economic *villa* system in this area. Intensification supposes a huge specialization towards wine's cash-crop production, the reorganization of the road network and the implantation of a new cadastral *system and landholding regime* that would allow a better exploitation of the territory and the implementation of wider winemaking factories equipped with a large number of processing facilities.[15] These changes are related at the same time to the urbanisation process of the central area of the Catalan coast, where the colony of *Barcino* is founded *ex novo*, and where the previous cities of *Baetulo* and *Iluro* become *municipia*. Thus, the urbanisation process is inseparable with the delimitation and reorganization of the territory attributed to each urban settlement. This set of

---

[13] Located on the axis of the Aude-Garona rivers, where it would be distributed inland towards Aquitania, Northern Gaul and the Atlantic coast. Later it reached points as far away as Germania and Britannia thru the axis of the Rhône-Loire-Seine-Marne-Rhine Rivers (Galliou 1984, 24-36; 1991, 99-105; Laubenheimer and Marlière 2010, 34-36; Laubenheimer 2015, 181-192).
[14] Miró 1988; Martinez 2015; for Laetanian wine trade routes see Martín i Oliveras 2015b, 196, fig.89.

[15] Such as the *ager centuriati* of *Barcino*. See Revilla 2008a, 99-123; Palet and Riera 2009, 131-140. For typology of wine processing facilities in this area, see Martin I Oliveras, Martin-Arroyo and Revilla 2016, 201, Fig. 2.

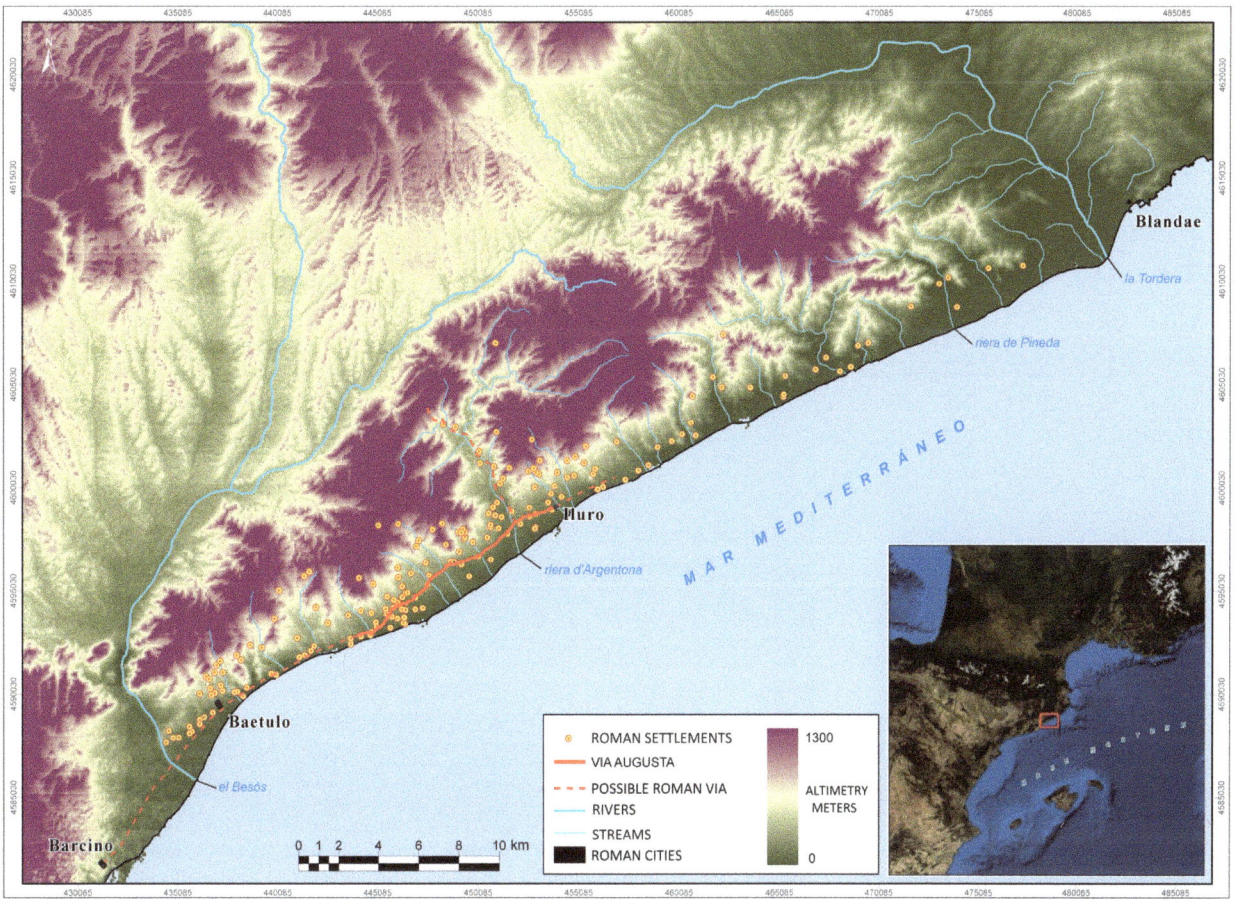

Figure 4.1. Roman settlements in the Laetanian central coast sample area between 1st century BC and 3rd century AD, according to Busquets, Moreno and Revilla (2013), p. 235, fig. 1.

factors allows us to understand the economic development of *Barcino*'s colony during the first and second centuries AD, as well as the expansion of the economic interests of its elites to nearby areas from an advanced moment of the first century AD.[16]

The fourth *phase of peak* is attested since the beginning of the 2nd century and keeps more or less continuous until the mid-third century AD. During this period the production of *amphorae* was clearly in recession, not so the production of wine as in some sites the archaeological record documented that the winemaking production and storage structures were reformed for producing much more surpluses. It implies that former little wine's cellars were progressively abandoned or reformed for being large transformation facilities. This fact could respond to two possibilities, not necessarily opposed. On one side, winemaking production and storage facilities suffered a process of concentration for being more efficient and competitive. On the other side, it seems that the demand for common wine substantially grows-up. This fact was probably connected with a significant increase of wine's exportation in containers of greater capacity such as *cupae* (barrels) and *culleii* (wineskins), and possibly as a consequence of a change in large-consumption market orientation, having to reduce costs when supplying these heavily-used and strongly competitive markets.

Finally, the fifth *phase of decline* seems to start at the end of the second century AD. The crisis could be due by different factors. On one hand, the concatenation of several external factors of negative nature such as plagues and barbarian incursions.[17] On the other hand and perhaps as a consequence of the formers, due to economic reasons that imply a loss of productive capacity and/or competitiveness front other producers with lower costs that supposes a new change in the market orientation. The viticultural centres of Laetanian region were gradually abandoned, reduced or restructured in resilience to carry out other agricultural activities and reoriented again towards self-sufficiency.

Related to wine-growing intensification and specialization process, over a hundred of agrarian establishments have so far been identified in the NE of *Citerior*'s province as having traces of pressing facilities or spaces for storing liquids, mainly wine as well. Over fourteen of them were

---

[16] To understand the expansion of the *Barcino* elites, see Carreras and Guitart 2009; Carreras *et al.* 2013; Olesti 2006, 175-200; 2009, 141-158; and also see Martin i Oliveras et al. 2007, 195-212; Rodà *et al.* 2005, 47-57; Olesti and Carreras 2012, 309-333; 2013, 147-189; 2015.

[17] As regards the Laetanian region historically we must to take into account overlapping events such as the Antonine Plague (165-180 AD); the Plague of Cyprian (250-266 & 270 AD); and the Franks incursion (Ca.259-276 AD).

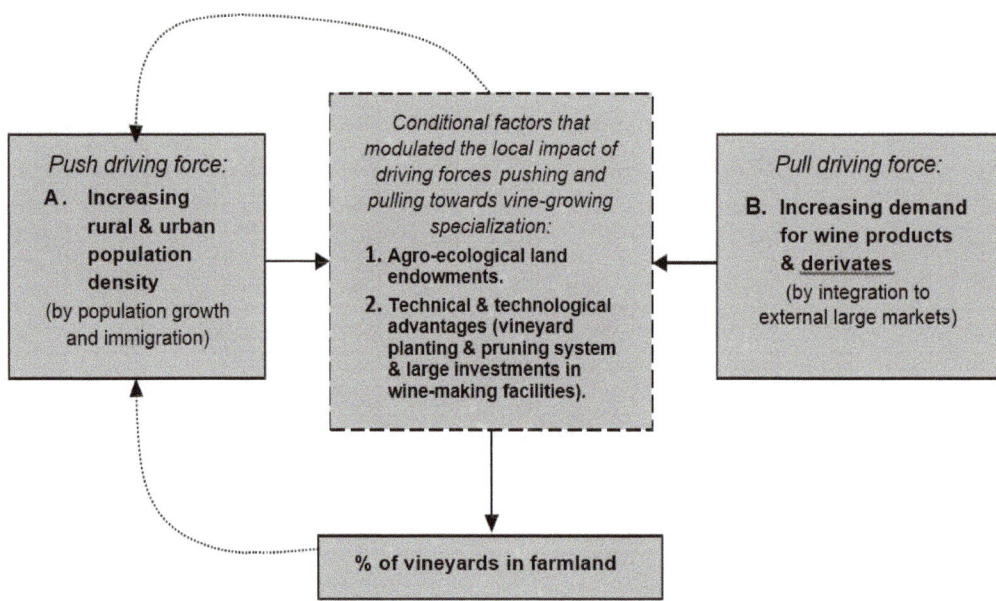

Figure 4.2. Pushing or pulling drivers and conditioning factors of Laetanian region's wine-growing specialization. Based on Badia-Miró & Tello (2013), p. 22, Fig.1.

Figure 4.3. Laetanian region with its Latin toponymical items and main roads (1st. century BC to 3rd. century AD).

concentrated in the Laetanian Region (Peña 2010). These facilities vary greatly in importance, from modestly-built facilities with a single press to large buildings with four or more presses, distinguish three different sizes[18]:

- *Large establishments 1500/2000 m2* with a complex spatial organization and a basically productive

---

[18] Settlement typology: Revilla 2010a, 35-42. In the absence of an overall study on the rural settlement in *Hispania Citerior* see: Prevosti 2005, 345-445; Revilla, Gonzalez and Prevosti (eds.) 2008-2011, 19-80. Facilities equipped with 5-6 presses must be highlighted in the Roman villa of Pont del Treball Digne-La Sagrera-Barcelona (Alcubierre and Hinojo 2015, 372-398; 2016). Facilities equipped with 4 presses must be highlighted in Veral de Vallmora, (Teià, Barcelona) (Martín i Oliveras 2009, 193-213), and in the Roman *villa* of Els Ametllers (Tossa de Mar, Girona) (Palahí and Nolla 2010). See also Sanchez *et al.* 1997; Martín i Oliveras 2009, 19-38; 2012, 59-98; 2015b; Peña 2011-2012, 37-57.

*Viticulture and demography in the Laetanian region (*Hispania Citerior Tarraconensis*), 1st c. BC – 3rd c. AD*

Figure 4.4a. General view of *Barcino* Roman colony hinterland plain in the 3rd century AD- Image from *Barcino* 3D http://ajuntament.barcelona.cat/arqueologiabarcelona/pla-barcino/barcino3d.

Figure 4.4b. *Ager Barcinonensis centuriato* proposal from Palet, Julià, Riera, Orengo *et al.* (2012), p.341-352.

function. These places would contain all infrastructure needed for making and storing wine on a certain scale. There are several *calcatoria* for treading the grapes, some pressing rooms or *torcularia* with from 4 to 6 presses, various tanks or *lacus* for collecting the must and different storing spaces set aside of between 100 and 200 *dolia* an aggregate, called *cellae vinariae*. Artisan activities as pottery, blacksmith and so on; have also been identified in most of these settlements.

- *Medium establishments 1000/1200 m2* with a wide range of buildings. Most were used for producing wine and had one or two presses, a collecting *lacus* and a *cella vinaria* of between 30 and 50 *dolia*.
- *Small establishments 400/500 m2* with a simpler spatial organization also dedicated to winemaking production with a single press and a *cella vinaria* of 5-10 *dolia*. These facilities often are integrated into small or medium size *villae* or urban *domus* probably aimed at self-consumption or for local trade.

In the mid or late 2nd or even the early 3rd century AD agricultural establishments would see the abandonment or gradual reduction of pressing facilities. In the case of the pottery workshops, some would disappear between the last third of the 1st and the beginnings of the 2nd century AD, while others would convert and diversify their production. This ensured their continuity during the 2nd and 3rd centuries AD (Revilla 1995, 104-113; Revilla 2004a, 185-188; Tremoleda 2008, 113-150).

The sample area: quantifying the population in the Laetanian central coast between the Besòs and Tordera rivers

Laetanian central coast is a territory located between those rivers in the north-east of the present province of Barcelona occupying a space of about 60 kilometres in length and about 7 kilometres in width, characterized by a special relief and peculiar natural conditions. It is oriented southwest-northeast and opens completely to the sea while a coastal mountainous range separates it from the internal Vallesian depression. The relief is structured around three basic elements, the mentioned coastal mountain range occupied by Mediterranean forest, where to the north some foothills reach the sea creating a more abrupt relief in certain places; the two main rivers Tordera (*Arnum*) and Besós (*Baetulo*) and a simple hydrological system formed by streams with torrential regime, called *"rieres"* flowing in the north-south direction connecting the strand and the interior layout of the *Via Augusta*; and a strip of coast formed by beaches and old marshes that ascends gently towards the mountain range through a succession of natural terraces suitable for agriculture. This geomorphological configuration facilitates communications and integrates spaces with different vegetation and resources. The cities that controlled and administered these lands are settled on the coast, near the main rivers or streams (Figure 4.1).

This sample area offers also special conditions of the study. In the first place, it is characterized by a dense rural setting with well differentiated situations and a very complex evolution between the end of the Republic and the 5th and 6th centuries AD. On the other hand, the configuration and particular evolution of the settlement patterns, between the late 2nd century and the end of the 1st century BC, provides an exceptional case study to analyse the general transformation process of the socioeconomic structures associated with the Roman conquest in this area of the Iberian Peninsula (Prevosti 1981a; 1981b; 1991, 135-141;

Revilla 2008a, 99-23). Finally, the existence of several cities makes it possible to appreciate Rome's specific use of urban foundations and the attribution of various legal statutes to control a large geographical area which was vital to ensure communications between the Pyrenees and *Tarraco*, as well as access to the inland.

In parallel, the available evidence allows us to explore the possible modalities and rhythms that followed the evolution of the urban phenomenon between the Late Republic and the Early Roman Empire in the north-east of *Hispania*. The central coast was initially organized around some indigenous cores, replaced, from the first third of the 1st century BC by foundations of new plant and regular urbanism such as *Baetulo*, *Iluro* and perhaps *Blanda*e.[19] These foundations, or at least some of them, should have received the status of *municipium* at the beginning of the imperial period.[20] The foundation of the *Barcino* colony, towards the change of era, had important consequences in the organization of the immediate territory, conditioning onwards the situation of the civic communities onwards.

The evolution of the Roman settlement pattern in the central coast of Catalonia has been the subject of some attention. This interest has generated a considerable bibliography, characterized by the diversity of objectives and methodological approaches. These results have been integrated, in turn, in the bibliography dedicated to the rural habitat and the economic structures of this sector of *Hispania Citerior* and the whole of Iberian Peninsula (Gorges 1979; Keay 1990, 119-150; Olesti 1997, 1-20; Járrega 2000, 271-301; Prevosti 2005, 291-480). However, the heterogeneity of efforts done by scholarship activities during the 19th and 20th centuries combined with amateur initiatives as far as the eighties and the preventive and emergency archaeological interventions and research programs executed during last four decades has generated an uneven quality of documentation obtained. Particularly noteworthy is the poor reliability of the documentation collected prior to the mid-twentieth century and the difficulties involved in its use and comparison with the data collected through the dissemination of a more rigorous archaeological methodology from the last quarter of this century. Nonetheless, this territory constitutes an exemplary case study.

Another serious problem is the radical transformation that the territory has experienced since the 1960's and 1970's, as a result of the combination of accelerated industrialization, urbanisation and infrastructure construction. The nature of the available documentation explains that the majority of interpretative proposals on the organization of Roman habitat have concentrated on the study of constructive typologies, in detriment of

---

[19] Mention apart would have the proto-urban Roman Republican settlement of *Ilturo* (Cabrera de Mar, Maresme), that would respond to another type of previous colonizing strategy (Garcia Roselló *et al.* 2000, 29-54; Olesti 2000, 55-86).

[20] For *Baetulo*: Guitart and Padrós 1991, 56; for *Iluro*: Revilla and Cela 2006, 91.

global approaches to the evolution of the territory and, in particular, on the relations between demographic aspects, such as density and distribution of rural and urban habitat and forms of exploitation (Revilla 2008b, 121; 2010a, 28).

However, the archaeological documentation currently available allows a more complex approach to the organization of this territory, integrating different types of evidence as typologies and functions of the habitat, road network, economic activities, forms of representation and status linked to the funerary world, hydraulic infrastructures as well as data on relief, soils, and so on. This set of data, largely collected in *the Inventory of the Archaeological Heritage of Catalonia* (IPAC), complemented by specialized bibliography and unpublished documentation, has been treated by GIS with the objective of defining some of the patterns of distribution of habitat and activities in rural areas (Busquets, Moreno and Revilla 2013, 239-249.). This analysis can contribute to defining the socio-economic and cultural strategies that explain, in the last instance, the particular configuration that assumes the territory of cities like *Baetulo* and *Iluro* and some aspects of its evolution. For this study, the corresponding spatial data have been selected the scope of study and the specific chronology of the IPAC database. The represented geographical entities as Roman archaeological sites have been linked to an alphanumeric database that can be exploited by performing simple or complex queries. Based on this information in the geo-referenced space, base mapping of diverse nature: Topographic maps, orthophotos, old aerial photographs and so on, has been used to generate, through a small edition, cartographic products or maps. These maps have a large capacity for displaying information and are essential for further geospatial studies.

Despite the lack of suitable documentation, it seems that the development of viticulture and activities related to the production and trade of the Laetanian wines and derivates had a significant impact on the demographic dynamics of the region.[21] However, the whole attempt to evaluate this impact has been limited by the absence of detailed demographic studies (Garcia Merino 1975; Carreras 1996, 95-122; Marzano 2011, 196-228; Carreras 2014, 53-82). There are some global proposals on the calculation of the population for all *Hispania* in those were differentiated percentages of urban and rural population. Attempts have also been made to define from the surface of an urban area and the densities of the population per hectare (Ha), the population that would have some Roman cities. These last calculations allow us to identify some categories of urban settlements, as we have seen in the section above (Table 4.2), which can be compared with its administrative function in certain cases. In the *Citerior* province, it would be the particular situation of *Tarraco* as administrative capital (Carreras 1996, 95-122; 2014, 53-82). Other cities could be included in the category of *agro-cities* as nuclei of the population with a basically agricultural economy (Lo Cascio 1994, 23-40; 1997, 3-76).

Macroscopic calculations made propose a total urban population of *Hispania* in the Early Roman Empire about 1,100,000 individuals. The amounts change between 1,000,170 and 1,250,213 individuals, depending on the value of density of inhabitants per hectare used (Carreras 2014, 56). Pliny the Elder (*NH* III.3.7-17; IV.4.18-30; IV.35.113-118), indicates that *Hispania*e had 399 urban entities between *civitates* and *populi*. Nowadays we know the approximate extension of 209 cities, which is a good sample of the total. The urban population may have represented approximately 25% of the total population of Roman *Hispaniae*. The total population of the Iberian Peninsula is estimated in 4,298,062 individuals, which can be distributed in 1,165,198 individuals of urban population from a density of 233 inhabitants per hectare and 3,132,864 individuals as the rural population.[22] To obtain this amount of rural population we ought to extrapolate the sparse data contributed by Pliny "The Elder" (III.4.28), for the three *Conventus iuridici* of the *Citerior* province: *Caesaraugustanus*, *Carthaginiensis* and *Tarraconensis*, resulting in a density of 7.12 persons/km2. In addition to the urban population, the total density of the Iberian Peninsula was 7.4 persons/km².[23] These values can be also compared with the rural population of the historical census documented from the fifteenth century (1497 to 1553), in which rural areas are recorded with an average of two "fireplaces", called "fogatges" in Catalonia, working as residence units with an average of 4-5 inhabitants for each one, therefore resulting and average of 9-10 people per km2 (Iglesias 1979).

Due to the scarcity of adequate literary and archaeological documentation for the Roman period, it is difficult to apply these calculations directly to a regional or local spatial analysis. Nevertheless, in our sample area of study, the existence of the three main urban settlements which *Pomponius Mela* (*De Chorographia*, 2.90), defines as *parva oppida civium Romanorum*: *Baetulo*, *Iluro* and *Blandae* are so well attested. Note that the last one is quite unknown archaeologically. The estimated dimensions for each of them are between 14 Ha for *Baetulo*, 7-8 Ha for *Iluro* and for the scarcely known *Blandae*, along with the 10 Ha of *Barcino* (Carreras 2014, 77-78). This value means a population for each one of them between the 3,262

---

[21] There are different "derivates" such as the *mulsum* -honey wine-, boiled musts or syrups such as the *defruntum*, the *sapa* and the *caroenum*,, subproducts like the *acetum* -vinegar-, weak wines like the *posca* -water mixed with vinegar- or plonks such as the *lora* and the *faex* for slaves comsumption. See Martín i Oliveras 2015c, 22-25.

[22] This is a conservative calculation with an average density of 233 hab./Ha. There are some archaeological empirical data to calculate it, as in the case of *Hermopolis* (Egypt), which has a population census of 37,100 inhabitants, associated to an urban extension of 120 Ha, which supposes a density of 309 hab./Ha. Other data are provided by the Colonial Italic foundations of which we have the number of settlers and their extension (Conventi 2004), with values such as *Augusta Praetoria* (288 hab./Ha), *Bononia* (240 hab./Ha), *Common* (452 hab./Ha), *Aquileia* (328 hab./Ha.), *Concordia* (300 hab./Ha) or *Augusta Taurinorum* (219 hab./Ha.).
[23] These calculations slightly correct the former value indicated in Carreras 1996, 107; see also Carreras 2014, 64.

inhabitants for *Baetulo* and 2,330 inhabitants for the rest if we accept the density of 233 persons/ha as right. The total of the urban population was thus established in 10,052 individuals that suppose an estimated rural population of 30,756 inhabitants if we accept the hypothesis that the urban population would be approximately 25% of the total. The overall population of the territory is therefore set at 40,808 people. The value of these calculations is limited, since it is not based on precise knowledge neither of the extension of some cities as *Blandae* nor of the rural habitat. A simple calculation applying the density of 7.12 persons/km2 for the whole *Citerior* province in the surface of approximately the present shires of the Maresme and Barcelonès territories (544,4 Km2), give a much smaller number of the rural population, about 3,876 rural inhabitants, that is almost 9 times less than expected.[24] This generates the paradox that the proposed hypothesis about the existence of economic intensification and specialization of winegrowing production and trade had been developing in a territory almost empty of inhabitants. Other possibilities can be there are different densities of population in different parts of the territory or there is a large number of floating population move in on during the diverse agrarian and artisan activities seasons. In this sense, it is also necessary to take into account the possibility that the small towns documented in the territory controlled a determinate agricultural area, functioning as residential agro-cities that concentrate a bigger percentage of the total population. It supposes that in the neighbourhoods of the cities, in a range of about 5 kilometres, there were few habitational sites due that the landowners and his main tenants would also reside in those towns (Lo Cascio 1997, 3-76).

Discussion

In any case, the main problem for the study of the demographic behaviour in the ancient times is the lack of suitable and contrasted information that allows us quantifying precise demographic calculations and the difficulty of establishing evolutionary dynamics from these calculations, which only provide an order of magnitude (Carreras 1996, 115). In order to improve these values, one must resort to the archaeological evidence documented in surveys and excavations, that could inform us about the dimensions of habitats and chronologies; to modern historical and ethnographical data inferred from *census* and from the study of the same activity in the same territory over the time; all of the resources with their respective methods that also presents some difficulties (Hin 2013, 298-341). But there are other problems related to the interpretation of the evidence and the use of the general calculations described above. The relationship between agricultural economics and demography has been analysed in the perspective, assumed, but not adequately demonstrated, of the capacity of the economy to increase both output and productivity and thus to sustain a parallel and direct increase in population (Blázquez 1978; Montenegro and Blázquez 1982). The apparent confirmation of this capacity would be provided by a large number of rural settlements documented in the Laetanian territory between the late Republican period and the Empire, as well as the urbanisation process. This has served to defend certain hypotheses about settlement pattern and the structure of local society in the central coast of Catalonia, apparently dominated by small and medium-sized landowners (Prevosti 1981a; 1981b; 1991, 135-141; 2005, 291-480). Implicitly, this increase in the number of settlements has also been interpreted as proof of a population increase in absolute terms.

These proposals are based on limited information, but often on incorrect theoretical and methodological assumptions.[25] One of these is the direct relationship between number of settlements and increase of population since the settlements increase can be explained initially by the spread of new forms of production supposing a hierarchy of the habitats and the functions assumed by the various types of rural settlements (Revilla 2015, 3, 8-10). Nevertheless, certain situations, involving mobility and the concentration of labour generated precisely by this new productive economy, can paradoxically cause the "invisibility" of a part of the population. Two extremes can be mentioned. On the one hand, the urban residence of rural landowners and labourers working in agrarian activities; on the other hand, and from a different perspective, the seasonal residence of the workforce in the rural settlements dedicated to the agrarian transformation, concerning the needs of the annual agricultural cycle and seasonal occupation of rural centres especially those specializing in viticulture in Laetanian region and surroundings (Burch *et al.* 2005). In this same context, one can include that part of the rural population were also engaged in artisanal activities that would work by contract as was the well-known case of Egypt (Cockle 1981, 87-97). This population could have migrated from one to another territory and could be relocated when finalized their contract. The organization of work in the large pottery workshops located in the coastal territory, also nearby of towns and the lower course of the Llobregat River may have generated forms of mobility like these (Revilla 2007, 1183-1192).

In the valuation of the various types of rural settlements, we must take into account the impossibility of quantifying the exact number of residents for each one (Carreras 2014, 65). The technology supposes concentration of labour organized strictly and hierarchically, assuming diverse and specialized functions. Part of these tasks would be met by fixed staff (probably slaves, but not only), dependent on a landowner. In this sense, the prescriptions given by

---

[24] In the modern census called "*fogatges*", the population densities for these territories, including cities, documented between 11-14 persons/km² in 1497 and between 16-20 persons/km² in 1553 for the Maresme shire.

[25] The problems in interpretation of archaeological survey's data are indicated by Carreras 1996; 2014, 65. The comparison of data values from different methodological sources could be an important tool to improve accuracy.

the agronomists and related to the tasks, facilities and the technology that had to be present in a *fundus* can be understood; another part of the labour would be hired. The productive processes may also imply a seasonal and eventual need for workforce, in particular related to certain phases of the agricultural cycle and the subsequent transformation of the product. This would imply the mobility of certain categories of the population between territories and regions and a particular concentration in the *agro-cities*, secondary settlements and villas, generating a differentiated occupation of the rural landscape. Consequently, any archaeo-demographic calculation should assess the impact of many important variables associated with the development of an agro-economic system and its specific forms of production. This impact, in the form of needs of labour and its organization, determines the spatial and temporal concentration of the population or its seasonal mobility in a given territory. Artisan and craft activities are also an important work-sharing factor to take into account.

Conclusion

The regional variability is one of the main points for understanding the changes in the rural settlement patterns, the demographic dynamics and the socioeconomic behaviours of any historic ancient period.

The specific interactions between the intra-regional and extra-regional economic networks were an important catalyst for the inception, development, and expansion of Roman viticulture phenomenon in the *regio Laeetana* between the 1st century BC and the 3rd century AD.

Archaeo-demographic analysis can also help us to calculate the amount of population, its distribution and its fluctuations over the time, trying to identify the causes of increases or decreases as regards the quantification of self-consumption needs and hand labour available.

The level of dependency of rural population from a given area to the regional markets and the urban centres and its later expansion to external markets, in our case of study western Europe, Italic peninsula and the city of Rome itself, are responding to a series of patterns and socioeconomic behaviour that could be analysed by economic and econometric models.

The wide utilization of mathematics, statistics and linear programming models in further studies will allows us to analyse, interpret and make predictions, regressions and reconstructions about the evolution of an ancient economic system, in relation with potential calculation of crop yields, the consumption level, the productive surplus susceptible of being traded in external markets and the study of several variables as the sales prices, the market reactions, the production, trade and transportation costs, the business tendency and the consequences of economic policies in the socio-political affairs.

Bibliography

Alcubierre, D., Hinojo, E., and Rigo, A. "Primers resultats de la intervenció a la vil·la romana del Pont del Treball a Barcelona." *Tribuna d'Arqueologia* 2012-2013 (2015): 372-398.

Alcubierre, D., Ardiaca, J., Artigues, P. L., and Llobet, S. "Resultats preliminars de la nova intervenció arqueològica a la vil·la romana del Pont del Treball a Barcelona.", *Tribuna d'Arqueologia* 2013-2014 (2016): 271-312.

Badia-Miró, M., and Tello, E. "An agency-oriented model to explain vine-growing specialization in the province of Barcelona (Catalonia, Spain) in the mid-nineteenth century." *AAWE-American Association of Wine Economics. Working paper 133* (2013): 1-31.

Badia-Miró, M., and Tello, E. "Vine-growing in Catalonia: the main agricultural change underlying the earliest industrialization in Mediterranean Europe (1720–1939)." *European Review of Economic History* 18 (2014): 203–226.

Beltrán de Heredia, J., and Comas, M. "Installacions vinícoles vinculades a domus: Els exemples de Barcino i Baetulo". In *El vi Tarraconense i Laietà ahir i avui. Actes del simpòsium*, Doumenta 7, edited by Prevosti, M., and A. Martín i Oliveras, 151-165. Tarragona: Institut Català d'Arqueologia Clàssica, 2009.

Berni, P., and Revilla, V. "Los sellos de las ánforas de producción tarraconense: representación y significado" In *La producció i el comerç de les àmfores de la provincia Hispania Tarraconensis. Homenatge a Ricard Pascual i Guasch (Barcelona, 17 i 18 de novembre de 2005)*, edited by A. López Mullor, and X. Aquilué, 95-111. Barcelona: Museu d'Arqueologia de Catalunya, 2008.

Blázquez, J. M. *Economía de la Hispania romana*, Bilbao: Nájera, 1978.

Boserup, E. *The Conditions of Agricultural Growth: The Economics of Agrarian Change Under Population Pressure*. London: G. Allen & Unwin, 1965.

Bowman, A., and Wilson, A. (eds.) *The Roman Agricultural Economy: Organization, Investment and Production*. Oxford Studies on the Roman Economy. Oxford: Oxford University Press, 2013.

Burch, J. *El Fundus de Turissa entre el segle I a.C. i l'I d.C.: arqueologia de dos establiments rurals: Mas Carbotí i Ses Alzines*. Girona: Aula de Prehistòria i Món Antic. Universitat de Girona, 2005.

Busquets, F., Moreno, A., and Revilla, V. "Hábitat, sistemas agrarios y organización del territorio en el litoral central de la Laietània". In *Paysages ruraux et territoires dans les cités de l'Occident romain: Gallia et Hispania*, edited by J.L. Fiches, R. Plana-Mallart and V. Revilla Calvo, 239-249. Montpellier: Presses Universitaires de la Méditerranée, 2013.

Carreras, C. "Una nueva perspectiva para el estudio demográfico de la Hispania romana." *Boletín del Seminario de estudios de Arte y Arqueología* 62 (1996): 95-122.

Carreras, C. "Del Mujal a Xanten: noves visions del comerç romà del vi de la Tarraconense", In *El vi Tarraconense i Laietà ahir i avui. Actes del simpòsium*, Documenta 7, edited by Prevosti, M., and A. Martín i Oliveras, 175-177. Tarragona: Institut Català d'Arqueologia Clàssica, 2009.

Carreras, C. "Nuevas tendencias y datos sobre la demografía romana en la Península Ibérica." *Boletín del Seminario de estudios de Arte y Arqueología* 80 (2014): 53-82.

Carreras, C. "New views on the wine imports from Hispania Tarraconensis". In *Amphorae from the Kops Plateau (Nijmegen): Trade and supply to the Lower-Rhineland from the Augustan period to AD 69/70*. Archaeopress Roman Archaeology 20, edited by C. Carreras and J. Van den Berg, 93-104. Oxford: Archaeopress, 2017.

Carreras, C., and Guitart, J. (eds.) *Barcino I. Marques i terrisseries d'àmfores al Pla de Barcelona*. Barcelona: Institut d'Estudis Catalans, 2009.

Carreras, C., López Mullor, A., and Guitart, J. (eds.) *Barcino II. Marques i terrisseries d'àmfores al Pla de Barcelona*, Corpus international des timbres amphoriques (Institut d'Estudis Catalans) 18. Tarragona: Institut Català d'Arqueologia Clàssica, 2013.

Cockle, H. "Pottery manufacture in Roman Egypt." *Journal of Roman Studies* 71 (1981): 87-97.

Comas, M. *Baetulo. Les àmfores*, Monografies Badalonines 8. Badalona: Museu de Badalona, 1985.

Conison, A. *The Organization of Rome's Wine Trade*, unpublished PhD dissertation University of Michigan. Ann Arbor, 2012.

Conventi, M. *Città romane di fondazione*, Studia archaeologica 130. Rome: L'Erma di Bretschneider, 2004.

Corsi-Sciallano, M., and Liou, B. *Les épaves de Tarraconaise à chargement d'amphores Dressel 2-4*, Archaeonautica 5. Paris: CNRS, 1985.

Cunliffe, B., (1988, French translation) *La Gaule et ses voisins. Le grand commerce dans l'Antiquité*, Antiquité synthèses 4. Paris: Picard, 1993.

De Soto, P. *Anàlisi de la xarxa de comunicacions i del transport a la Catalunya romana: estudis de distribució i mobilitat*, unpublished PhD dissertation Universitat Autònoma de Barcelona. Barcelona, 2010.

Dewar, R.E. "Environmental productivity, population regulation and carrying capacity." *American Anthropologist Journal* 86 (1984): 601-614.

Duncan-Jones, R. *The Economy of the Roman Empire*. Cambridge: Cambridge University Press, 1974.

Dyson, S. L. *The Roman Countryside*, Duckworth debates in archaeology. London: Duckworth, 2003.

Fiches, J.-L., Plana, R., and Revilla, V. (eds.) *Paysages ruraux et territoires dans les cités de l'Occident romain: Gallia et Hispania*. Montpellier: Presses Universitaires de la Méditerranée, 2013.

Flórez, M. *Dinàmica del poblament i estructuració del territori a la Laietània interior. Estudi del Vallès Oriental de l'Època Ibèrica fins a l'alta Edat Mitjana*, unpublished PhD dissertation Universitat Autònoma de Barcelona. Barcelona, 2011.

Galliou, P. "Days of wine and roses? Early Armorica and the Atlantic wine trade." In *Cross-channel trade between Gaul and Britain in the pre-Roman Iron Age*, Occasional paper. Society of antiquaries of London NS 4, edited by S. Macready, and F.H. Thompson, 24-36. London: Society of antiquaries of London, 1984.

Galliou, P. "Les amphores Pascual 1 et Dressel 2-4 de Tarraconaise découvertes dans le Nord-Ouest de la Gaule et les importations de vins espagnols au Haut Empire." *Laietania* 6 (1991) 99-105.

García Merino, C. *Población y poblamiento en Hispania romana: el Conventus Cluniensis*, Estudia romana 1,2. Valladolid: Dep. de Prehistoria y Arqueologia, Univ. de Valladolid, 1975.

García Roselló J., Martín Menéndez, A., and Cela Espín, X. "Nuevas aportaciones sobre la romanización en el territorio de Iluro (Hispania Tarraconensis)." *Empúries* 52 (2000): 29-54.

Gliessman, S.R., Engles, E., and Krieger, R. (1998, 2nd edition) *Agroecology: Ecological Processes in Sustainable Agriculture*. Boca Ratón: CRC Press, 2007.

González Cesteros, H. "Hallazgos de productos tarraconenses en la frontera germana. Un mercado secundario." In *La difusión comercial de las ánforas vinarias de Hispania Citerior-Tarraconensis (siglos I a.C. -I d.C.)*, Archaeopress Roman Archaeology 4, edited by V. Martínez, 205-220. Oxford: Archaeopress, 2015.

Gorges J.-G. *Les villas hispano-romaines. Inventaire et problématiques archeologiques*, Publications du Centre Pierre Paris 4. Paris: De Boccard, 1979.

Guitart, J., and Padrós, P "La ciutat Romana de *Baetulo* (Badalona). Història i urbanisme." *Revista Espais* 31 (1991): 50-56.

Heckscher, E. "The Effect of Foreign Trade on the Distribution of Income." *Ekonomisk Tidskrift* 21 (1919): 497–512.

Hin, S. *The demography of Roman Italy. Population Dynamics in an Ancient Conquest Society 201 BCE–14 CE*. Cambridge: Cambridge University Press, 2013.

Hopkins, K. "Taxes and Trade in the Roman Empire (200 B.C.-A.D. 400)." *Journal of Roman Studies* 70 (1980): 101-125.

Iglesias, J. *El fogatge de 1553*. Barcelona: Fundació Salvador Vives Casajuana, 1979.

IPAC. *Inventari del Patrimoni Arqueològic i Paleontològic de Catalunya*. Barcelona: Generalitat de Catalunya, Department de Cultura, 1987.

Isoardi, D. "Demographic analysis of Pre-roman populations near the Greek colony of Massalia (southern France)." In *From the Pillars of Hercules to the Footsteps of the Argonauts*, Colloquia Antiqua 4, Mnemosyne Supplementum, edited by G. R.Tsetskhladze, J.-P. Morel, and A. Hermary, 37-59. Leuven-Paris-Walpole: Peeters, 2012.

Járrega R. "El poblament rural i l'origen de les *villae* al nord-est d'*Hispania* durant l'època romana republicana (segles II-I aC)." *Quaderns de prehistòria i arqueologia de Castelló* 21 (2000) 271-301.

Jonhston, D. E. (1994, 5th ed.) *Roman Villas*. Aylesbury: Shire Archaeology, 2004.

Keay, S. J. "Processes in the Development of the Coastal Communities of Hispania Citerior in the Republican Period." In *The Early Roman Empire in the West*, edited by T. Blagg and M. Millet, 119-150. Oxford: Oxbow, 1990.

Kehoe, D. "The State and Production in the Roman Agrarian Economy." In *The Roman Agricultural Economy: Organization, Investment and Production*, Oxford Studies on the Roman Economy, edited by A. Bowman and A. Wilson, 33-53. Oxford: Oxford University Press, 2013.

Laubenheimer, F., and Marlière, E. *Échanges et vie économique dans le Nord-Ouest des Gaules (Nord/Pas-de-Calais, Picardie, Haute-Normandie). Le témoignage du IIe s. av J.-C au IVe s. ap. J.-C. Volume I*, Institut des sciences et techniques de l'antiquité (Séries). Besançon: Presses Universitaires de Franche-Comté, 2010.

Laubenheimer, F. "Les circuits d'exportation des vins de Tarraconaise en Gaule." In *La difusión comercial de las ánforas vinarias de Hispania Citerior-Tarraconensis (siglos I a.C. -I d.C.)*, Archaeopress Roman Archaeology 4, edited by V. Martínez, 181-192. Oxford: Archaeopress, 2015.

Lo Cascio, E. "The Size of the Roman Population: Beloch and the Meaning of the Republican Census Figures." *Journal of Roman Studies* 84 (1994): 23-40.

Lo Cascio, E. "Le procedure di recensus dalla tarda repubblica al tardo antico e il calcolo della popolazione di Roma." In *La Rome impériale: démographie et logistique. Actes de la Table ronde (Rome, 25 mars 1994)*, Collection de l'École française de Rome 230, edited by the École française de Rome, 3-76. Rome: École française de Rome, 1997.

López Mullor, A., and Martín Menéndez, A. "Las anforas de la Tarraconense." In *Cerámicas hipanorromanas. Un estado de la cuestión*, edited by D. Bernal, and A. Ribera, 689-724. Cádiz: Universidad de Cádiz, 2008a.

López Mullor, A., and Martín Menéndez, A. "Tipologia i Datació de les Àmfores Tarraconenses produïdes a Catalunya." In *La producció i el comerç de les àmfores de la província Hispania Tarraconensis. Homenatge a Ricard Pascual i Guasch (Barcelona, 17 i 18 de novembre de 2005)*, edited by A. López Mullor, and X. Aquilué, 33-94. Barcelona: Museu d'Arqueologia de Catalunya, 2008b.

Lowe, B. *Roman Iberia. Economy, Society and Culture*. London: Duckworth, 2009.

Malthus, T.R. (1798, 4th ed.) *An essay on the principle of population*. London: J. Johnson, 1872.

Martín i Oliveras, A. "Parc Arqueològic *Cella Vinaria* (Teià-Maresme-Barcelona) Descobrint el celler romà de Vallmora." In *El vi Tarraconense i Laietà ahir i avui. Actes del simpòsium*, Documenta 7, edited by Prevosti, M., and A. Martín i Oliveras, 193-213. Tarragona: Institut Català d'Arqueologia Clàssica, 2009.

Martin i Oliveras, A. "Arquelogia del vino en época romana: El proyecto Cella Vinaria y el complejo vitivinicola de Vallmora (Teià – Maresme - Barcelona). Nuevas aportaciones a la investigación." In *De vino et oleo Hispaniae. Areas de produccion y procesos tecnologicos del vino y el aceite en la Hispania romana. Coloquio Internacional (Murcia, 5-7 de mayo de 2010)*, AnMurcia 27-28, edited by J. M. Noguera, and J. A. Antolinos, 113-140. Murcia: Universidad de Murcia, 2011-2012.

Martín i Oliveras, A. "Anàlisi tecnofuncional d'estructures productives i vitivinícoles d'època romana. Identificació i localització a Catalunya de fosses de maniobra de premses de biga amb contrapès tipus arca lapidum." *Pyrenae* 43, vol. 2 (2012): 59-98.

Martín i Oliveras, A. "Arqueología del vino en época romana: teoría económica, lógica productiva y comercial aplicada al envasado, la expedición, el transporte y la distribución de ánforas vinarias del noreste peninsular (s. I a.C.-Id.C.)." In *La difusión comercial de las ánforas vinarias de Hispania Citerior-Tarraconensis (siglos I a.C. -I d.C.)*, Archaeopress Roman Archaeology 4, edited by V. Martínez, 19-38. Oxford: Archaeopress, 2015a.

Martín i Oliveras, A. *Arqueologia del Vi a l'Época Romana. Del Cultiu al Consum. Marc Teòric i Epistemològic*. Barcelona: Societat Catalana d'Arqueologia, 2015b.

Martín i Oliveras, A. "Arqueologia del Vi a l' Época Romana. Del Cultiu al Consum. Aspectes Ideològics i Qualitatius". *Auriga*, 79: 22-25. 2015c.

Martín i Oliveras, A., Rodà de Llanza, I., and Velasco Felipe, C. "*Cella Vinaria* de Vallmora (Teià, Barcelona). Un modelo de explotación vitivinícola intensiva en la Layetania, Hispania Citerior (s. I a.C.- s. V d.C.)." *Histria Antiqua* 15 (2007): 195-212.

Martín i Oliveras, A., Martín-Arroyo, D., and Revilla, V. "The Wine Economy in Roman Hispania. Archaeological data and modellization." In *Economía romana. Nuevas perspectivas. The Roman Economy. New perspectives*, Col·lecció Instrumenta 55, edited by J. Remesal, 189-236. Barcelona: Edicions Universitat de Barcelona, 2017.

Martín-Arroyo, D. J. "Modelización de la *ratio riparia/uinea*: El emparrado romano entre Hasta Regia y Gades." In *Lacus autem idem et stagnus, ubi inmensa aqua convenit*, Estudios Históricos sobre Humedales en la Bética (II), edited by L. G. Lagóstena, 105-124. Cádiz: Seminario Agustín de Horozco de Estudios Económicos de Historia Antigua y Medieval, 2016.

Martínez, V. (ed.) *La difusión comercial de las ánforas vinarias de Hispania Citerior-Tarraconensis (siglos I a.C. -I d.C.)*, Archeopress Roman Archaeology 4. Oxford: Archaeopress, 2015.

Marzano, A. "Rank-size analysis and the Roman cities of the Iberian Peninsula and Britain: some considerations." In *Settlement, Urbanization, and Population*, Oxford studies on the Roman economy, edited by A. Bowman, and A. Wilson, 196-228. Oxford: Oxford University Press, 2011.

Miret, M., Sanmartí, J., and Santacana, J. "From indigenous structures to the Roman world: Models for the occupation of central coastal Catalonia." In *Roman Landscapes. Archaeological survey in the Mediterranean region*, Archaeological monographs of the British school at Rome 2, edited by G. Barker, and J. Lloyd, 47-53. London: British School at Rome, 1991.

Miró, J. *La producción de ánforas romanas en Catalunya. Un estudio sobre el comercio del vino de la Tarraconense (siglos I a.C.-I d.C.)*, BAR International Series 488. Oxford: BAR Publishing, 1988

Montenegro, A., and Blázquez, J. M. *Historia de España, t. II, España romana (218 a. De J.C.-414 de J.C.), vol. I, La conquista y la explotación económica*. Madrid: Espasa-Calpe S.A., 1982.

Morley, N. *Metropolis and hinterland: the city of Rome and the Italian economy 200 B.C. – A.D.200*. Cambridge: Cambridge University Press, 1996.

Morrison, K. D. "The intensification of Production: Archaeological Approaches." *Journal of Archaeological Method and Theory* 1, vol. 2 (1994):111-159.

Nolla, J.M. "Una nova àmfora catalana. La Tarraconense 1". In *El vi a l'antiguitat. Economia, producció i comerç al Mediterrani occidental. Actes del I Col·loqui Internacional d'Arqueologia romana. (Badalona, 1985)*, Monografies Badalonines 9, edited by J. Mayné, 217-223. Badalona: Museu de Badalona, 1987.

Nolla, J.M., and Solías, J.M. "L'àmfora tarraconense 1. Característiques, procedència, àrees de producció, cronología." *Butlletí Arqueològic de Tarragona* 6-7 (1984-1985): 107-144.

Olesti, O. *El territori del Maresme en època republicana (s. III-I aC): Estudi d'arqueomorfologia i història*. Mataró: Caixa d'Estalvis Laietana, 1995.

Olesti O. "El origen de las *villae* romanas en Cataluña." *Archivo Español de Arqueología* 70 (1997):1-20.

Olesti O. "Integració i transformació de les comunitats ibèriques del Maresme durant el s. II-I aC: un model de romanització per a la Catalunya litoral i prelitoral." *Empúries* 52 (2000): 55-86.

Olesti, O. "Propiedad de la tierra y élites locales. El ejemplo del *ager Barcinonensis*." In *Histoire, Espaces et Marges de l'Antiquité: Hommages à Monique Clavel-Lévêque Vol. 4*, edited by A. Gonzalès, and M. Garrido-Hory, 175-200. Besançon: Presses universitaires de Franche-Comté, 2006.

Olesti, O. " Propietat i riquesa a l'*ager Barcinonensis*." In *Barcino I. Marques i terrisseries d'àmfores al Pla de Barcelona*, edited by C. Carreras, and J. Guitart, 141-158. Barcelona: Institut d'Estudis Catalans, 2009.

Olesti, O., and Carreras, C. "Esclavos y libertos en la producción vinícola y alfarera en el *Ager Barcinonensis*: de la marginalidad al éxito económico." In *Dipendenza ed emarginazione nel mondo antico e moderno. Atti del 33 Convegno G.I.R.E.A.*, Le vie del diritto 2, edited by F. Reduzzi Merola, and M.V. Bramante, 309-333. Rome: Aracné, 2012.

Olesti, O., and Carreras, C. "Le paysage social de la production vitivinicole dans l'ager Barcinonensis." *Dialogues d'Historie Ancienne* 39, vol. 2 (2013): 147-189.

Olesti, O., and Carreras, C. 2015: "De servus a propietario agrícola: el esclavo en el mundo de la producción anfórica en el Ager Barcinonensis — From Servus to Landowner: the slave in the Amphora Production System in the Ager Barcinonensis." In *Los espacios de la esclavitud y la Dependencia desde la Antigüedad Homenaje a Domingo Plácido. Actas del XXXV Coloquio del GIREA (2012)*, edited by A. Beltrán, I. Sastre, and M. Valdés, 561-587. Besançon: Institut des Sciences et Techniques de l'Antiquité, Presses universitaires de Franche-Comté, 2015.

Oller, J. *El territorio y poblamiento de la Layetania interior en época antigua (ss. IV a.C-I dC)*, Col·lecció Instrumenta 51. Barcelona: Universitat de Barcelona, 2015.

Ohlin, B. *Interregional and International Trade*, Harvard Economic Studies 39. Cambridge, Mass.: Harvard University Press, 1967.

Palahí, L., and Nolla, J.M. *Felix Turissa. La vil·la romana dels Ametllers i el seu fundus (Tossa de Mar, la Selva)*, Documenta 12. Tarragona: Institut Català d'Arqueologia Clàssica, 2010.

Palet, J.M. *Estudi territorial del Pla de Barcelona. Estructuració i evolució del territori entre l'època ibero-romana i l'altmedieval. Segles II/I aC-IX/X dC*, unpublished PhD dissertation Universitat de Barcelona. Barcelona, 1997.

Palet, J.M., Fiz, J.L., and Orengo, H.A. "Centuriació i estructuració de l'ager de la Colònia de Barcino: Anàlisi arqueomorfològica i modelació del paisatge." *Quaderns d'Història i Arqueologia de la Ciutat de Barcelona* 5 (2009): 106-123.

Palet, J.M., Orengo, H.A., and Riera, S. "Centuriación del territorio y modelación del paisaje en los llanos litorales de Barcino y Tarraco. Una investigación Interdisciplinar a través de la integración de datos arqueomorfológicos y paleoambientales." In *Sistemi Centuriali e Opere di Assetto Agrario tra Età Romana e primo Medioevo. Atti del convegno borgoricco (Padova)-Lugo (Ravenna)*, Agri Centuriati 6-7, edited by P.L. Dall'aglio, and G. Rosada, 113-129. Pisa – Rome: F. Serra, 2011.

Palet, J.M., Julià, R., Riera, S., Orengo, H.A., Picornell, L., and Llergo, Y. "The role of the Montjuïc promontory (Barcelona) in landscape change: Human impact during Roman times." In *Variabilités environnementales, mutations sociales: Nature, intensités, échelles et temporalités des changements*, edited by F. Bertoncello, and F. Braemer, 341-352. Antibes: Éditions APDCA, CNRS, 2012.

Palet, J. M. & Riera, S. 2009: "Modelació antrópica del paisatge i activitats agropecuàries en el territori de la colònia de Barcino: aproximació des de l'arqueomorfologia i la palinología." In *Barcino I. Marques i terrisseries d'àmfores al Pla de Barcelona*, edited by C. Carreras, and J. Guitart, 131-140. Barcelona: Institut d'Estudis Catalans, 2009.

Peña, Y. *Torcularia. La producción de vino y aceite en Hispania*, Documenta 14. Tarragona: Institut Català d'Arqueologia Clàssica, 2010.

Peña, Y. "Variantes tecnológicas hispanas en los procesos de elaboración de vino y aceite en época romana." In *De vino et oleo Hispaniae. Areas de produccion y procesos tecnologicos del vino y el aceite en la Hispania romana. Coloquio Internacional (Murcia, 5-7 de mayo de 2010)*, AnMurcia 27-28, edited by J. M. Noguera, and J. A. Antolinos, 37-57. Murcia: Universidad de Murcia, 2011-2012.

Perkins, P. *Etruscan settlement, society and material culture in central coastal Etruria*, BAR International Series 788. Oxford: BAR Publishing, 1999.

Pons Pujol, L. "La annona militaris en la Tingitana: observaciones sobre la organización y el abastecimiento del dispositivo militar romano." In *L'Africa Romana. XIV Convegno Internazionale di Studi. Ai confini dell impero : contatti, scambi, conflitti. Tozeur, 12-15 dicembre 2002*, edited by M. Khanoussi, P. Ruggeri, and C. Vismara, 1663-1680. Rome: Carocci, 2004.

Prevosti, M. *Cronologia i poblament a l'àrea rural de Baetulo*, Monografies badalonines 3. Badalona: Oikostau, 1981a.

Prevosti, M. *Cronologia i poblament a l'àrea rural d'Iluro*. Mataró: Caixa d'Estalvis Laietana, 1981b.

Prevosti, M. "The establishment of the villa system in the Maresme (Catalonia) and its development in the roman period." In *Roman Landscapes. Archaeological survey in the Mediterranean region*, Archaeological monographs of the British school at Rome 2, edited by G. Barker, and J. Lloyd, 135-141. London: British School at Rome, 1991.

Prevosti M. "L'època romana." In *Història agrària dels Països Catalans, vol. I, L'antiguitat,* edited by E. Giralt, 291-480. Barcelona: Fundació catalana per a la recerca i la innovació, Universitat de Barcelona, 2005.

Remesal, J. "El sistema annonario como base de la evolución económica del Imperio romano." In *Le commerce maritime romain en Méditerranée occidentale, colloque international tenu à Barcelone du 16 au 18 mai 1988*, PACT 27, edited by T. Hackens, and M. Miró, 355-367. Strasbourg: Conseil de l'Europe, 1990.

Remesal, J. "El abastecimiento militar durante el Alto Imperio Romano. Un modo de entender la economía antigua." *Boletim do CPA* 17 (2004): 163-182.

Revilla, V. *Producción cerámica, viticultura y propiedad rural en Hispania Tarraconensis (siglos I a.C.-III d.C.)*, Cuadernos de arqueología 8. Barcelona: L'Estaquirot, 1995.

Revilla, V. "Ánforas y epigrafía anfórica en Hispania Tarraconensis." In *Epigrafía anfórica*, Col·lecció Instrumenta 17, edited by J. Remesal, 159-196. Barcelona: Universitat de Barcelona, 2004a.

Revilla V. "El poblamiento rural en el noreste de Hispania entre los siglos II a.C. y I d.C.: organización y dinámicas culturales y socioeconómicas." In *Torres, atalayas y casas fortificadas. Explotación y control del Territorio en Hispania (S. III a. de C. – S. I d. de C.)*, edited by P. Moret, and T. Chapa, 175-202. Jaén: Universidad de Jaén, 2004b.

Revilla, V. "Onomástica en epigrafía anfórica de la Hispania Tarraconense: algunas consideraciones sobre su significado y métodos de análisis." In *Acta XII Congressus internationalis epigraphiae graecae et latinae, Barcelona, provinciae Imperii Romani inscriptionibus descriptae, Barcelona, 3-8 septembris 2002*, Monografies de la Secció historico-arqueològica (Institut d'Estudis Catalans) 10, edited by M. Mayer, G. Baratta, and A. Guzmán, 1183-1192. Barcelona: Institut d'Estudis Catalans, 2007.

Revilla V. "La villa y la organización del espacio rural en el litoral central de Cataluña: implantación y evolución de un sistema de poblamiento." In *Actes del Simposi Les vil·les romanes a la Tarraconense. Implantació, evolució i transformació. Estat actual de la investigació del món rural en època romana (Lleida, 2007), vol. I*, Monografies (Museu d'Arqueologia de Catalunya-Barcelona) 10-11, edited by V. Revilla, J.-R. González, and M. Prevosti, 99-123. Barcelona: Museu d'Arqueologia de Catalunya, 2008a.

Revilla, V. "Agrarian Systems in Roman Spain: archaeological approaches." In *New Perspectives on the Ancient World: Modern perceptions, ancient representations*, BAR International Series 1782, edited by P. P. Funari, R. S. Garrafoni, and B. Letalien, 117-129. Oxford: BAR Publishing, 2008b.

Revilla, V. "La producción anfórica en el sector meridional de Cataluña: prácticas artesanales, viticultura y representaciones culturales". In *La producció i el comerç de les àmfores de la provincia Hispania Tarraconensis. Homenatge a Ricard Pascual i Guasch (Barcelona, 17 i 18 de novembre de 2005)*, Monografies del Museu d'Arqueologia de Catalunya 8, edited by A. López Mullor, and X. Aquilué, 189-226. Barcelona: Museu d'Arqueologia de Catalunya, 2008c.

Revilla, V. "Hábitat rural y territorio en el litoral oriental de Hispania Citerior: perspectivas de análisis." In *El poblamiento rural romano en el sureste de Hispania. 15 años después*, edited by J. M. Noguera, 20-75. Murcia: Universidad de Murcia, 2010a.

Revilla V. "Rural Settlement in the central littoral area and the interior regions of Catalonia in the 1st and 2nd centuries BC, in Time of Changes. The beginning of the Romanization." *Studies on the Rural World in the Roman period* 5 (2010b) 139-159.

Revilla, V. "Viticultura, territorio y hábitat en el litoral nororiental de Hispania Citerior durante el Alto Imperio." In *De vino et oleo Hispaniae. Areas de produccion y procesos tecnologicos del vino y el aceite en la Hispania romana. Coloquio Internacional (Murcia, 5-7 de mayo de 2010)*, AnMurcia 27-28, edited by J. M. Noguera, and J. A. Antolinos, 67-83. Murcia: Universidad de Murcia, 2011-2012.

Revilla, V. "Agricultura, artesanado rural y territorio en el noreste de Hispania Citerior: estructuras y dinámicas." In *La difusión comercial de las ánforas vinarias de Hispania Citerior-Tarraconensis (siglos I a.C. -I d.C.)*, Archaeopress Roman Archaeology 4, edited by V. Martínez, 1-17. Oxford: Archaeopress, 2015.

Revilla, V. "La economía en las ciudades romanas del noreste de la Hispania Citerior." In *Coloquio Internacional In Roma nata, per Italiam fusa, in provincias manat. La ciudad romana en el Noroeste: nuevas perspectivas (Lugo, 24-25 de octubre de 2016)*, edited by M.D. Dopico, and M. Villanueva. Lugo, in press.

Revilla V., Cela X. "La transformación material e ideológica de una ciudad de Hispania: Iluro (Mataró) entre los siglos I y VII d.C." *Archivo Español de Arqueología* 79 (2006): 89-114.

Revilla, V., González, J. R., and Prevosti, M (eds) *Actes del Simposi Les vil·les romanes a la Tarraconense. Implantació, evolució i transformació. Estat actual de la investigació del món rural en època romana". (Lleida, 28-30 novembre 2007)*, Museu d'Arqueologia de Catalunya, Monografíes 10-11. Barcelona: Museu d'Arqueologia de Catalunya, 2008-2011.

Ricardo, D. (1817, 3rd ed.) *On the Principles of Political Economy and Taxation"*. Ontario: Batoche Books Kitchener, 2001.

Riera, S. *Evolució del paisatge vegetal Holocè al pla de Barcelona a partir de les dades pol·liníques*, unpublished PhD dissertation Universitat de Barcelona. Barcelona, 1994.

Rodà, I., Martín i Oliveras, A., and Velasco Felipe, C. "Personatges de Barcino i el vi laietà. Localització d'un fundus dels PedaniiClementes a Teià (El Maresme) a partir de la troballa d'un signaculum de plom amb inscripció (segle II dC)." *Quaderns d'Arqueologia i Història de la ciutat de Barcelona* 1 (2005): 47-57.

Rubin, J. "Expulsion from the garden (Boserup and Malthus)." *Peasant Studies Newsletter* 1 (1972): 35-38.

Ruestes, C. *El poblament antic a la Laietània Litoral (del Besòs a la riera de Caldes): l'aplicació de un GIS (Sistema d'Informació Geogràfica), a l'estudi de la seva evolució i les seves relacions espacials*, unpublished PhD dissertation Universitat Autònoma de Barcelona. Bellaterra, 2002.

Saller, R. P. "Framing the Debate over Growth in the Ancient Economy." In *The Ancient Economy*, Edinburgh readings on the ancient world, edited by W. Scheidel, and S. von Reden, 251-269. Edinburgh: Edinburgh University Press, 2002.

Samuelson, P.A. "International Trade and the Equalisation of Factor Prices." *The Economic Journal* 58, vol. 230 (1948): 163–184.

Samuelson, P.A. "Prices of Goods and Factors in General Equilibrium." *Review of Economic Studies* 21, vol. 1 (1953-1954): 1–20.

Sánchez, E., and Esther Gurri i C. *El jaciment romà del Morè. Sant Pol de Mar, Maresme*, Excavacions Arqueològiques a Catalunya, 13. Barcelona: Generalitat de Catalunya, Departament de Cultura, 1997.

Scheidel, W. *Death on the Nile: disease and the demography of Roman Egypt*, Mnemosyne, bibliotheca classica Batava. Supplementum 228. Leiden: Brill, 2001.

Scheidel, W. "Creating a metropolis: a comparative demographic perspective." In *Ancient Alexandria between Egypt and Greece*, Columbia studies in the classical tradition 26, edited by W. V. Harris, and G. Ruffini, 1-31. Leiden - Boston: Brill, 2004a.

Scheidel, W. "Human mobility in Roman Italy, I: The free population." *Journal of Roman Studies* 94 (2004b): 1-26.

Scheidel, W. "Human mobility in Roman Italy, II: The slave population." *Journal of Roman Studies* 95 (2005): 64-80.

Scheidel, W. "The Roman slave supply." *Princeton/Stanford Working papers in Classics* (2007a): 1-22.

Scheidel, W. 2007b: "Demography". In The Cambridge Economic History of the Greco-Roman World, edited by W. Scheidel, I. Morris, and R. P. Saller, 38-86. Cambridge: Cambridge University Press, 2007.

Scheidel, W., and Friesen, S.J. "The size of the economy and the distribution of income in the Roman Empire." *Princeton/Stanford Working papers in Classics* (2009): 1-34.

Smith, A., (1776, digital ed.) *An Inquiry into the Nature and Causes of the Wealth of Nations*, Metalibri The Digital Edition. Sao Paulo, 2007. (URL: http://metalibri.incubadora.fapesp.br)

Tchernia, A. *Le vin de l'Italie romaine. Essai d'histoire économique d'après les amphores*, Bibliothèque des Écoles françaises d'Athènes et de Rome 261. Rome: École Française de Rome, 1986.

Tello, E., Badia-Miró, M., Cussó, X., Garrabou, R., and Valls. F. "Explaining vineyard specialization in the province of Barcelona (Spain) in the mid-19th century." *Working Paper E08/201 of the Faculty of Economics and Business at the University of Barcelona* (2008) 1-42.

Tello, E., and Badía-Miró, M. "Land-use profiles of agrarian income and land ownership inequality in the province of Barcelona in mid-nineteenth century." *Working Paper of the Spanish Agricultural Society DT-SEHA 11–01* (2011): 1-30.

Tremoleda, J. "Les instal·lacions productives d'àmfores tarraconenses." In *La producció i el comerç de les àmfores de la província Hispania Tarraconensis. Homenatge a Ricard Pascual i Guasch (Barcelona, 17 i 18 de novembre de 2005)*, Monografies del Museu d'Arqueologia de Catalunya 8, edited by A. López Mullor, and X. Aquilué, 113-150. Barcelona: Museu d'Arqueologia de Catalunya, 2008.

Van Minnen, P. "Agriculture and the 'Taxes-and-Trade' Model in Roman Egypt." *Zeitschrift für Papyrologie und Epigraphik* 133 (1998): 205-220.

Witcher, R. "The Extended Metropolis: Urbs, Suburbium and Population." *Journal of Roman Archaeology* 18 (2005): 120-138.

5

Growing grapes in populous landscapes: demography, food, land and vine agroforestry in central Adriatic Italy

Dimitri Van Limbergen

Ghent University

Abstract: The Early Imperial period was a time of significant urban expansion in central Adriatic Italy (*Picenum et Ager Gallicus*), sustained by the proliferation of an elite class that took control over both town and country from the Augustan era onwards. With the presence of some 40 Roman towns, the region was not only one of the more urbanised areas of Roman Italy – surpassed only by *Latium et Campania* – but also the most urbanised tout court along the entire Adriatic coast (3.5-4.9 towns/1000 km²). Also, during the last 15 years, systematic field surveys have indicated that the territories of these towns were among the most densely populated rural areas of the peninsula in Imperial times (76-109 persons/km²). In the course of this 200-year period, feeding such a growing population against the background of natural constraints was likely to be an increasing challenge. Given the ubiquity of cereals in the Roman diet, a rising demand for this food product in particular must have given way to important changes in land availability and organisation. With this paper, I would like to discuss the potential consequences of such a process with regard to how local viticulture practices may have evolved in the area, bound by environmental restrictions. The focus is hereby on the *arbustum*, that is, an extensive agrarian technique in which vines were trained upon rows of fruit trees in combination with the inter-cultivation of cereals and other crops. It is argued that, through the increasing adoption of arbustum fields, the Romans found an albeit temporary solution for successfully combining a high demand for cereals with an equally high demand for wine, in this way holding off – at least for a while – a Malthusian doom scenario.

Keywords: Urban and rural demography, Viticulture, Arbustum, Wine, Agroforestry

Introduction

Recent archaeological work in central Adriatic Italy – an area of roughly 11,600 km² including the Marche region and the northern part of Abruzzo – has identified the Early to Mid-Imperial period (ca. 25 BC – AD 200) as a time of significant urban and rural prosperity (Vermeulen et al. 2017). What remains less clear, however, is how these developments impacted local agriculture in general, and viticulture in particular. The era marks significant changes in the wine amphora trade – apparently pointing to ta time of diminishing exports (Van Limbergen 2018a) – but at the same time the (wine) press record hints at a phase of vitality until at least the beginning of the 2nd century AD (Van Limbergen 2019). It thus seems that the decrease of central Adriatic amphorae does not directly mirror a crisis in local viticulture, but rather reflects the reduction of a particular sector of the wine trade, one oriented towards overseas rather than domestic markets. This also implies that the Late Republican wine boost as a result of favourable conditions outside Italy is only part of the story (Van Limbergen et al. 2017). Indeed, the 40 towns that expanded across central Adriatic Italy throughout the 2nd-1st century BC could also incite wine farmers to ramp up production and invest in costly infrastructure (Morley 1996; De Ligt 2012; Tchernia 2016). Local towns were primary markets for country produce in the Roman world (Vera 1994), but their effect is much less detectable in archaeology, as land transport mostly made use of perishable wooden barrels and skins (Marzano 2013). This bias is thus likely to have distorted our diachronic reconstructions of viticulture, and so urges us to fully take into account domestic demographic trajectories when trying to explicate Early to Mid-Imperial changes in local viticulture (and agriculture). Furthermore, such changes must be framed better within the possibilities and constraints of the local environment.

This chapter explores this issue for central Adriatic Italy by exploring the significance of the *arbustum* – a type of vine agroforestry with vines, trees and cereals (or other crops) cultivated together – as a workable response to changing population and market conditions. The argument is divided into three main parts. The first part offers a novel outline of the urban and rural population developments in central Adriatic Italy on the basis of recent excavation and field survey data. Fieldwork by the PVS in town and country now allows for a finer reconstruction of these processes, and for a more detailed calculation of urban and rural population densities in Late Republican and Early/

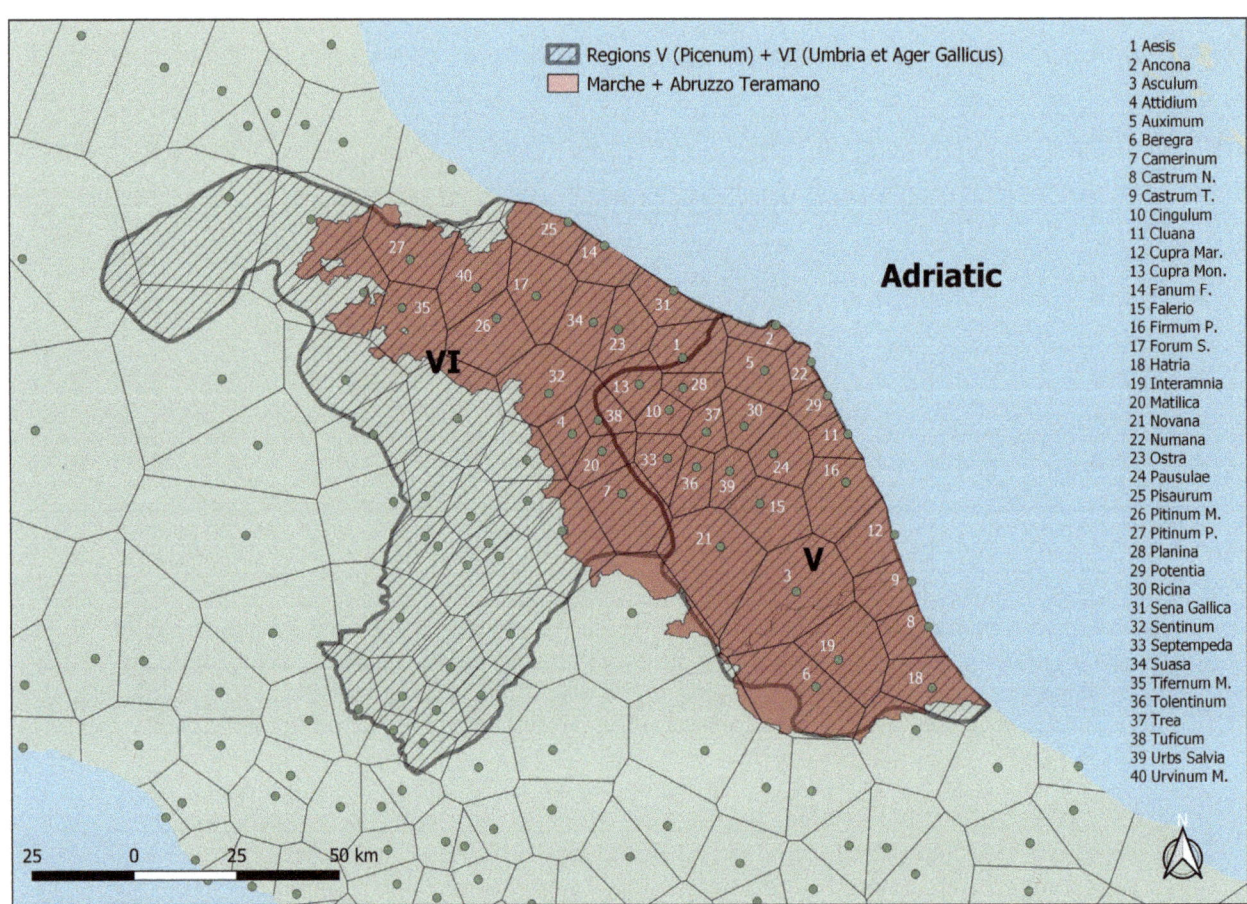

Figure 5.1: Roman towns in central Adriatic Italy with their hypothetical territories (Map by author).

High Imperial times (Van Limbergen and Vermeulen 2020). The second part then elaborates a methodology on how to determine the caloric, food and land needs of these populations. These models are essentially probabilistic, and thus not meant to provide us with real figures, but rather with a credible range of estimates, and thus a useful order of magnitude. The third part explores the relation between population (growth) and land (constraints); and the potential of the *arbustum* as a (temporary) sustainable solution in the region in Early and Mid-Imperial times. To this purpose, I first sketch the theory behind the argument and stress the diverse and multifaceted nature of agricultural intensification processes in pre-industrial societies. I then make use of a selection of art historical, literary and archaeological data to elucidate the principal features of the *arbustum*, and the application of this system in Roman Italy. Finally, I integrate these data with comparative evidence from later periods to assess the suitability of the *arbustum* for sustainable viticulture in central Adriatic Italy. The concluding part reflects on the implications of my observations and calculations for the development of Imperial viticulture and agriculture in central Adriatic Italy and beyond.

Urban demography

Urbanisation was a relatively late and gradual phenomenon in central Adriatic Italy. The first colonies were only founded in the 3rd to mid-2nd century BC, and then mostly along the coast: *Hatria* (L) and *Sena Gallica* (R) somewhere between 290/289 and 285/283 BC; *Castrum Novum* (R) and *Firmum Picenum* (L) in 264 BC; *Aesis* in 247 BC (R), *Pisaurum* and *Potentia* in 184 BC; and finally *Auximum* in either 174 or 157/156 BC.[1] More inland, the Roman conquest merely triggered the development of a series of road stations and *praefecturae* along the newly established *Via Flaminia* and its many side roads. Only after the Social War (91-88 BC) were these settlements gradually converted into proper administrative towns (*municipia*). From the mid-1st century BC onwards, this process intensified and became accompanied by an intense program of veteran colonization and land distribution. With 40 towns, the region so evolved into one of the more urbanised regions (in terms of urban density) of Early/High Imperial Italy – only surpassed by *Regio* I *Latium et Campania* along the Tyrrhenian coast – and the most urbanised *tout court* along the entire Adriatic coast (Figure 5.1). Unsurprisingly, recent archaeological research in the central Adriatic area has also identified this period – and in particular the time between the reigns of Augustus and Hadrian (27 BC – AD 138) – as an era of significant urban structural vitality and prosperity.[2]

---

[1] (L) = Latin; (R) = Roman; the Roman colony of Pollentia (later Urbs Salvia) was probably founded at the time of the Gracchian land reforms (133-123 BC).

[2] For a more detailed account of these developments, in particular in relation to the towns in the Potenza valley, see Van Limbergen et al. 2017a, 147-150; 2017b, 353-356; for a wider framing, see Vermeulen 2017; and also Van Limbergen 2015, 171-340.

What remained less clear until now, is how precisely these towns developed demographically. Textual accounts do not provide us with useful information on this matter. We know that in the 3rd century BC, Latin colonies were generally much larger than Roman colonies – 2,500 to 6,000 vs. 300 families – because of their different foundational background (Salmon 1969; Delplace 1993; Rosenstein 2012). The contrast became less pronounced after the Second Punic War (218-201 BC), when Latin colonies received on average 3,000-4,000 settlers vs. 2,000 in Roman colonies (Pelgrom 2013). But there are no specific references to the actual number of settlers that were sent to the colonies in the central Adriatic area. The references to the military forces provided by Rome's allies in the region – like *Camerinum*, which was able to furnish a cohort of 600 men for the Second Punic War (Livy XXVIII.45) and two cohorts (1,000-2,000 men) for the battle of *Vercellae* (101 BC) (Cic. *Balbo* XX.46) – are neither of much help. While the ability of the allied town to provide that many troops hints at a noteworthy population – especially towards the end of the 2nd century BC – we remain clueless about how many of them actually came from the settlement itself, or from its territory.[3] Duncan-Jones tried to estimate the population of some Italian towns based on epigraphic evidence for large-scale gifts and distributions. For *Pisaurum* in AD 120-180, he so estimated the presence of a free population of 9,800-18,550 people (or a total population of 12,600-23,800 people); for *Sentinum* in AD 88/96 this resulted in a free population of 2,725-5,550 people (3,500-7,200 in total). He then used the lower figure for the proper urban centre, and the higher figure when including their territories (Duncan-Jones 1982, 268-269, 273, Table 7). All these figures certainly provide valuable orders of magnitude, but offer little ground to make reliable demographic reconstructions.

The best – even if by no means flawless – way to establish workable urban demographic models for Roman Italy is by extrapolating population figures from the archaeological data at one's disposal. This is usually done by multiplying room, house and town surfaces by informed average numbers of people (Hanson 2016). It should come as no surprise that most literature concerns the well-preserved and intensively studied towns of *Pompeii* (Beloch 1898; Fiorelli 1873; Nissen 1877; Storey 1997; Wallace-Hadrill 1991; 1994), *Ostia* (Calza 1926; Meiggs 1960; Nibby 1829; Packer 1971; Storey 1997), Rome (Bairoch 1989; Brunt 1971; Carcopino 1940; Hopkins 1978; Lo Cascio 1999; Robinson 1992; Stambaugh 1988; Storey 1997) and *Cosa* (Fentress et al. 2003). In a recent paper, however, we developed a variant of the method to derive population figures from urban contexts known only by (field-, aerial-, geophysical-) survey; and applied this method to the towns of *Potentia* and *Trea* in the Potenza River Valley in central Marche.[4]

This approach essentially consisted of three steps:

- Step 1: Make an inventory of all available data (as collected from archaeology, literature, epigraphy) on a) the extent and internal plan of the site; b) the area covered by public, religious and commercial space; c) the size of the *insulae* and the street grid; and d) the existence of suburban quarters;[5]
- Step 2: Use this inventory to establish a) the total area occupied by urban infrastructure and public buildings; and b) the potential area available for residential buildings;
- Step 3: Collect all available archaeological data on the size and nature of private architecture in the wider study area (in this case the 40 towns in the Central Adriatic region); and use this information to a) fill-in the residential area of each town; and b) calculate the number of inhabitants.

For this chapter, I took this approach one step further and expanded our efforts to all 40 towns in the study area. This entailed, of course, a significant bias in data availability, as both site preservation and research history varied greatly from one town to another. Many of them developed on hilltops or strategic points along the coast and rivers. Hence, these settlements continued to be occupied long after the Roman occupation, and are now covered by their medieval and modern successors. This often hinders large-scale and in-depth archaeological work, thus allowing at best for a fragmentary reconstruction of the original town plan and architecture (21/40).[6] Even if not (completely) built over in later periods, some towns have simply received little scientific attention (7/40).[7] For others, their exact location is still up in the air (2/40).[8] The least problematic contexts obviously are those towns that remained (largely) unoccupied after their abandonment, and whose traces and structures have been systematically studied over the years (10/40). *Potentia*, *Trea*, *Suasa*, *Ostra* and *Sentinum* are the best-documented examples in this category, but *Ricina*, *Septempeda*, *Urbs Salvia*, *Tifernum Mataurense* and *Forum Sempronii* offered much useful information as well. This latter group of ten towns was thus used as the basis for our analysis, with their data taken to inform and/or complement the reconstructions in the other towns. The full inventory with ample references can be found elsewhere; the most relevant results are summarized below.[9]

---

[3] The same goes for *Sena Gallica*, whose inability to deliver troops for Rome's second armed conflict with Carthage in 209 BC (Livy XXVII.38.3-7), suggests a low population level, but not much more.
[4] Van Limbergen and Vermeulen 2020, for a full discussion of the methodology and the results, as well as a review of the main issues and guidelines in urban demographic modelling in archaeology.
[5] For *Potentia* and *Trea*, see Van Limbergen and Vermeulen 2020.
[6] These is the situation for the towns of *Pisaurum* (Pesaro), *Urvinum Mataurense* (Urbino), *Fanum Fortunae* (Fano), *Sena Gallica* (Senigallia), *Aesis* (Jesi), Ancona, *Auximum* (Osimo), *Cingulum* (Cingoli), *Matilica* (Matelica), *Tolentinum* (Tolentino), *Tuficum* (Borgo Tufico), *Camerinum* (Camerino), Numana, *Attidium* (Attiggio), *Cupra Maritima* (Cupra Marittima), *Firmum Picenum* (Fermo), *Falerio* (Pian di Falerone), *Asculum* (Ascoli Piceno), *Interamnia Praetuttiorum* (Teramo), *Castrum Novum* (Giulianova) and *Hatria* (Atri).
[7] This holds true for Pitinum Mergens, Pitinum Pisaurense, Planina, Cluana, Pausulae, Castrum Truentinum, and Cupra Montana.
[8] This is so for *Beregra* and *Novana*.
[9] For the original version of this inventory (with maps), see Van Limbergen 2015; it was published as a gazetteer (with references) in Vermeulen 2017 (Van Limbergen and Vermeulen 2017).

Dimitri Van Limbergen

Urban surface area (USA)

For 15 out of 40 towns in the study area, it was possible to deduce the approximate urban surface area delimited by the town walls. Together, these towns cover an area of 369.6 ha. For 20 out of 40 towns, it was only possible to estimate the urban area from circumstantial evidence, such as natural barriers or indirect spatial indications like the position of extramural cemeteries and suburban villas, or later medieval town areas. These towns cover an additional estimated area of 354.7 ha. For the remaining five towns, it was not possible to determine their physical extent on the basis of the available evidence, and here an average size of 21 ha (obtained from the sum of the other areas) was assigned. This gave another total area of 105 ha. The total urban surface area for the Early/High Imperial period can so be estimated at ca. 830 ha (Table 5.1). This tentative reconstruction suggests that most towns in central Adriatic Italy were small to medium urban centres; that is, of between 10-20 ha (17/40)[10] and 20-30 ha (14/40)[11] in size. Some of them were no more than agglomerations with urban status (<10 ha) (4/40)[12], but a few bigger towns of ≥40 ha (5/40)[13] existed as well.

Street grid (SG)

The proportional space taken up by the street grid could only be assessed in four towns: *Pisaurum* (9%), *Sena Gallica* (17.2%), *Potentia* (28.1%) and *Asculum* (9.6%). Despite being of roughly the same size, *Potentia*, *Pisaurum* and *Sena Gallica* display noticeable differences in the area reserved for the street grid (5.2, 1.8, 3.1 ha respectively). This is mostly linked to the size of the *insulae*, and to the number of streets and their average width. At *Pisaurum*, streets were between 3 and 4 m wide, while the town area itself was subdivided into 35 *insulae*, measuring mostly 2x2 *actus* (18), but also 2x2.5 (6), 2.5x1 (1) and 2x1 (3). *Potentia*, on the other hand, had wider streets of between 5 and 6 m, and consisted of 60 smaller *insulae* measuring 1x2 (24), 1x2.5 (12), 1x1.5 (12) and 1x1 (12) *actus*. This explains why the street grid took up ca. 10% at *Pisaurum*, but almost 30% at *Potentia*. The reasoning for the other two towns is similar. The street width at *Sena Gallica* was about 3.30 m (cf. *Pisaurum*), but the town was organized into 21 *insulae* of 1x2 *actus* and 21 irregularly-shaped *insulae*; thus explaining the ca. 17% share of the street grid. Finally, the town plan of *Asculum* has been reconstructed as a grid of 75 *insulae* of 2x2 *actus*, while recorded street widths measure 2.61 m (*cardines*), 3.90 m (*decumani*) and 5.40 m (*decumanus maximus*); thus resulting in a share of ca. 4.2 ha (ca. 10%) for the street grid (comparable to *Pisaurum*).

Table 5.1: USA, SG, PS and RS of all central Adriatic towns.

Town	USA (ha)	SG (ha)	PS (ha)	RS (ha)	RS (%)
Aesis	10	1.5	1.8	6.7	67
Ancona	11.6	1.7	2.5	7.4	63.8
Asculum	43.7	4.2	3.8	35.7	81.7
Attidium	10	1.5	1.8	6.7	67
Auximum	16	4.8	2.4	8.8	55
Beregra	21	3.2	1.8	16	76.2
Camerinum	20	3	2.4	14.6	73
Castrum N.	13.8	2.1	1.8	9.9	71.7
Castrum T.	25	3.8	1.8	19.4	77.6
Cingulum	15	2.3	1.8	10.9	72.7
Cluana	21	3.2	1.8	16	76.2
Cupra Mar.	16.7	2.5	2	12.2	73.1
Cupra Mon.	23	3.5	1.8	17.7	77
Fanum F.	25	5	4.3	15.8	63.2
Falerio	9	0.9	2.4	5.7	63.3
Firmum P.	18.6	2.8	1.7	14.1	75.8
Forum S.	25	3.8	3.9	17.3	69.2
Hatria	70	7	2.2	60.8	86.9
Interamnia	26	3.9	2.1	20	76.9
Matilica	12	1.2	1.8	9	75
Novana	21	3.2	1.8	16	76.2
Numana	16	2.4	2.4	11.2	70
Ostra	39	5.9	2.2	31	79.5
Pausulae	20	3	1.8	15.2	76
Pisaurum	18.9	1.8	2.4	14.6	77.2
Pitinum M.	9	1.4	1.8	5.8	64.4
Pitinum P.	5	0.8	1.5	2.7	54
Planina	21	3.2	1.8	16	76.2
Potentia	18.4	5.2	2.2	11.1	60.3
Ricina	25	3.8	2.8	18.5	74
Sena Gallica	18	3.1	1.8	13.1	72.8
Sentinum	15	2.3	2.9	9.9	66
Septempeda	15	2.3	1.8	11	73.3
Suasa	42	6.3	2.4	33.3	79.3
Tifernum M.	12	3.6	1.8	6.6	55
Tolentinum	21	3.2	1.8	16	76.2
Trea	11	1.1	2.8	7.1	64.5
Tuficum	20	3	2.4	14.6	73
Urbs Salvia	42	6.3	3	32.7	77.9
Urvinum M.	8	1.2	1.9	4.9	61.3
TOTAL	829.7	125	89.2	616	72.2

This sample size served for estimating the share of the street grid in towns with less data. At *Tifernum Mataurense*, for example, we know of street widths of 4.70 m and 6.20 m for the *cardines*, and of 6.40 m for the *decumani*, possibly following a grid of *insulae* of 1x2 *actus*. This means that both the street widths and the *insulae* approach

---

[10] Pisaurum, Tifernum M., Sena Gallica, Aesis, Sentinum, Ancona, Auximum, Cingulum, Numana, Potentia, Trea, Septempeda, Matilica, Attidium, Firmum P., Cupra Mar., Castrum N.
[11] Fanum F., Forum S., Planina, Tuficum, Ricina, Cluana, Pausulae, Tolentinum, Camerinum, Castrum T., Interamnia, Cupra Mont., Beregra, Novana.
[12] Pitinum P., Pitinum M., Urvinum M., Falerio.
[13] Suasa, Ostra, Urbs Salvia, Asculum, Hatria.

the dimensions at *Potentia*; thus suggesting a share closer to 30% than to 15% (*Sena Gallica*) or 10% (*Asculum, Pisaurum*) for the street grid. The same goes for *Auximum*, whose *insulae* seem similar to *Potentia*'s in both size and number. *Fanum Fortunae* had street widths similar to *Potentia*'s (5-6 m), but larger *insulae* (2.5x2.5, 2.5x1.25 *actus*), thus implying a somewhat smaller share of the street grid (perhaps 20%). Then again, *Sentinum* seems to approach *Sena Gallica*'s numbers, opting instead for a 15% share. A similar scenario can be envisaged for *Forum Sempronii*, whose streets were generally wider, but the *insulae* larger. Based on the same reasoning, smaller shares can be proposed for *Trea, Matilica, Falerio* and *Hatria* (10%). Unfortunately, it was not possible to estimate the share of the street grid for the other 26 towns, so an average share of 15% was assigned; an educated guess based on the examples here analysed (10-30%), and on the results from Marta Conventi's study on the organisation of urban space in Roman Italy (10-50% on the whole, but 12-30% for towns of ≤20 ha) (Conventi 2004) (Table 5.1).

Public space (PS)

Forum

For many towns, the position and/or the size of the *forum* remain unclear. For *Pisaurum, Cingulum, Matilica* and *Falerio*, its dimensions have been hypothesized on the basis of either its supposed connection with the modern main square of the town, or its theoretical location inside the reconstructed street grid (2,450 to 4,900 m²). These are thus hypothetical figures unfit for further analysis. Precise data on its dimensions were only recovered from seven towns, ranging from 1,800 m² (*Sentinum*) to 5,400 m² (*Cupra Maritima*), but mostly between 3,200 and 3,600 m². This gives an average size of ca. 3,323 m². This means that for the towns for which the size of the walled area is known – that is, *Sentinum, Potentia, Trea* and *Urbs Salvia* – the *forum* took up between 0.5% (*Urbs Salvia*) and 3.2% (*Trea*). The average was 1.7%. Logically, the proportional difference in surface share is linked to the size of the settlement; that is, higher in smaller towns and lower in larger towns. It therefore seemed better to use the average of 3,323 m² (rather than a fixed percentage) for the towns for which the size of the *forum* was unknown. This resulted in the *forum* occupying between 0.5 and 3.3% in all towns, except for *Pitinum Pisaurense* (6.6%), *Urvinum Mataurense* (4.2%), *Pitinum Mergens* (3.7%), *Falerio* (3.7%) and *Cupra Marittima* (5.4%) – all smaller settlements with an estimated surface area of ≤10 ha.

Theatre

The presence of a theatre was confirmed archaeologically in 16 towns, but the size of the building could only be determined in 13 of them.[14] The surface area ranged from 1,500 m² (*Suasa, Firmum*) to 6,650 m² (*Urbs Salvia*). The average is ca. 2,870 m². The share of the theatre could be determined for eight towns, and ranged from 0.4% (*Hatria*) to 4.7% (*Urvinum Mataurense*), but mostly from 0.8% to 1.6%. The derived average was 1.5%. The average of 2,870 m² was used for the remaining towns. As such, the theatre took up between 0.4% (*Suasa, Ostra*) and 5.7% (*Pitinum Pisaurense*) of the total urban space, but mostly between 1.1 and 3.2%. Still, we cannot be certain that a theatre was included in every town.

Amphitheatre

An amphitheatre was identified in ten towns.[15] The existence of this spectacle building is supposed on the basis of epigraphy in *Pisaurum, Auximum* and *Tuficum*. 19th century testimonies mention the remains of amphitheatres at *Numana* and *Camerinum*, but their exact location remains unknown. *Cupra Montana* perhaps had an amphitheatre as well. Except for *Urbs Salvia*, these amphitheatres were all inserted into the main urban area, occupying at least 4,144 m² (*Interamnia*) and as much as 9,350 m² (*Fanum Fortunae*). The average was 6,390 m². Percentage-wise, the proportions ranged from 0.6% (*Hatria*) to 5.9% (*Ancona*); the average was 2.8%. Applying again the average of 6,390 m² to the other towns in the area, the building could occupy a share of as little as 1.6% (*Ostra, Interamnia*) and as much as 12.8% (*Pitinum Pisaurense*). In most cases this was no more 5.3%. We should however not expect every town to have had an amphitheatre.

Other public architecture

There is much less information on other public buildings in the central Adriatic towns, such as the main temples, the *tabernae*, the *basilica*, the *macellum*, the public baths and cisterns.[16] Only a few bath houses have been completely excavated.[17] A recurrent scheme seems to be the combination of a large bath complex close to the *forum* – known sizes lie between 2,000 and 5,200 m², with an average of 3,666 m² - with a smaller thermal building located elsewhere within the settlement (300-1,208 m²; average of 633 m²). Sizes of the basilica are known from *Trea* (700 m²) and *Cupra Marittima* (1,403 m²) (1,050 m² average); *macella* are known in *Forum Sempronii* (400 m²), *Potentia* (512 m²), *Castrum Truentinum* (600 m²) and *Trea* (900 m²) (603 m² average); *tabernae* in their entirety have been registered at *Suasa* (4,000 m²), *Trea* (4,500 m²) and *Potentia* (4,800 m²) (4,433 m² average). A town also

---

[14] Theatres are currently known at Urvinum Mataurense, Fanum Fortunae, Suasa, Ostra, Aesis, Potentia, Ricina, Urbs Salvia, Firmum Picenum, Falerio, Asculum, Interamnia Praetuttiorum and Hatria. At Forum Sempronii, Pitinum Mergens and Sentinum, theatres have been noted in the past, but their position is unknown at the moment.

[15] Fanum Fortunae, Forum Sempronii, Suasa, Ancona, Ricina, Urbs Salvia, Falerio, Asculum, Interamnia Praetuttiorum, Hatria.

[16] Cisterns have not been incorporated in these calculations, as these often occupied higher areas of the town, either immediately inside or outside the town walls, or served as platforms for the construction of other buildings and squares, such as the *forum*.

[17] These include the baths unearthed at Tifernum Mataurense, Forum Sempronii, Ostra, Sentinum, Septempeda, Matilica, Falerio and Interamnia Praetuttiorum.

housed a number of temples. *Pompeii*, for example, had nine temples spread out over 66 ha. Based on archaeology and epigraphy, it seems that central Adriatic towns typically housed one large main temple and between two and five smaller temples within their urban area.[18] The larger temples varied from 384 to 613 m² (540 m² average), while the smaller temples mostly occupied 117 to 300 m² (202 m² average), with also 2 temples below 100 m² known (74 m² average). In order to acknowledge the presence of temples within the urban fabric of all towns, it was decided to assign one temple of each category to the towns without information.

The overall size of public space and its implications for the size of residential space

There are only a handful of towns for which we have an approximate idea of the total space taken up by public architecture. One of the best documented towns is *Trea*, where the central zone with all identified public buildings occupied a total area of ca. 11,465 m² (ca. 1.2 ha), or 10.6% of the entire urban area. The most recent town plan easily allows for a total central public area of ca. 23,150 m² (2.3 ha) or ca. 21% of the urban surface, including as such the temples for Minerva and Vittoria (whose existence is attested in epigraphy), a *Serapeum* for the goddess Isis (known through 18th century excavations) and a bath house in the southern section of the town. By adding the guestimated averages for thermal buildings and a small temple – but not those for theatres and amphitheatres, as these have not been attested despite intense geophysical and aerial photography – the total public area may have occupied 23,985 m² (2.4 ha) or ca. 22% of the urban area.

Another well-documented example is *Potentia*, where the known public buildings cover an area of ca. 12,123 m² (1.2 ha) or ca. 6.6% of the walled town area. The most recently published plan makes a central public space totalling ca. 19,200 m² (1.9 ha) – perhaps including a basilica, a thermal complex and a few temples – entirely plausible. At least one other thermal building and a structure resembling a small theatre (1,838 m²) have been attested within the walled area. There is no proof for an amphitheatre. The public space may thus have covered an area of ca. 21,671 m² (2.2 ha) or 11.8%. The same reasoning may be followed for *Sentinum*, where the known public buildings around the central *forum* cover an area of ca. 5,399 m² (0.5 ha) or 3.6% of the walled area. But the published archaeological map – indicating further public buildings west and south of the *forum* – suggests the existence of a town organized around a large central public area of ca. 24,894 m² (16.6%). Add to this the thermal complex (3,800 m²) to the east of the *forum*, and the public space probably covered ca. 28,694 m² (2.9 ha) or 19.1% of the town.

Another good example is *Ostra*. The known public buildings and open zones cover an area of ca. 14,800 m² (1.5 ha) or about 3.8% of the estimated urban area. Add to this the averages for the other probable public buildings – possibly including *tabernae* around the *forum*, a basilica, a *macellum*, two other temples and a second bath house – and the total public area may have covered ca. 22,131 m² (2.2 ha) or 5.7% of the town. Finally, some useful data also come from *Suasa* and *Forum Sempronii*. The known public buildings at *Suasa* amount to ca. 16,550 m² (1.7 ha) or 3.9% of the total urban area; the estimated total area (based on the averages for a basilica, a *macellum*, a large bath complex and temples) perhaps covered ca. 25,215 m² (2.5 ha) or 6%. At *Forum Sempronii*, the principal public area (with the *forum*, the *Augusteum* and other temples, the *macellum*, and the main thermal complex) covered a likely area of ca. 29,400 m² (2.9 ha) or 11.8% of the town. Adding a second bath (1,208 m²), the amphitheatre (5,560 m²) and the supposed theatre enlarges this area to ca. 39,038 m² (3.9 ha) or 15.6%.

For the remaining towns, the total public space was guestimated by combining the known public buildings with the averages for the other buildings. An amphitheatre, however, was only added when attested in epigraphy or archaeology. This made the share of public space ranging from 3.1-5.7% to 23.7-30.5%. Logically, the lower shares come from the larger towns (*Suasa*, *Ostra*, *Urbs Salvia*, *Castrum Truentinum*, *Hatria*), while the higher ones come from the smaller towns (*Pitinum Pisaurense*, *Urvinum Mataurense*, *Trea*, *Falerio*). For 21 out of 40 towns, the share ranged between 10.1 and 21.6%; that is, for settlements of between 10 and 25 ha. The public space covered between 7.8 and 9.3% of the urban area in the 10 remaining towns. This means that the availability of residential space ranged from 54.5% at *Pitinum Pisaurense* (the smallest town) to 86.9% at *Hatria* (the largest town). For most towns the residential space seems to have had a share of 60-70% (Table 5.1).

Residential space (RS) and the size of the total urban population in the Early/High Empire

Putting an approximate figure on the number of people that occupied these residential areas is another exercise full of uncertainties. Our knowledge on urban housing in central Adriatic Italy is fragmentary to start with. We mostly have scattered remains of walls and floors that fit together to form rooms and partial house plans at best.[19] What seems clear, however, is that they all refer to houses of the *domus*-type, organised around an *atrium* and/or a

---

[18] Temples are known at *Pisaurum* (2), *Pitinum Pisaurense* (2), *Tifernum Mataurense* (1), *Fanum Fortunae* (1), *Forum Sempronii* (6, of which 3 attested archaeologically), *Suasa* (1), *Ostra* (2), *Ancona* (2), *Tuficum* (3, only attested epigraphically), *Numana* (1), *Potentia* (1), *Ricina* (1), *Trea* (4), *Attidium* (1), *Pausulae* (1), *Urbs Salvia* (3), *Tolentinum* (1), *Camerinum* (1), *Firmum Picenum* (1), *Falerio* (3), *Cupra Marittima* (1), *Asculum* (4), *Interamnia* (2).

[19] This holds true for almost all towns in the area: Pitinum Pisaurense, Urvinum Mataurense, Fanum Fortunae, Pitinum Mergens, Aesis, Auximum, Cingulum, Planina, Tuficum, Numana, Septempeda, Attidium, Cluana, Pausulae, Tolentinum, Camerinum, Firmum Picenum, Falerio, Cupra Maritima, Castrum Truentinum, Castrum Novum, Hatria, Cupra Montana, Beregra, Novana, Interamnia, Ancona.

peristylium. In the towns built on a hilltop, we can also see that these *domus* were often organised in terraces that followed the morphology of the hill. In terms of chronology, it seems that *domus* are present in these towns from the Late Republican period onwards. For some of the oldest colonies in the area, they appeared already in the 3rd-2nd century BC. In most cases, however, private architecture was long confined to simpler buildings with walls made out of pebbles and clay, and rudimentary floors. *Domus* only began to characterize the urban area of most settlements between the mid-2nd century BC and the mid-1st century BC, often still in co-existence with simpler housing structures. The towns of *Suasa* and *Urbs Salvia* are exemplary in this case. *Domus* especially started dominating the entire urban fabric from Augustan times onwards, and peaked in both number and richness in the course of the 1st-2nd century AD. Well-documented examples are *Pisaurum* and *Interamnia*; clear signs of their continuing prosperity in the High Imperial period come from *Tifernum Mataurense, Forum Sempronii, Suasa, Ricina* and *Matelica*. But these processes have been registered in most of the central Adriatic towns. One apparent feature of many houses is their longevity, well-attested by a number of interventions in the 1st-2nd century AD pointing to restorations, transformations, enlargements or reductions, or also to an overall re-organization of existing spaces. Another noteworthy evolution in the same period is the tendency to build very large *domus*, often enlarging and/or incorporating pre-existing smaller houses. This process is best visible in the well-studied *Domus dei Coiedii* at *Suasa* (1st century AD), and the *Domus di Europa* at *Forum Sempronii* (mid-2nd century AD). But the trend is noticeable in the *domus* of the *Piazza della Cattedrale* at *Hatria*, and the *domus* of the Via Mazzolari at *Pisaurum* (mid-1st century AD) as well. All these houses seem to have maintained their grandeur and vitality until the beginning of the 3rd century AD, but afterwards signs of decay and abandonment usually began to kick in. Towards the 5th-6th century AD, all houses seem to either have been abandoned completely, or converted in part into graveyards.

Based on current knowledge, we may reasonably assume that the residential layout of the central Adriatic towns in the Early/High Imperial period resembled the typical mixture of houses of varying sizes that we know so well from contemporary towns like *Pompeii* and *Herculaneum*. In order to tentatively recreate this layout, the original study used the data on house sizes and plans (number of rooms) from the best-documented contexts, collected by excavation (*Pisaurum*; *Forum Sempronii*; *Matilica*; *Sena Gallica*; *Urbs Salvia*; *Tifernum Mataurense*; *Suasa*: Domus dei Coiedii, Casa ad Atrio, Casa del Primo Stile), aerial photography (*Tifernum Mataurense, Ostra, Trea*) and geophysical survey (*Sentinum*). This resulted in a total of 27 houses, of which 20 provided data on size, and 7 on both size and the number of living rooms.[20] For the

Table 5.2: Estimated occupation densities for 19 houses in central Adriatic Italy.

Town	House size (m²)	Occupation density (based on 34-39 m²/p)
Tifernum M.	360	9-10
	390	10-11
	470	12-13
	480	12-14
	560	14-16
	925	23-27
	950	24-27
Sentinum	300	7-8
	460	11-13
Ostra	400	10-11
	540	13-15
	800	20-23
Pisaurum	625	16-18
	875	22-25
Forum S.	1,200	30-35
Matilica	600	15-17
	1,200	30-35
Trea	1,300	33-38
	1,600 (x2)	41-47

Table 5.3: Estimated densities for 7 houses in central Adriatic Italy (OD1*=34-39 m²/p; OD2**= 1 room/p).

Town	House size (m²)	N° of rooms	OD1*	OD2**
Suasa	189	3	4-5	3
Urbs Salvia	392	7	10-11	7
Suasa	455	7	11-13	7
Sena Gallica	459	10	11-13	10
Suasa	480	7	12-14	7
Tifernum M.	925	12	23-27	12
Suasa	3,300	15	38-44	15

first group of houses (20/27), the number of inhabitants was estimated by using Wallace Hadrill's 34-39m²/person ratio (Table 5.2); for the second group (7/27) an informed ratio of one person/room was applied as well (Table 5.3) (Wallace Hadrill 1994). These houses were then grouped

---

[20] Excluding as such entrances, passageways, storage- and commercial rooms. For a full discussion of these houses (with their detailed plans), see Van Limbergen and Vermeulen 2020; also Van Limbergen 2015, 260-264.

Table 5.4: House sizes in central Adriatic Italy ranked according to 200m² surface bands (all sizes are in m²).

300–500	500–700	800–1,000	1,000–2,000
300	540	800	1,200 (x2)
360	560	875	1,300
390	600	925	1,500
392	625	955	1,600 (x2)
400 (x2)			
455			
459			
460			
470			
480 (x2)			
12	4	4	6
420	580	890	1,400

and ranked according to surface bands of 200 m² (Table 5.4).[21]

This provided us with:

1. an arrangement of houses that could be used to fill in the residential areas (for each house of >1,000m², there were now 2 houses of 300-500m², and ca. 0.7 house of 500-700m² and 800-1,000m² each) (Table 5.4);
2. usable average sizes for each of the subgroups: 420m², 580m², 890m², 1,400m² (Table 5.4);
3. informed occupation densities for estimating their number of inhabitants, based on the ratios of both surface area and room numbers: 7-12 persons (420m²), 10-15 p. (580m²), 12-17 p. (890m²), 15-20 p. (1,400m²).[22]

This calculation method was then applied to all 40 towns in the study area. For example, for *Potentia*, I so recreated a residential area of ca. 11.1 ha consisting of 68 (420m²), 24 (580m²), 24 (890m²) and 34 (1,400m²) houses; thus inhabited by ca. 1,515 to 2,265 people. All results are listed below (Table 5.5). This gave a total urban population of between ca. 82,640 and 123,645 people in the region in the Early/High Imperial Empire. Urban densities were 134-201 persons/ha for residential areas only (R), and 97-143 p/ha for total town areas (T).

Urban structural and demographic growth?

Together with the apparent structural peak of these towns in the first two centuries of the Empire, there are reasons to believe that these settlements were also less populated in the Late Republican period.[23] Many towns used a smaller surface area in the two centuries preceding the Augustan era. This evolution is most obvious in towns where a more

Table 5.5: Urban populations and urban densities in central Adriatic Italy in the Early/High Empire.

Town	Population	P/ha (R)	P/ha (T)
Aesis	890-1,330	133-199	89-133
Ancona	970-1,450	131-196	84-125
Asculum	4,805-7,185	135-201	110-164
Attidium	890-1,330	133-199	89-133
Auximum	1,200-1,800	136-205	75-113
Beregra	2,170-3,245	136-203	103-155
Camerinum	1,960-2,930	134-201	98-147
Castrum N.	1,330-2,000	134-202	97-145
Castrum T.	2,615-3,910	135-202	105-156
Cingulum	1,465-2,190	134-201	98-146
Cluana	2,170-3,245	136-203	103-155
Cupra Mar.	1,650-2,460	135-202	99-147
Cupra Mon.	2,400-3,590	136-203	104-156
Fanum F.	2,140-3,200	135-203	86-128
Falerio	760-1,130	133-198	84-126
Firmum P.	1,910-2,850	135-202	103-153
Forum S.	2,350-3,515	136-203	94-141
Hatria	8,225-12,300	135-202	118-176
Interamnia	2,665-3,985	133-199	103-153
Matilica	1,200-1,795	133-199	100-150
Novana	2,170-3,245	136-203	103-155
Numana	1,560-2,330	139-208	98-146
Ostra	3,760-5,690	121-184	96-146
Pausulae	2,040-3,050	134-201	102-153
Pisaurum	2,010-3,005	138-206	106-159
Pitinum M.	810-1,210	139-208	90-134
Pitinum P.	365-545	135-201	73-109
Planina	2,170-3,240	136-203	103-154
Potentia	1,515-2,265	137-204	82-123
Ricina	2,480-3,710	134-201	99-148
Sena Gallica	1,775-2,655	136-203	99-148
Sentinum	1,330-2,000	134-202	89-133
Septempeda	1,465-2,190	133-199	98-146
Suasa	4,490-6,715	135-202	107-160
Tifernum M.	890-1,330	135-201	74-111
Tolentinum	2,170-3,245	136-203	103-155
Trea	845-1,260	119-177	77-115
Tuficum	1,960-2,930	134-201	98-147
Urbs Salvia	4,390-6,565	134-201	105-156
Urvinum M.	680-1,015	138-207	85-127
Total/Average	82,640-123,645	134-201	97-143

---

[21] In this final ranking, the 189m² house at *Suasa* was omitted, as it was quickly transformed into a 455m² house (leaving us with 26 houses); and only the 1,500m² living space of the huge *Domus dei Coiedii* at *Suasa* (3,300m²) was taken into account (Van Limbergen and Vermeulen 2020).

[22] The selection of these criteria obviously involved some trial and error, and the reasoning is fully explained in Van Limbergen and Vermeulen 2020.

[23] The data (with references) to support the following statements can be found in Van Limbergen and Vermeulen 2017; and Van Limbergen 2015.

restricted circuit wall preceded the Early Imperial one (*Sentinum, Ancona, Potentia, Firmum*). In other cases, the archaeological evidence on the whole suggests that a significant part of the urban area was unoccupied in Late Republican times (*Pisaurum, Fanum Fortunae, Sena Gallica, Urbs Salvia, Interamnia*). Indications for the border zones of these towns only becoming fully occupied in the Early Imperial period especially comes from *Pisaurum*, where the 1st century AD *domus* buildings in the southern part of the settlement were often constructed in previously unoccupied areas. The same goes for Ancona, where many houses in the south-eastern section were only built in the Augustan period or later, in one case at the expense of an earlier artisanal quarter. Another development which suggests that large portions of the marginal zones within the walled urban area remained free from buildings prior to the Augustan period, is the repeated insertion here of massive spectacle buildings such as an amphitheatre. Estimating the size of these Late Republican settlements is an impossible task, but based on the evidence at hand, reductions of 35% (Firmum), 41% (Ancona), and 63% (Fanum F.) seem entirely plausible. The data for *Potentia* hint at only 13% (based on changes in the circuit wall), but the recent surveys by the PVS strongly suggest that most of the colony's intramural area was only extensively occupied in the 2nd century BC, and then built-up more intensively in the course of the 1st century AD (Vermeulen et al. 2017).

Along with the expansion of the inhabited areas towards the border zones of the settlements, there are indications for a contemporary expansion of public space in their central zones. Indeed, it seems that the monumentalisation of the central Adriatic towns from the Augustan period onwards went hand in hand with a complete reorganisation of their urban layout. In some towns, these developments led to the reconversion of central residential space into public space. This was the case for *Ostra* and *Hatria*, where the Imperial baths close to the *forum* replaced Late Republican housing quarters. At *Urbs Salvia*, the Imperial temple near the *forum* was built on the place of a former housing area, as attested by the remains of a Late Republican house beneath the sacral building. Remains of Late Republican houses were also found at Ancona, in an area that was next included in the central *forum* complex with its adjacent public buildings. A similar scenario seems plausible for *Trea*, where the development of the *forum* complex deeply altered the layout of the central section of the town. The same is true for Septempeda, where the Early Imperial baths near the forum replaced a Late Republican artisanal building. These developments suggest that not only were the public areas more restricted in these towns before the Early/High Imperial period, but their private and artisanal zones were also more confined and located closer to the centre.

Another telling argument is that many towns show traces of the development of suburban quarters from the Augustan period onwards, often intensifying in the course of the 1st century AD. The process is most visible at *Pisaurum*, but extramural expansions have been noted for *Urvinum M., Sentinum, Auximum* and *Camerinum* as well. Work by the PVS in the Potenza Valley also points to the presence of some form of suburban residential occupation at *Potentia, Trea* and *Ricina* in the first two centuries AD. Such developments are usually interpreted as signs for a lack of space within the original enclosed area (De Ligt 2012). This would imply that these areas only began to be exploited after the space within the town walls had been completely built-up, thus creating a shortage in residential space. In turn, this makes it unlikely that the whole walled area of these towns was densely built-up, or had already reached its maximum occupation, in Late Republican times.

Finally, mostly smaller houses characterised the layout of the Late Republican residential areas. Indeed, the larger houses all seem to date to the Imperial period – or at least to do not appear before the mid-1st century BC – often enlarging and/or incorporating pre-existing smaller houses. In sum, the evidence suggests that most towns were smaller – or less densely occupied – in the Late Republic, with often less space reserved for public buildings, and smaller houses in residential areas. Therefore, I would argue that these towns peaked both structurally and demographically in the Early/High Empire.

Rural demography

Field surveys by the PVS in the surroundings of *Potentia* (lower valley) and *Trea* (middle valley), as well as in the area between *Matilica* and *Camerinum* (upper valley), have also shed light on the rural occupation in the Potenza valley. In fact, during more than a decade of fieldwalking, the PVS recorded a total of 125 rural habitation sites (Figure 5.2).[24] A recent evaluation of the artefact data assigned these sites to one or more phases from Iron Age to Late Antiquity (Figure 5.3).[25] Taking into account the estimated size of the surface scatters (fully reliable size data for most rural sites are lacking) and the nature of the material evidence, the 89 sites that were identified as Roman and/or late antique were then grouped into the following six settlement types:[26]

1. small rural units (concentrations of ≤ 300 m²);
2. small farms (concentrations of 300-1,200m²);
3. medium-sized farms (concentrations of 1,200-2,400m²);
4. large farms or small/simple villas (concentrations of 2,400-4,000m²);
5. (large) villas (concentrations of 4,500-9,500m²);
6. village-like or roadside nucleated settlements (concentrations of 3,000-10,000m²).

---

[24] For the full site catalogue, see Van Limbergen et al. 2017b.
[25] The diachronic data distribution method – which also allowed for fixing thresholds to distinguish between certain and possible phases of occupation (not visualised in Table 5.6) in 50-year periods – used to translate the recorded artefact scatters into diachronic site patterns is discussed at length in Van Limbergen et al. 2017a, 112-115.
[26] A full discussion of these categories can be found in Van Limbergen et al. 2017, 120-122.

Dimitri Van Limbergen

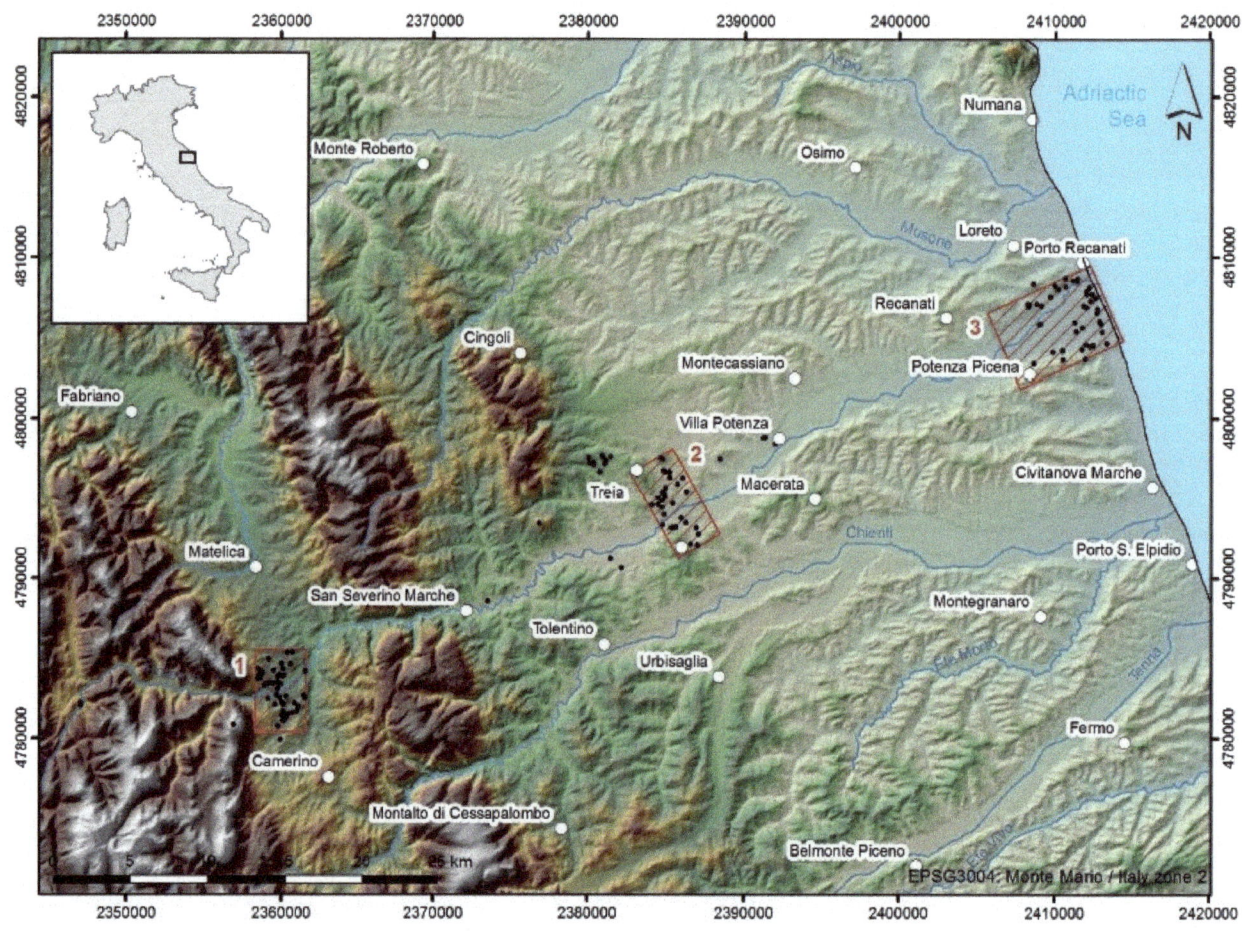

Figure 5.2: Topographic map of central Marche with the three main areas of artefact survey and location of sites intensively surveyed by the PVS team (After Van Limbergen et al. 2017c, 11, fig. 3).

This in turn allowed for establishing diachronic settlement patterns in the valley from Late Republican to late antique times (Figure 5.4). I am aware that building settlement typologies and changes on the basis of field survey data involves incorporating a series of assumptions. We cannot be certain, for example, that the extent of the surface scatter directly reflects the size of the underlying site (Scheidel 2008, 49-54; Rathbone 2008; De Ligt 2012, 249-254; Launaro 2011, 69). The classification of sites into farms and villas (small and large) also remains a largely arbitrary, and thus potentially erroneous, exercise. There is notable disagreement on what defines a farm surface area-wise in survey scatters. Robert Witcher, for example, in his study on Rome's suburbium, applies the category 'farm' to surface scatters of <1,000 m² that display a limited range of material culture (Witcher 2005). Then again, for the Greek world, upper area limits for farmsteads are sometimes placed around 1,400-2,000 m² (Osborne 2004). Similar concerns apply to what constitutes a villa.[27]

The application of possible recovery rates – that is, the extent to which the number of registered sites reflects the number of sites that actually occupied the landscape – is a tricky matter as well (Witcher 2011). For this exercise, I assumed a 100% recovery rate for all sites, but surely this is too optimistic. This seems especially so for the lower valley transect close to the river mouth of the Potenza river, where a succession of anthropogenic and natural processes has significantly changed the landscape since Roman times; suggesting that the survey results may not accurately reflect the original situation (Taelman et al. 2017; Van Limbergen et al. 2017, 124, note 40). Recovery rates may have varied according to site type too, and this should be incorporated into future models as well.[28] Finally, the survey transects of the PVS are all situated close to Roman nucleated settlements in the valley. It is entirely possible that more remote areas were less densely occupied; thus implying that the recorded site densities might not be representative for the entire study area. This is another bias that should be addressed in the future.

The extrapolation of population figures from these tentative site patterns introduces yet another uncertainty; that is, the assignment of household sizes to each of the site categories. There is significant discussion on the number of inhabitants that one should assign to different

---

[27] Robert Witcher considers villas to refer to those scatters of >1,000 m² that have a greater variety in, and a higher quality of material culture (Witcher 2005).

[28] Possible recovery rates may be 30% for farms, 60% for villas, and 80% for villages (Witcher 2011).

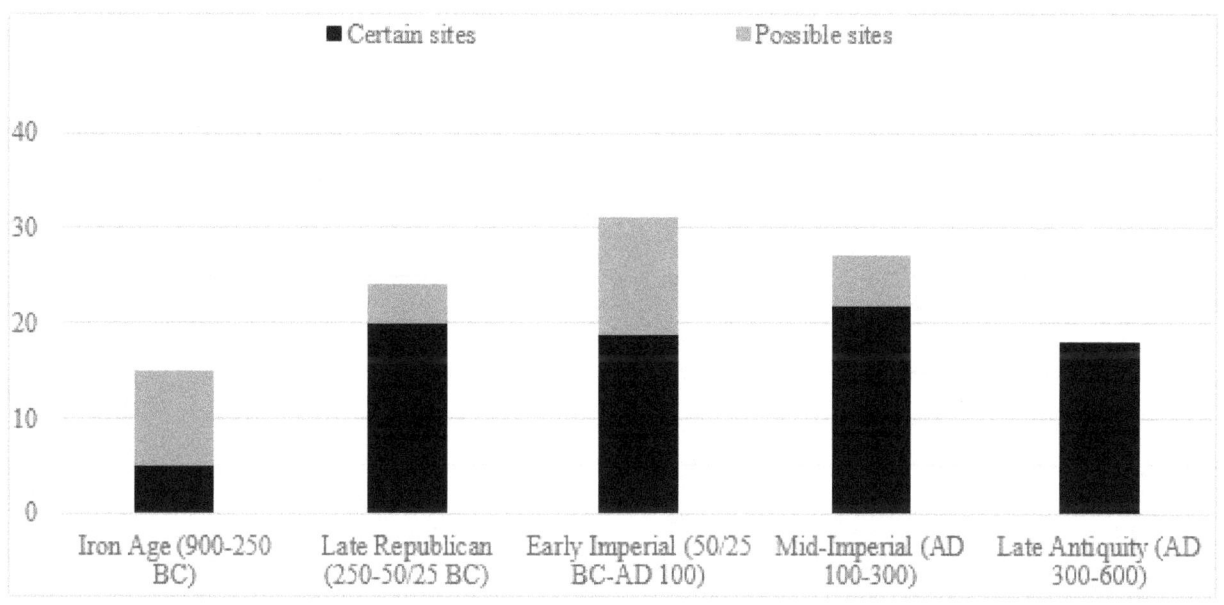

A

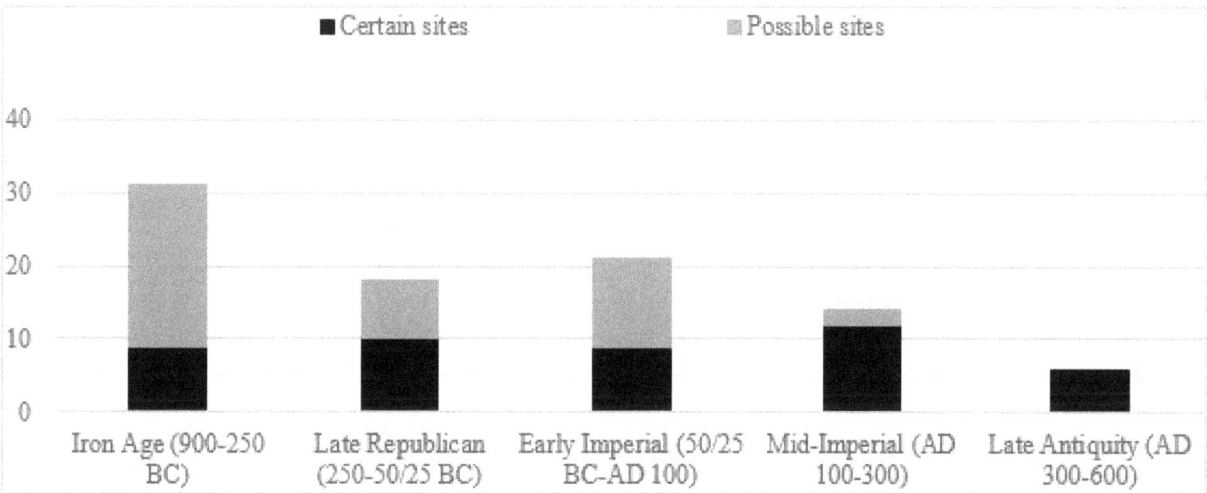

B

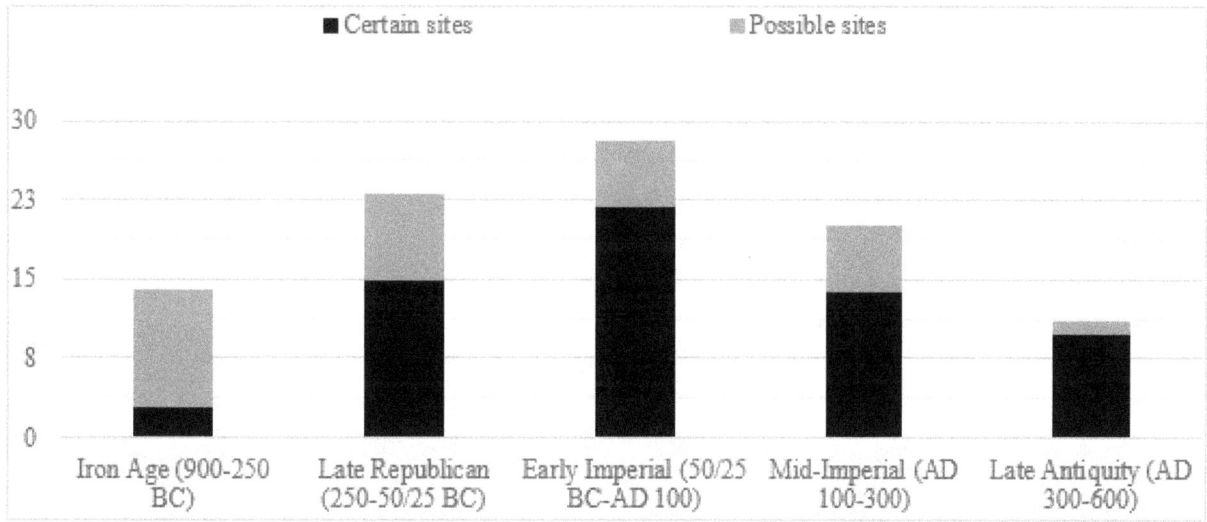

C

Figure 5.3: Rural site dynamics in the Potenza Valley (A: upper valley; B: middle valley; C: lower valley) between Iron Age and Late Antiquity (After Van Limbergen et al. 2017b, 122-123, Tables 6-8).

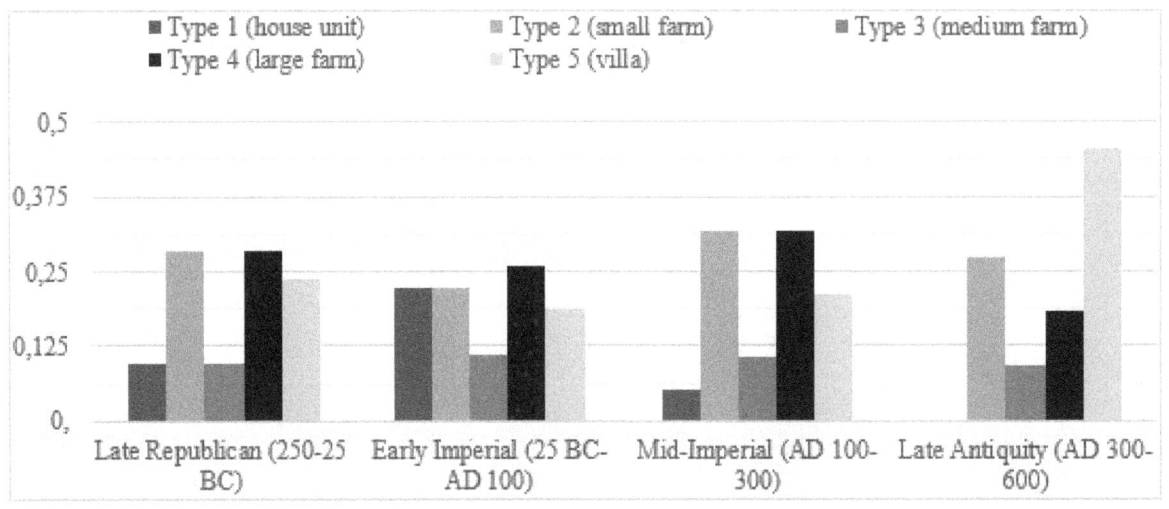

A

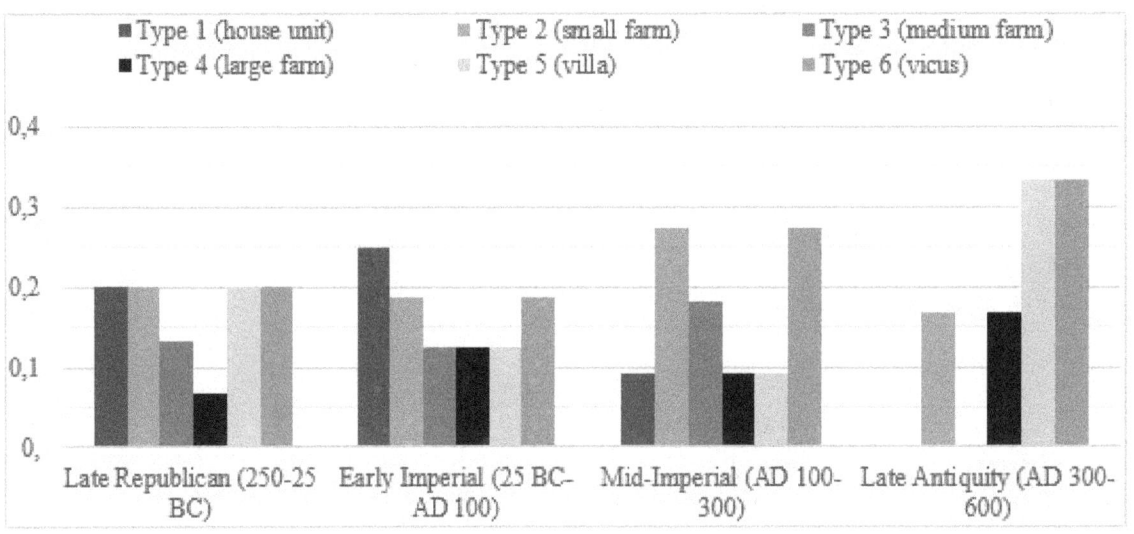

B

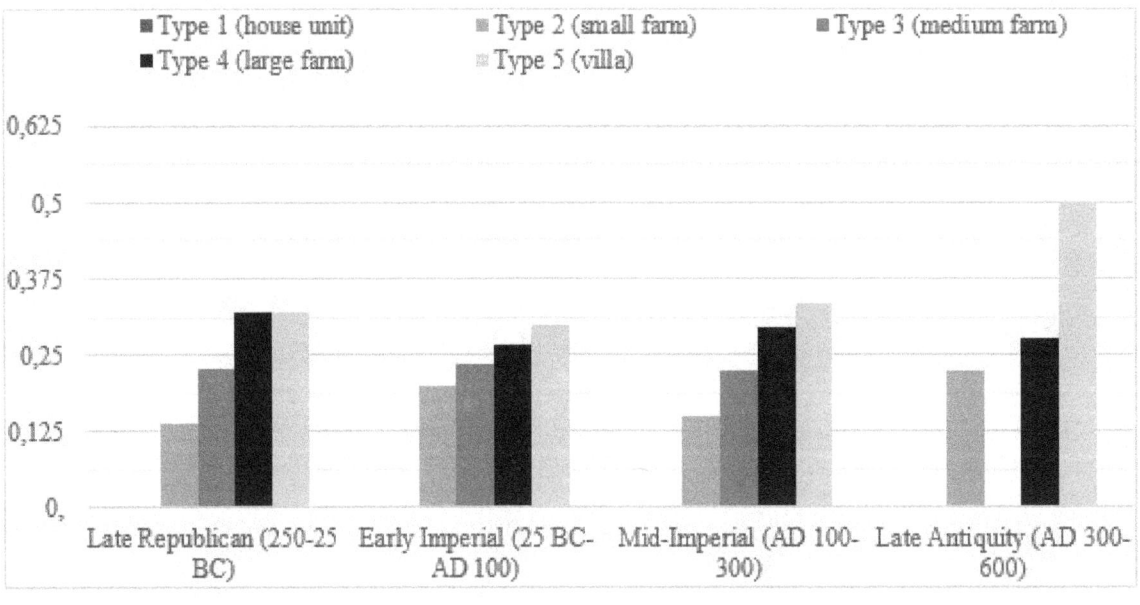

C

Figure 5.4: Diachronic rural settlement patterns in the Potenza Valley from the Late Republic to Late Antiquity (A: upper valley; B: middle valley; C: lower valley) (After Van Limbergen et al. 2017b, 145, Tables 15-17).

site types, but some consensus in the literature exists on the use of a nucleated family of five persons per Roman farm (even if there is some scepticism on the viability of such small families in high mortality pre-industrial societies) (Witcher 2011).[29] Informed estimates for villas generally range between 15 (Witcher 2005) and 35 persons (Perkins 1999, 167), with recurrent averages of 20 (De Graaf 2012) and 25 people (Attema and De Haas 2011). Similarly, for villages or small nucleated settlements, suggested occupation densities range from 50 (Perkins 1999, 166) to 100 people per site (Attema and De Haas 2011; Osborne 2004). In sum, assessing the size of the rural population at any moment in time based on field survey data is a highly contested matter, but this kind of evidence remains our best source to analyse population changes in the Roman world (Sbonias 1999; Ghisleni et al. 2011; Witcher 2011).

Taking into account all these remarks and considerations, the extrapolation of rural population figures on the basis of site typology and household size followed an established approach:

- The data from the three individual intensive survey transects in the Potenza Valley (upper valley: 3.15 km²; middle valley: 3.69 km²; lower valley: 3.88 km²) were combined to establish densities per site type (number of sites/km²);
- These densities were applied to the territories of the four towns in the valley (*Potentia, Ricina, Trea, Septempeda*) (Figure 5.1) to arrive at tentative total site numbers per type for each territory;[30]
- These site numbers were multiplied with an informed range of inhabitants according to site type; this gave an estimated population range for each territory;
- I performed this calculation for both the Late Republican and the Early Imperial period.

I hereby tested three different reconstructions. In a first scenario, I kept all six original site types and assigned the following occupation densities to them: 2-5 people for a small rural unit, 5-10 for a mall farm, 10-15 for a medium-sized farm, 15-20 for a large farm/simple villa, 20-25 for a (large) villa, and 100-250 people for a villages or roadside settlement (A).[31] In a second scenario, I grouped the rural units, small farms and medium farms (types 1 tot 3) into the general category 'farm' (1a) (5-10 people), and the simple villa and large villa (types 4 and 5) in the category 'villa' (2a) (20-25 people) (B). In a third scenario, I repeated this exercise with only the rural units and small farms (type 1, 2) counted as 'farms' (1b), and the medium farms, simple and large villas (types 3 to 5) counted as 'villas' (2b) (C). As an example, the three scenarios are outlined for the territory of *Potentia* (121 km²) in the Early Empire in Table 5.6. For this chapter, however, only the first scenario (A) was used to calculate the rural population for all four territories, as the results between the three models were not dramatically different.

This gave a total rural population for the territories in the Potenza Valley of between ca. 47,274 and 68,235 people in the Early Empire, and ca. 39,336 to 55,875 people in the Late Republic (Table 5.7). Rural population densities were thus 76-109 (EI) to 63-90 (LR) persons/km².

Table 5.6: The size of the rural population for the territory of *Potentia* (121km²) in the Early Empire (T = site type; U = upper valley; M = middle valley; L = lower valley; Tot = Total; POT == Potentia).

A							
T	U	M	L	Tot	/km²	POT	Population
1	0	4	6	10	0.9	109	218-545
2	6	3	6	15	1.4	169	845-1,690
3	7	2	3	12	1.1	133	1,330-1,995
4	8	2	7	17	1.6	194	2,910-3,880
5	9	2	5	16	1.5	182	3,640-4,550
6	0	3	0	3	0.3	3	300-750
Tot	30	16	27	73	\	790	9,243-13,410
B							
1a	13	9	15	37	3.5	424	2,120-4,240
2a	17	4	12	33	3.1	375	7,500-9,375
6	0	3	0	3	0.3	3	300-750
Tot	30	16	27	73	\	802	9,920-14,365
C							
1b	6	7	12	25	2.3	278	1,390-2,780
2b	24	6	15	45	4.2	508	10,160-12,700
6	0	3	0	3	0.3	3	300-750
Tot	30	16	27	73	\	789	11,850-16,230

Table 5.7: The estimated size of the rural population for the territories (expressed in km²) of the four towns in the Potenza Valley in the Late Republic (LR) and the Early Empire (EI).

Town	Ter.	LR Pop	EI Pop
Potentia	121	7,792-11,205	9,243-13,410
Ricina	184	11,724-16,685	13,872-19,970
Septempeda	186	11,281-15,740	14,034-20,195
Trea	133	8,539-12,245	10,125-14,660
Total	624	39,336-55,875	47,274-68,235

---

[29] Alternatives have been proposed for e.g. Roman Britain (20 people for the smaller sites; Millett 1990, 185), the Ager Caeretanus (10 people/farm; Enei 2001, 72) and the Ager Cosanus (10 people or 2 families per site of <1,000m²; Perkins 1999, 167).

[30] These town territories are entirely hypothetical and were defined by Voronoi (a.k.a. Thiessen) polygons in QGIS: *Potentia* (121 km²), *Ricina* (184 km²), *Trea* (133 km²), *Septempeda* (186 km²). The Voronoi polygon creates boundaries around a set of points in the landscape – in this case the 40 central Adriatic towns – as such dividing the latter into regions based on distance to points. Admittedly, this is a very theoretical and simple subdivision that does not take into account parameters such as the size of the town (larger towns may have had larger territories and vice versa) and the nature of the surrounding landscape, but it suffices for the purpose of this exercise.

[31] I decided not to apply the calculated densities for villages and roadside settlements to the entire territories, and instead use the detected number of villages, as there is little information on how frequent these villages were in the landscape, and their inclusion seemed to result in disproportionate total rural population numbers.

Based on this reconstruction, it seems that the countryside in the Potenza Valley was relatively densely populated from the 2nd century BC onwards, but most intensely between ca. 50 BC and AD 100. This impression seems in line with the aforementioned influx of veterans in the central Adriatic area from the mid-1st century BC onwards. If such densities characterized all 40 town territories – a yet unverifiable assumption – a total area of 10,525 km² would have been populated by **663,075-947,250** (LR) or **799,900-1,147,225** (EI) people at one point.[32] The corresponding urbanisation rate in the Early Empire would have been ca. 10% (ranging from 7 to 13%); low but perfectly acceptable according to pre-industrial standards. This scenario would also give some substance to the overall impression one gets from the urban archaeological record in the Early/High Empire; that is, most towns being service centres with a considerable amount of space reserved for public activities, and a rather modest habitation component, catering for the many people that lived in their immediate surroundings or further afield.

Food and land requirements

Daily caloric intakes

How much food did these people need? It remains difficult to estimate how many calories a person consumed on a daily basis in antiquity.[33] It goes without saying that the number of calories required per day is not constant, but fluctuates according to a number of parameters. The chief variables that influence daily caloric intakes are five: stature, weight, age, gender and the level of physical activity. Logically, taller and heavier persons need more energy than smaller and leaner ones; children and elderly people need less than adults, men more than women, and more active persons require more than less active ones.[34] To illustrate, according to the FAO, a male of 35 years of age, weighing 65 kg and measuring 172 cm needs about 3,490 calories when performing heavy work. An adult male of 25 years of age with the same weight and height needs only 2,580 calories when involved in light activity work, such as office tasks. The same goes for adult females, with housewives needing fewer calories (1,990) than rural women (2,235) (FAO 1985).

In practice, this makes it easier to calculate the needs of a particular group of people; that is, individuals who tend to have roughly the same height, weight, age and sex, or perform similar occupational activities. A good example are military men. According to the Recommended Daily Allowance of the United States Army, a soldier aged 16-19 with average weight and height (175.2 cm) neds 3,600 cal. per day. Soldiers aged 25 with the same weight and height need 3,200 cal. More than actual consumption rates of modern American soldiers, however, these are amounts that the army considers adequate for recruits to efficiently perform their duties on an annual basis, including periods of intensive activity (U.S. Army 1961).

Jonathan P. Roth recently used these U.S. Army guidelines to establish the food requirements of Roman soldiers. Roth took three parameters into account: body height, age of recruitment and the length of active service; all based on literature and archaeology. He thus argued that the average Roman soldier aged 30, measuring 170 cm and weighing 66 kg needed 3,000 cal. per day (Roth 1999, 9-13). This is in line with the amount that Polybius' infantrymen would have received (2,906 cal.) from a daily ration of four *modii* of wheat (Foxhall and Forbes 1982).

Matters tend to complicate when one tries to model the needs of groups with a more heterogeneous character. For antiquity, peasant families logically come to mind.[35] The daily physical labour of farmers was clearly different from those of soldiers, but households typically also included women, children and elderly people; each having particular caloric needs. Nathan Rosenstein has used the 1985 FAO statistics to estimate the minimum and maximum caloric requirements of a hypothetical household of a father aged 50, a mother of 40, one son of 15, another one of 20, and a daughter of 10 in Mid-Republican Italy. According to the FAO, depending on their level of physical activity (heavy to moderate), adult males aged 18 to 30 years and weighing 65 kg need 2,990 to 3,530 cal. per day. Adult males aged 30 to 60 with the same weight need 2,900 to 3,380 cal. Adult females aged 30 to 60 and weighing 55 kg only need 2,170-2,400 cal. per day.[36] Rosenstein thus used these values for the 20-year old son, the father and the mother. For the 15-year old son, he considered an intake of 2,650-2,900 cal.; for the daughter 1,950-2,350 cal. All these values took into account that the family members carried out moderate to heavy work on a daily basis; that is, levels of occupational activities characteristic for subsistence farming (Rosenstein 2004, 66-71).[37]

---

[32] The total area refers to the territories of the towns combined, as defined by Voronoi polygons in QGIS: *Aesis* (268 km²), *Ancona* (102 km²), *Asculum* (756 km²), *Attidium* (148 km²), *Auximum* (267 km²), *Beregra* (614 km²), *Camerinum* (401 km²), *Castrum N.* (188 km²), *Castrum T.* (208 km²), *Cingulum* (128 km²), *Cluana* (114 km²), *Cupra Mar.* (223 km²), *Cupra Mon.* (201 km²), *Fanum F.* (175 km²), *Falerio* (319 km²), *Firmum P.* (361 km²), *Forum S.* (338 km²), *Hatria* (280 km²), *Interamnia* (422 km²), *Matilica* (180 km²), *Novana* (623 km²), *Numana* (82 km²), *Ostra* (222 km²), *Pausulae* (249 km²), *Pisaurum* (228 km²), *Pitinum M.* (354 km²), *Pitinum P.* (375 km²), *Planina* (132 km²), *Sena Gallica* (187 km²), *Sentinum* (340 km²), *Suasa* (265 km²), *Tifernum M.* (340 km²), *Tolentinum* (169 km²), *Tuficum* (169 km²), *Urbs Salvia* (149 km²), *Urvinum M.* (324 km²).
[33] For a discussion on food and calories in antiquity, see Van Limbergen 2018b.
[34] In the case of infants, additional energy is required for growth, while women need supplementary energy in periods of pregnancy, and lactation (FAO 1973; 1985; 2004).

[35] Previous estimates for the food needs of peasant households in antiquity, mostly estimated in equivalents of wheat, include Brunt 1971; Hopkins 1978; 1980; 1983; Kehoe 1988, 14-15; and Erdkamp 2005, 48-49).
[36] Energy and Protein Requirements: Report of a Joint FAO/WHO/UNU/ Expert Consultation; see also Rosenstein 2004, 66-67.
[37] Rosenstein's approach is similar to the one used by Lin Foxhall and Hamish Forbes in their study on the role of grain as a staple food in antiquity. In order to define the number of cal. per day needed by human beings in antiquity, they used the FAO 1973 guidelines on Energy and Protein Requirements for 'moderately active' (2,852 cal.), 'very active' (3,337 cal.) and 'exceptionally active' (3,822 cal.) adult males (based on

One potential bias in Rosenstein's model is his reliance on energy requirements based on modern individuals with possibly different anthropometric features than Roman farmers. In a more recent attempt at outlining the needs of both a poorly and well-fed rural population in the Middle Tiber Valley, Helen Goodchild started from the skeletal evidence at *Herculaneum*, *Pompeii* and *Isola Sacra* to estimate the height, weight and BMR (Basal Metabolic Rate) of Roman agricultural workers.[38] She then used these figures in conjunction with the FAO statistics to estimate the daily nutritional requirements of a hypothetical Roman farmer. She hereby assumed both a light, moderate and heavy workload within a nine-hour working day (the latter was derived from ethnographic parallels). Next, she recalibrated these figures according to the annual workload outlined by Columella in his agricultural handbook. She so arrived at an average daily energy requirement of 3,283 cal. for a Roman farmer. She repeated the calculation for female adults, assuming a workload identical to those used for male adults. Finally, by combining the figures for males and females, and modelling a variety of height, weight and age ranges, she set the daily food intake for Roman farmers at 1,951-3,798 cal. (Goodchild 2007).

Goodchild's approach to combine the FAO guidelines with ancient skeletal and textual data can be further elaborated on to explore the potential food needs of the urban and rural population in central Adriatic Italy. Human energy needs are defined by the FAO as *"the amount of food energy needed to balance energy expenditure in order to maintain body size, body composition and a level of necessary and desirable physical activity consistent with long-term good health"* (FAO 2004, 4).[39] In other words, human beings need a certain amount of calories in order to compensate for the energy expended on the maintenance of their vital bodily functions and the performance of physical activities in the day. As mentioned above, the main parameters influencing these needs are body height, weight, age, sex and the level of activity. Together, the first four variables influence the so-called Basic Metabolic Rate or BMR of a person; that is, the minimal amount of energy required in order to keep going the basal metabolism processes of the human body. This includes the energy needed to maintain a series of functions essential to life, such as cell functions, the upkeep of body temperature, brain functions, etc. In doing so, the BMR is responsible for 45-70% of all energy that is expended by a person in 24 hours (FAO 2004). BMR is calculated by using the Harris-Benedict equations for men and women, and is expressed in either Kilocalories (kcal) or Megajoules (MJ) (Harris and Benedict 1918; 1919):

BMR (men) = 66.473 + (13.7516 x weight in kg) + (5.0033 x height in cm) – (6.7550 x age in years)

BMR (women) = 655.0955 + (9.5634 x weight in kg) + (1.8496 x height in cm) – (4.6765 x age in years)

To illustrate, a male adult aged 25, weighing 55 kg and measuring 160 cm would have a BMR of 1,454.76 kcal or 6.09 MJ; a female adult of the same age, height and weight would have a BMR of 1,427.11 kcal or 5.98 MJ.

Information on heights in antiquity can be collected from skeletal evidence. Roman measurable skeletal material is still relatively scarce in Italy, but some progress has been made in recent years. For example, a recent study by Geoffrey Kron on 49 bone studies from Italy between 500 BC and AD 500 focused on evidence regarding the stature of adult males. Kron's analysis revealed mean heights of between 163.5 and 172.7 cm for individual sites; that is, a mean height of 168.3 cm (Kron 2005, 73-74; 2008, 80-81). Another study by Monica Giannecchini and Jacopo Moggi-Cecchi focused on stature evidence from bone samples in central Italy between Iron Age and Medieval times. Five out of 17 contexts were Roman burial sites in Lazio, Molise and Marche. For these sites, mean heights of 164.2-165.4 cm for men, and 149.5-153.2 for women, were established. The data from Roman *Suasa* in the Marche region gave mean heights of 164.7 and 153.2 cm for men and women respectively (Giannecchini and Moggi-Cecchi 2008). Further data for the study area come from cemeteries in five other towns: *Potentia* (Porto Recanati), *Urvinum M.* (Urbino), *Fanum F.* (Fano), *Septempeda* (Civitanova Marche and *Forum S.* (Fossombrone). For adult men, the mean heights ranged between 163.3 (*Potentia*) and 168.5 cm (*Fanum F.*); for women they ranged from 150.0 (*Urvinum M.*) to 157.4 cm (*Potentia*).[40]

Next, I derived the possible weight of these individuals from their height by using their BMI or Body Mass Index. Obviously, it is impossible to measure the BMI of people in antiquity directly, but values of 18.9 and 24.5 are nowadays considered to be the lower and upper limits for a healthy BMI. The FAO has recently put forward a mean value of 21 as an acceptable value on the level of the population as a whole (FAO 2004, 40). Of course, we

---

a man aged 20-39, weighing 62 kg). In a second model, they also used these guidelines to set the needs of a hypothetical household of six at 15,495 cal. This family consisted of one very active female of 52 kg and 60-69 years of age (1,947 cal.), one very active adult male of 62 kg and 20-39 years (3,337 cal.), one very active adult female of 52 kg and 20-39 years of age (2,434 cal.), one very active male child of 13-15 years of age (3,237 cal.), one female child of 10-12 years (2,350 cal.), and one child of 7-9 years of age (2, 190 cal.) (Forbes/Foxhall 1982, 48-49, based on FAO 1973).

[38] Weight was here calculated from height by using the Body Mass Index (in this case = 20-25) through the formula: weight in kg = (height³/10,000) x Body Mass Index; the BMR was calculated by the standard Harris-Benedict formula for male adults: BMR (per day) = 66.4730 + (13.7516 x weight in kilos) + (5.0033 x height in cm) – (6.7550 x age in years) (Goodchild 2007, 317-319).

[39] It is of course another matter entirely to which extent people in antiquity were able to achieve this on a daily basis.

[40] The data for *Forum Sempronii* were collected from a preliminary excavation report of a late antique cemetery positioned within the circuit wall of the Roman town, but the height ranges were too general (and very high) in order to use them in this chapter (Mei/Gobbi 2014, 944). There has been some discussion on the methodology used to derive these heights, so it is possible that these heights will change in the future (e.g. Killgrove 2019; Flohr 2019).

cannot know whether or not people effectively maintained a healthy BMI in Roman times – and in many cases they probably did not – but by including both the lower and the higher end of the acceptable spectrum, it becomes at least possible to outline the implication for both a poorly and a well-fed population. Therefore, BMI values of 18.9, 21 and 24.5 were used in order to outline a range of possible BMR in antiquity. I hereby applied the following formula:

Weight in kg = (height^2/10,000) x BMI (= Body Mass Index)

To illustrate, an adult male of 163.3 cm (*Potentia*) – with a BMI value of 18.5, 21 or 24.9 – might have had a body weight of 49, 56 or 66 kg. Assuming that he was 25 years old, his BMR would have ranged from 1,388.47 to 1,622.24 kcal. Based on the central Adriatic skeletal data, models were then created for a minimum male height of 163.3 cm (*Potentia*) and a maximum height of 168.5 cm (*Fanum F.*); for females this was 150.0 (*Urvinum M.*) and 157.4 cm (*Potentia*). Three age groups were considered: 18-29 years (models for 18-25-29), 30-59 years (30-45-59), and 60+ years (60-65-70).[41]

After BMR, the second most influential parameter on daily energy expenditure is the Physical Activity Level or PAL. This is especially so for adults, as their BMR is relatively constant and no more energy is required for growth. The FAO 1973/1985 reports simply classified the PAL of adult persons according to their involvement in light, moderate or heavy occupational activities (FAO 1973; 1985). The 2004 report modified this classification and instead opted to rank the physical activity levels according to certain 'lifestyles', so combining 'obligatory' and 'discretionary' activities carried out during the day.[42] They classified these lifestyles into three main categories:[43]

1. **Sedentary or light activity lifestyle**: persons who fall under this category usually have occupations that do not demand much physical effort. In addition, they are not required to walk long distances, generally use motor vehicles for transportation, do not exercise or participate in sports regularly, and spend most of their leisure time sitting or standing, all this with little body displacement (e.g. talking, reading, watching television, listening to the radio, using computers). A good example is male office workers in urban areas, who only occasionally engage in physically demanding activities during or outside working hours. Another example may be rural women living in villages that have some facilities, like electricity, piped water and nearby paved roads. These women spend most of their time selling produce at home or in the marketplace, or doing light household tasks and take care for their children in or around their houses.

2. **Active or moderately active lifestyle**: this category of people has occupations that are not strenuous in terms of energy demands, but do involve more energy expenditure than those described for sedentary lifestyles. For example, they can be persons with sedentary occupations who regularly spend a certain amount of time in moderate to vigorous physical activities, during either the obligatory or the discretionary part of their daily routine. This may be the daily performance of one hour (either continuous or in several sessions throughout the day) of moderate to vigorous exercise such as jogging, running, cycling, aerobic dancing or various sports activities. In fact, all these activities can raise a person's average PAL from 1.55 (corresponding to the sedentary category) to 1.75 (the moderately active category). Moderately active lifestyles are also connected with certain 'physical' jobs, like masons and construction workers, or rural women in villages that are less developed than in the previous category, where performing agricultural tasks or walking long distance to fetch water and fuelwood are a fixed part of the daily routine.

3. **Vigorous or vigorously active lifestyle**: persons that attain this type of lifestyle engage regularly in strenuous work or in strenuous leisure activities for several hours. Good examples are women with non-sedentary occupations, or that swim or dance an average of two hours a day, or non-mechanised agricultural workers who handle a machete, hoe or axe for several hours daily and walk long distance over rugged terrains, often carrying heavy loads.

For each of these three categories, the FAO assigned a range of daily PAL values: 1.40-1.69 for the sedentary lifestyle, 1.70-1.99 for the active lifestyle, and 2.00-2.40 for the vigorous lifestyle.[44] For each lifestyle, the FAO also outlined a list of typical activities associated with a certain style and their corresponding average PAL values attained by a person living according to that particular style: 1.53 for the sedentary level, 1.76 for the active, and 2.25 for the vigorous one (Table 5.8). Obviously, the determining factor for ascribing a certain person's lifestyle to one of the three categories is the intensity of this person's habitual physical activity in the course of a 24-hour period; that is, the more intense it is, the higher the PAL value.

We may use these lists as a basis for estimating the mean hypothetical PAL value of Roman town dwellers. Of course, the daily activities listed by the FAO are a reflection of modern living practices and cannot be extrapolated directly to the pre-industrial Roman era. What we can

---

[41] For the details of all these calculations, see Van Limbergen 2015, Appendix IV.

[42] With the term 'obligatory', reference is made to activities such as work, school and other demands made on people by their economic, social and cultural environment; 'discretionary' activities, on the other hand, include personal tasks, household affairs, leisure time, etc. (FAO 2004, 7).

[43] The following definitions are taken in part from the FAO 2004 report, 39.

[44] Pal values higher than 2.40 cannot be maintained over longer periods of time, but may be manageable during shorter periods of intense physical activity. An illustration may be cycling games that last for several weeks (FAO 2004, 39).

imagine, however, is that Roman towns were populated by individuals – or groups of individuals – that had lifestyles comparable to the three FAO classes. Roman towns were typically 'melting-pots', with a range of human beings present, either free or slaved, young and old, belonging to lower or higher social classes, or involved in different occupational activities. For central Adriatic Italy, the latter matter is well-illustrated by a recent study of Alessandro Cristofori on the epigraphic record of *Picenum*. These textual sources mention persons involved in a variety of occupations, mostly administrative in nature (managers, magistrates, public notaries, delegates, overseers, financial controllers, secretaries, clerks) – and thus comparable to the living conditions falling under the sedentary or light activity lifestyle – but also artisans and other workshop personnel, various traders and dealers, sportsmen and musicians (gladiators, ball players, trumpeters), soldiers, farmers, servants and bailiffs, or people involved in the construction industry (architects, builders, carpenters, plaster workers) and the medical sector (surgeons, nurses) (Cristofori 2004). These jobs reflect social positions and levels of physical activity that may be associated with either moderately or vigorously active lifestyles, as these people spent different amounts of energy in the course of their daily activities, and thus had varying energy needs.

Still, in order for the FAO lists to be applicable, they first need to be adapted to the circumstances in antiquity. For instance, the original tables attribute a one hour time-period to activities such as 'driving car to/from work' or 'commuting to/from work on the bus', while reserving only one hour for 'walking at varying paces without a load'. These were replaced by three hours of 'walking' in order to reflect the fact that people displaced themselves mostly on foot within urban contexts in antiquity. Other modifications include the omitting of 'general household work' in favour of 'non-mechanised domestic tasks', and the reducing of leisure time from 3-4 hours in modern times to only one hour in antiquity. In the end, this resulted in higher mean PAL values of 1.63 for the sedentary lifestyle (1.53 for modern), 1.87 for the active lifestyle (1.76 for modern), and 2.35 for the vigorous lifestyle (2.25 for modern) (Table 5.9; 4.10; 4.11). This was considered fit as people in antiquity had an overall harsher and physically more demanding lifestyle than modern individuals.

As such, for adults, the total amount of energy required for each day (TEE = Total Energy Expenditure) can be calculated by multiplying the BMR with the daily PAL value:

TEE = BMR x PAL

For instance, a male adult of 25 years, weighing 56 kg and measuring 163.3 cm (*Potentia*), with a BMR of 1,484.72 kcal. would have had a daily energy requirement between 2,078.61 and 2,509.18 cal. when involved in a sedentary lifestyle, but 2,524.02-2,954.59 cal. with an active lifestyle, or 2,969.44-3,563.33 cal. with a vigorous lifestyle. Based on the mean PAL values calculated for the Roman period, the average caloric requirements of this person would have been 2,420.09 (sedentary), 2,776.43 (moderately active), or 3,489.09 cal. (vigorously active).

When taking into consideration the whole range of PAL values – that is, from a minimum of 1.40 up to a maximum of 2.40 – the caloric requirements for Roman male urban are 1,518 to 4,324 kcal.; for female persons they are 1,410-3,492 kcal. If we take into account only the mean values – which are most likely to reflect the living circumstances of urban dwellers in Roman Italy – the caloric needs range

Table 5.8: PAL ranges and mean PAL values according to different lifestyles (After FAO 2004, Table 5.3).

Lifestyle	PAL range	PAL mean
Sedentary or light activity	1.40-1.69	1.53
Active or moderately active	1.70-1.99	1.76
Vigorous or vigorously active	2.00-2.40	2.25

Table 5.9: Estimated daily energy expenditure of Roman town dwellers (R) compared with modern figures (M) with a sedentary or light activity lifestyle (Based on FA0 2004, 36, Table 5.1).

Main daily activities	Time (hours)		Energy Cost (PAR)		Time x Energy Cost	
	M	R	M	R	M	R
Sleeping	8	8	1	1	8.0	8.0
Personal care	1	1	2.3	2.3	2.3	2.3
Eating	1	1	1.5	1.5	1.5	1.5
Cooking	1	1	2.1	2.1	2.1	2.1
Work (sitting)	8	8	1.5	1.5	12.0	12
Non-mechanised domestic tasks	\	1	\	2.3	\	2.3
Household work	1	\	2.8	\	2.8	\
Driving car to/from work	1	\	2.0	\	2.0	\
Walking at various paces without a load	1	3	3.2	3.2	3.2	9.6
Light leisure activities	2	1	1.4	1.4	2.8	1.4
Totals	**24**	**24**			**36.7**	**39.2**

Table 5.10: Estimated daily energy expenditure of Roman town dwellers (R) compared with modern figures (M) with an active or moderately active lifestyle (Based on FA0 2004, 36, Table 5.1).

Main daily activities	Time (hours)		Energy Cost (PAR)		Time x Energy Cost	
	M	R	M	R	M	R
Sleeping	8	8	1	1	8.0	8.0
Personal care	1	1	2.3	2.3	2.3	2.3
Eating	1	1	1.5	1.5	1.5	1.5
Cooking	\	1	\	2.1	\	2.1
Work (standing, carrying light loads)	8	8	2.2	1.5	17.6	17.6
Non-mechanised domestic tasks	\	1	\	2.3	\	2.3
Commuting to/from work on the bus	1	\	1.2	\	1.2	\
Walking at various paces without a load	1	3	3.2	3.2	3.2	9.6
Low intensity aerobic exercise	1	\	4.2	\	4.2	\
Light leisure activities	3	1	1.4	1.4	4.2	1.4
Totals	24	24			42.2	44.8

Table 5.11: Estimated daily energy expenditure of Roman town dwellers (R) compared with modern figures (M) with a vigorous or vigorously active lifestyle (Based on FA0 2004, 36, Table 5.1).

Main daily activities	Time (hours)		Energy Cost (PAR)		Time x Energy Cost	
	M	R	M	R	M	R
Sleeping	8	8	1	1	8.0	8.0
Personal care	1	1	2.3	2.3	2.3	2.3
Eating	1	1	1.4	1.4	1.4	1.4
Cooking	1	1	2.1	2.1	2.1	2.1
Work (non-mechanized, agricultural)	6	6	4.1	4.1	24.6	24.6
Collecting water/wood	1	\	4.4	\	4.4	\
Non-mechanised domestic tasks	1	3	1	2.3	2.3	6.9
Walking at various paces without a load	1	3	3.2	3.2	3.2	9.6
Light leisure activities	4	1	1.4	1.4	5.6	1.4
Totals	24	24			53.9	56.3

from 1,768 to 4,146 kcal. for men, and from 1,641 to 3,419 kcal. for women. Combined, this means that urban dwellers in central Adriatic Italy were likely to require between **1,705 and 3,783** calories per day (Table 5.12). These are the values on which to base the urban food and land model. For the rural population, only PAL values associated with moderately or vigorously active lifestyles should be taken into account, as farmers most likely had an overall more burdening lifestyle than town dwellers (**1,956-3,783** kcal.).

Table 5.12: Minimum and maximum caloric needs for Roman men and women between 18 and 70 years of age in central Adriatic Italy.

PAL	Men		Women		Combined	
	Min	Max	Min	Max	Min	Max
1.40 - 2.40	1,518	4,234	1,410	3,492	1,464	3,863
1.63 1.87 2.35	1,768	4,146	1,641	3,419	**1,705**	**3,783**

Annual food and land needs

The impact of the amount of land that was required to feed the urban and rural population of central Adriatic Italy needs to be assessed on the basis of dietary composition and main crop yields. There is no room here to discuss in depth Roman eating habits and dietary proportions, but it is commonly assumed that in antiquity between 60 and 75% of the daily calories came from cereals. The remaining 25 to 40% mostly consisted of wine and olive oil, vegetables and fruit, and possibly some meat or fish.[45] For this paper, I modelled shares of 60% cereals, 15% wine, 10% oil and 15% other foodstuffs (e.g. vegetables, fruit, meat, fish) for urban dwellers; for the rural population, I set the shares at 75%, 5%, 5% and 15% respectively. These proportions are inspired by the assumption that the diets of peasants relied more on cheaper calories, and were thus less varied, than those

---

[45] For a recent discussion of the issue, see Van Limbergen 2018b.

Table 5.13: Estimated annual food needs of the Early Imperial population in central Adriatic Italy (wheat, wine, olive oil).

POP	Wheat (kg)	Wine (l)	Oil (l)
82,640 (UrbLow)	9,049,080-20,511,248	10,858,896-24,436,648	603,272-1,206,544
123,645 (UrbHigh)	13,539,128-30,688,689	16,246,953-36,561,827	902,609-1,805,217
779,900 (RurLow)	128,463,940-219,012,620	35,035,620-78,870,140	2,959,630-6,159,230
1,147,225 (RurHigh)	184,244,335-314,110,205	50,248,455-113,116,385	4,244,733-8,833,633

of town residents.[46] In the face of assessing the amount of land required for meeting these diets, I only modelled the data for the so-called 'Mediterranean triad': grains, grapes (wine) and olives (oil).

Assuming such diets, a town dweller in antiquity (men and women combined) consumed ca. 109.5-248.2 kg of wheat, 131.4-295.7 l of wine, and 7.3-14.6 l of oil per year. A countryman had an annual per capita consumption of ca. 160.6-273.8 kg of wheat, 43.8-98.6 l of wine, and 3.7-7.7 l of oil. We should hereby include extra amounts of oil for purposes other than food, such as bathing and lighting. It is impossible to determine how much olive oil a person in Roman Italy 'consumed' in the form of soaps, perfumes, medicines and lamps, so I decided not to model this variable here. A useful order of magnitude, however, may come from ancient oil lamps, which needed a good 10 litres of fuel per year when used on a daily basis (Wunderlich 2003, 256; Crnobrnja 2008, 411; Van Limbergen 2018b, 1065). Most urban households probably had more than one lamp, so an additional 10 l of non-food oil per urban person seems entirely possible. The case was probably different for rural households, as people in the countryside likely had a less luxurious lifestyle than those living in a town.

We can thus get a sense of what was consumed by an urban population of 82,640-123,645 people, and a contemporary rural population of 799,900-1,147,225 people, in the Early Empire. For the lowest combined population, 137,513,020 to 239,523,868 kg of wheat, 45,894,516 to 103,306,788 l of wine, and 3,562,902 to 7,365,774 l of oil was needed (Scenario 1); for the highest population, total consumption was 197,783,463 to 344,798,894 kg (wheat), 66,495,408 to 148,678,212 l (wine), and 5,147,342 to 10,638,850 l (olive oil) (Scenario 2) (Table 5.13).

For an assessment of the theoretical amount of land needed to produce these quantities of food, I used a net yield of 400 kg of wheat, 2,000 l of wine and 450 l of olive oil per hectare; all considered realistic averages for most lands of Roman Italy when cultivated intensively (i.e. if the piece of land was used exclusively for that particular crop).[47] For Scenario 1 this meant an aggregate surface of between 374,647.5 and 666,831.5 ha (3,746.5-6,668.3 km²); for Scenario 2 this was 539,144.9 to 959,978.2 ha (5,391.4-9,960 km²). The decisive factor in terms of land usage was clearly the amount of cereals, as these crops occupied about 90% of all cultivated land ((3,437.8-5,988 km² in Scenario 1; 4,944.6-8,620 km² in Scenario 2).

As recalled, the total area covered by the 40 town territories (calculated by Voronoi polygons) was ca. 10,525 km² (Figure 5.1). Even if these territories are largely hypothetical, they do allow to get a sense of how much land was available to the people living in central Adriatic Italy. It is impossible to know precisely how much of this land was suitable to agriculture, or how much of it was used for pasture, or which zones were covered by woodlands. But this model can provide us with a useful framework to explore under which circumstances population and food production were likely to impact land availability and land use in the area. From this point of view, it seems telling that under Scenario 2 – with the highest urban and rural population, and assuming that all people received adequate amounts of calories – ca. 94.5% of all land (9,960 out of 10,525 km²) had to be used for agriculture. Under Scenario 1 – with low populations and minimum caloric intakes – only 35.5% of all land (3,746.5 km²) was needed. In other words, the more people lived in the area, and the more prosperous and healthy they were, the more land use constraints manifested themselves. All evidence points to the Early/High Empire as the most likely period in which such developments may have taken place. Logically, this process would have touched in the first place cereal farming, thus straining the most suitable lands – that is, (alluvial) plains and valley bottoms – and expanding cultivation into less favourable areas (lower and higher hills). In fact, this may have become an issue rather quickly in the dominantly hilly and mountainous landscape of central Adriatic Italy (Table 5.14). To illustrate, we may confront the sizes of some of the coastal plains with the cereal land requirements of the overlapping town territories. Together, these territories cover an area of 1,502 km²; the relevant coastal plains amount to an aggregate 341 km².[48] To provide all people living in town and country with cereals under Scenario 2, ca. 699.1 to 1,215.4 km² of land was needed; that is, 46.5-81% of their territories, and largely exceeding the amount of plain land available. Under these conditions, good cereal land would

---

[46] Various other dietary scenarios are of course possible by adjusting the shares of the different foodstuffs. For a slightly different model, see Van Limbergen et al. 2017a.

[47] Spurr 1986; Jongman 2003, 115; Kehoe 2007, 551; even if grape and olive harvests in particular varied considerably from year to year (Kron 2012, 159; Mattingly 1994); see also Van Limbergen et al. 2017a, 359, note 59). In practice, however, fields were often cultivated following a 'mixed' regime, with different crops combined on the same plot (see *infra*).

[48] Data on the sizes of coastal plains from Cencini and Varani 1991, 34, fig. 1.

Table 5.14: Landscape zones with their respective share for the Marche region.

Landscape zone	Area (km²)	Share (%)
High mountains	2,350	24.16
Medium high hills	2,817.25	28.96
Internal plains	419.44	4.31
Low hills	3,069.37	31.56
Coastal alluvial plains	587.74	6.05
Internal alluvial plains	482.81	4.97
Total	**9,727.17**	**100**

have become scarce rather quickly, arguably resulting in land pressure and diminishing returns. Comparable issues in land availability can be envisaged in the upper valley areas as well, as major parts of some of the town territories consisted of high mountains unsuitable for arable cultivation. Again, this remains mostly a theoretical exercise, but helps in framing the potentially far-reaching implications of rising population and living standards in the area on agricultural developments in the 1st-2nd century AD.[49]

Turning to viticulture, we might easily imagine how these processes impacted the use of wine land as well. Basically, low count scenarios left vintners with more and better land to grow their vines, more possibilities to intensify and/or expand these lands, and thus more favourable conditions to optimize yields and maximize their profits. Some of this is arguably visible in the regional amphora evidence, which attests to a full participation of the central Adriatic area in the Late Republican overseas wine trade.[50] Under high count scenarios, however, things were bound to be rather different. (Good) wine land became scarcer, increasingly forcing vine growers to exploit less ideal areas and/or change the way in which they organized their fields.[51] More wine was needed by the local population as well, thus creating significant limitations for the export of local wine. Perhaps this is one of the reasons why wine amphora export clearly diminishes in this period, with what remains now being shipped almost exclusively to Northern Italy and Rome?

In any case, despite the strong indications for a populous landscape in the 1st century AD, clear and immediate signs of profound stress seem absent. This suggests that these communities were able – at least for a while – to keep going on the land available to them. Perhaps it is no coincidence either that the regional (wine) press record – contrary to the amphora evidence – now shows signs of considerable vitality?[52] This also raises the question of how vine growers in the area responded to the challenge of feeding a growing population while coping with ever larger limits in land availability.

In the final part of this paper, I would like to try and answer this question by exploring the significance of the *arbustum* as an ingenious response to land pressure in Early/High Imperial central Adriatic Italy. In particular, I will review the qualities of this ancient vine agroforestry system as a sustainable agricultural strategy by adopting a multidisciplinary approach that combines the ancient source material on the '*arbustum*' with comparative historical analysis and ethnographic analogy. We start with setting out the theoretical framework that substantiates our reasoning.

Possible responses to land pressure and the versatility of agricultural intensification

In her seminal 1965 work on the conditions of agricultural growth, Ester Boserup saw population pressure as an independent variable that drove agro-technological progress and environmental change. In her view, this progress or change in land use as a response to population growth followed an extensive-intensive continuum, along which people adapted ever more intensive labour- and capital demanding agricultural strategies and technologies essentially aimed at increasing the cropping frequency of the land (Boserup 1965). The appeal of this reasoning is its simple and unitary course. Indeed, the model sees all pre-industrial societies responding to population pressure in an identical way; that is, by intensifying their cropping systems. While this does acknowledge the power of man to alter the productive capacity of his environment, it considers these interventions in a too unilateral and universal way. In other words, the model displays a profound lack in variability and diversity when it comes to human productive- and intensification strategies.

Contrary to Boserup's narrow approach, agricultural regimes can actually follow different pathways towards intensification. The variability with which these processes manifest themselves is linked to spatial, environmental and historical determinants (e.g. tenurial or other socio-economic factors) that are either regionally or nationally defined. In order to model our reasoning on these matters, it is useful to briefly take a look at how Timothy Kaiser and Barbara Voytek in 1983 – and later also Kathleen Morrison in 1994 – have (re)defined the concept of intensification in agrarian societies (Kaiser and Voytek 1983; Morrison 1994). They see three principal pathways through which agricultural intensification may be achieved:

- Intensification proper: Intensification in itself already comprises a wide range of operations that transcend Boserup's model of increased cropping frequency. These not only include an upgrade in labour and capital for working a plot of land more intensively, but also a change in the frequency of ploughing, manuring, sowing and other farming operations. The construction

---

[49] Admittedly, this model does not take into account the possibility of grain imports, but it also excludes the amounts of grain that might have been shipped to Rome, which we know was a destination for central Adriatic wine in this period (Van Limbergen 2018a).
[50] Van Limbergen 2018a.
[51] There is plenty of evidence from pre-industrial Europe to support such claims (Grigg 1980; Van Leeuwen and Seguin 2006).
[52] Van Limbergen 2011; 2019.

of soil and water controlling facilities – whether or not permanent – or other interventions that can increase the amount of suitable farm land also fall under this category.
- Specialization: The second pathway entails all strategies in which labour and resources are channelled towards achieving one particular outcome. This may be the focusing on a single crop or the use of one particular production strategy. The strategy is often applied for producing food crops for a certain target market, rather than for responding to overall increased local demand. Still, the increased production of a certain crop or crop variety may also be stimulated by local environments.
- Diversification: The third and final pathway is often considered the opposite of intensification and therefore a strategy that we do not often – or at least not immediately – associate with scenarios of increasing agricultural productivity. It can, however, be seen as 'intensifying' in the sense that resources, crops and labour are combined to produce an overall higher yield. At first, such productive strategies might seem extensive, rather than intensive, but what is increased here is the number of components of a particular productive system, while at the same time labour is diversified as well. A distinction can be made between two types of diversification: temporal and spatial. Temporal diversification refers to systems that combine different planting and/or harvesting times on the same plot. Spatial diversification mixes crops or crop varieties with varying growth characteristics.

It is important to stress that these three subdivisions are not mutually exclusive, but most of the time are operating simultaneously in various combinations within a society in order to achieve intensification. Once again, how these strategies were applied and combined – and to which extent one strategy was preferred over the other to reach or maintain certain production levels – depended mostly on environmental and socio-economic interactions. It follows that pathways of intensification could differ substantially from one region to the other, or even from one society to the other. Kaiser and Voytek illustrated this concept by focusing on how the Neolithic Vinca culture in southeast Europe responded to population growth; that is, by shifting from transhumant caprines to sedentary bovines and swines combined with cereals (intensification and specialization), while at the same time making fuller and wider use of local resources through hunting, horticulture and fruticulture (diversification) (Kaiser and Voytek 1983, 330-336). A more recent illustration is the Classic Maya civilization, which – before its collapse in the 9th century AD – adopted a wide range of intensive strategies for sustaining its high-density population in a wet lowland area: terracing, drainage, irrigation and raised field cultivation (landesque capital intensification), combined with a focus on maize culture (specialization), grown within a cultivation lengthening regime (crop cycle intensification); that is, the lengthening of the period in which fields are cultivated on an annual basis before fallowing in order to increase long-term productivity (Johnston 2003; Beach et al. 2009; 2010).

Closer to our study area, intensification took yet another form in sixteenth and seventeenth century Holland, where population growth encouraged land reclamation in coastal areas and around inland lakes, combined with convertible husbandry – that is, alternating arable and pasture on a field – and horticulture. Then again, in eighteenth and nineteenth century Scandinavia, the increase of land under cultivation went hand in hand with a reduced fallow strategy and the widespread distribution of the potato (Grigg 1980).

These examples should suffice to show that intensification and sustainability strategies were multifaceted phenomena, whose particularities were regionally, timely and socially defined, always adapted to the surrounding environment and channelled according to the needs of the local population. This understanding should then encourage us to adopt an open-minded view on how farmers in the Roman world looked for sustainable agricultural practices in response to population growth and increasing land needs. Indeed, it is from this perspective that I now wish to explore how vine growing in central Adriatic Italy may have evolved within a wider process of agrarian change in the Early Empire, triggered by an ever growing demand for cereals.

Vinea, porculetum or arbustum?

There is still a widespread view that – with Italy's gradual turn towards commercial viticulture from the mid-3rd century BC onwards – intensive modern-style vineyards largely replaced traditional mixed vine plantations in the Late Republic. An attentive look at our literary sources, however, reveals little similarities with today's typical vine fields of Tuscany and Piemonte, with their dense pattern of long parallel rows of vines laid out over sloping fields. Most of the vine growing systems described by the ancient agronomists have a marked mixed aspect, with vines distributed between other crops. This is unsurprising, as agriculture in the ancient Mediterranean was primarily polyculture in nature (Barker/Rasmussen 1998). Even if many vine training techniques are reminiscent of modern practices – that is, vines trained with or without props and organized into long trenches or rows of pits within a field (referred to as '*vinea*' by the ancient agronomists)[53] – archaeology amply attests how viticulture in Roman Italy was mostly extensive in scheme (Boissinot 2009; Forni 2002, 133-148). Indeed, traces of vineyards suggest a widespread occurrence of so-called '*coltura promiscua*', with very wide intervals between vine rows used for the cultivation of other crops; a system which Pliny calls the '*porculetum*' and considers characteristic for Umbria (HN 17.35.171) (Broise and Jolivet 1995, 112; Calci and Sorella 1995, 122; Quilici 1992, 117-129). Some late antique mosaics also illustrate this matter for northern Africa. One of the famous mosaics from the villa at Tabarka in Tunisia, for example, depicts a farm whose gardens are filled with palisaded vines (*vitis characata*) plotted between

---

[53] E.g. Varro, *Rust.* 1.8; Columella, *Rust.* 4.12; 16.3-4; 17; 19; 26; Pliny, *HN* 17.35.164-166).

other trees. Another example is the harvest mosaic from Cherchell, where vines are grown on a wooden *pergola* (pergola) framework, as such allowing space for other crops to be cultivated beneath and between the vines (Balmelle and Brun 2005).

By far the most discussed mixed vine growing practice in Roman Italy, however, is the *arbustum* or tree-wedded vine. The '*arbustum*' was conceived as an orchard in which crops were cultivated between rows of trees that had vines attached to and running between them. The practice is a recurring theme in Imperial iconography. Harvest scenes of *arbusta* are frequently part of Dionysian and bucolic scenes on sarcophagi, mosaics and frescoes. Despite their clearly religious and festive connotations, these representations sketch a fairly truthful picture of this vine growing practice (Figure 5.6; 5.7). In essence, we can reconstruct the system as follows: vines were guided vertically along trees and stretched out between them in the form of festoons (Figure 5.5, A). In-between the rows of trees, cereals or other crops were grown (Figure 5.5, C). During the harvest, grapes were hand-picked by ladders and collected in small cone-shaped baskets (Figure 5.6; 5.7) Afterwards, they were transferred into carts for transport.

Pliny and Columella sum up the kind of trees that were best suited for this purpose. Columella makes a distinction between two types of *arbustum*: a native Italian one called '*arbustum Italicum*' and a Gallic type from northern Italy (Cisalpine Gaul) named '*arbustum Gallicum*'. For the first type, Columella mentions the use of the poplar tree, the elm and the ash tree (*Rust.* 5.6.5); the second type made use of smaller trees like the cranberry, the cornel, the horn-beam, the mountain-ash and the willow (*Rust.* 5.7.1). Pliny adds the fig and the olive tree to this list, as well as the lime, the maple and the rowan for the area to the north of the Po River (*HN* 17.35.200-201). When laid out thoughtfully, the choice of trees seem to have depended on two main criteria: height and foliage density. The former ensured the protection from snooping animals and the maximization of vine growth; the latter's consideration took into account the seeking of balance in sun exposure and the farmer's need to feed his animals. This explains the popularity of the poplar and the elm in Roman Italy; that is, two trees that grow to great heights fast and have a high foliage density.

Both authors also give advice on the most suitable grape varieties for the *arbustum*. Pliny mentions that the Albuelis vine produces more fruit at the top of trees, while the Visulla vine does the same on the bottom branches (*HN* 14.4.31). Of the Aminian cultivar family, the smaller twin sister variety (*Gemellarum minor*) only flourishes when it is trained on trees, while the elder sister variety (*Aminaea maior*) is also less likely to suffer damage when trained on trees than on wooden frames (*HN* 14.4.22). Other varieties that do well on an *arbustum* are the three Apian vines (*Apianae*), the Helvolan (*Helvolae*) and the Nomentan vines (*Nomentanae*) (Col. *Rust.* 3.2.17-24), as well as the Dactylides (Plin. *HN* 14.4). From the descriptions of the agronomists, it can be deduced that these were mostly prolific (highly productive), hard-berried and early ripening varieties. Both authors make further interesting comments on the relationship between quantity and quality in these kind of plantations. Columella advices farmers who eye the obtainment of high yields to concentrate their efforts on the lower sections of the tree; those who desire high quality should try to mount the vines as high as possible on trees (*Rust.* 5.6.24). The issue is even firmer expressed by Pliny, who is of opinion that not only are the better wines made from grapes at the top of the trees and the mass wines from those lowest down, but also that high-class wines *tout court* can only be obtained from vines on trees because of the beneficial effects of height and better sun exposure (*HN* 17.35.1999-200).

There is a clear link between the geographical distribution of the *arbustum* within Italy as indicated by the ancient agronomists, their practical guidelines and certain environmental circumstances. Indeed, it seems no coincidence that most references to this technique concern northern Italy (*Gallia Cisalpina*) – in particular the Po plain (Milan, Novara, Piacenza, Ravenna) and the Veneto/Trento area – and to a lesser extent central (Rieti, *Ager Caecubus* in today's Pontine Marshes in southern *Latium*) and southern Italy (*Campania*, Canosa di Puglia) (Figure 5.8); that is, mostly in lower and flatter areas with a high level of humidity and rainy and foggy weather conditions, especially in the bottom valley floors (Plin. *HN* 14.3.34). The damaging effects of excessive rainfall, fog or high soil humidity on grapes are well-known today. The main risk is the incidence of fungal infections that ultimately kill the vine (Jackson 2008). Even if the ancients did not understand the precise biochemical link between humidity and grape rot, they too were familiar with the problem, as is illustrated by Strabo's account of viticulture in the marshy environments around Ravenna, where vines yielded fruit quickly and in great quantities, but did not live beyond the age of 4-5 years (*Geogr.* 5.1.7). Varro's comments are particularly insightful to this purpose:

"In Italy, the people of Reate (Rieti) practice this custom. This variation in culture is caused chiefly by the fact that the nature of the soil makes a great difference; where this is naturally humid the vine must be trained higher, because while the wine is forming and ripening it does not need water, as it does in the cup, but sun. And that is the chief reason, I think, that the vines climb up trees." (Rust. 1.8.6)

Virgil expresses a similar opinion regarding the areas in which vines needed to be trained much higher – preferably with the aid of trees – in his Georgics:

"But if a soil exhales thin mists and curling vapours, if it drinks in moisture and throws it off again at will, if it always clothes itself in the verdure of its own grass, and harms not the steel with scurf and salt rust, that is the one to wreath your elms in joyous vines." (G. 2.217-221)

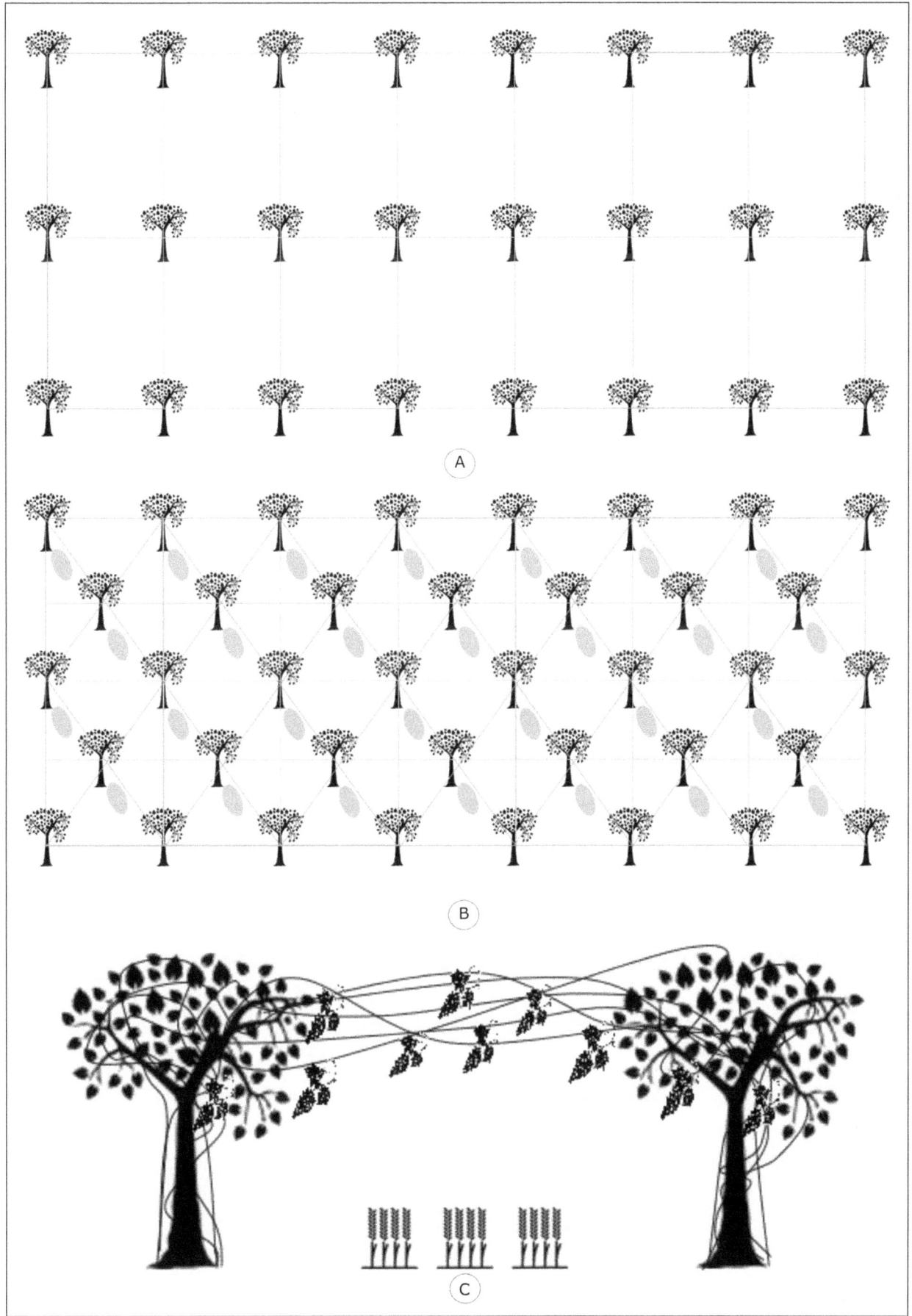

Figure 5.5: Schematic reconstruction of an arbustum: A) traditional layout; B) layout according to quincunxes; C) cereals cultivated between the rows of trees (Figure by author).

Figure 5.6: Sarcophagus representing a Dionysiac vintage festival, AD 290-300; on the left erotes are picking grapes with the aid of ladders from vines on trees (The J. Paul Getty Museum, Villa Collection, Malibu, California, Digital Image courtesy of the Getty's Open Content Program).

Figure 5.7: Detail of a sarcophagus of a child with vintage scenes, depicting erotes picking grapes from vines linked to trees, 2nd century AD (Photo © DAI Rome Photo Archives, neg. D-DAI-ROM-71.224, now in the National Museum of Palazzo Venezia, Rome).

Other than meeting well with the plant's natural climbing tendency, the training of vines high in trees in these areas was thus considered an excellent solution for protecting the grapes from these harmful environmental conditions. This also explains the agronomists' advice to use either grapes with hard skins or those that ripe early in the season (see *supra*), as the former are more resistant for bad weather conditions, while the latter avoid these by maturing before their arrival. This is not to say that lower-trained vines could not thrive in these areas; they were just confined to places where drainage, humidity and fog were less of a problem, like on hill slopes and higher plateaus. They were also more common in the warmer and dryer areas of central and southern Italy, where vines needed to be kept low and their growth limited in order to keep more groundwater available for the development of the grapes (Braconi 2008).

The popularity of the *arbustum* in Roman Italy, however, cannot just be explained by it being suitable for certain soils and climates. In fact, some of our textual and iconographic data suggest that its use within the peninsula was deeply rooted in pre-Roman peasant farming traditions. In a classical work on the history of the Italian agricultural landscape, Emilio Sereni has argued for a link between the diffusion of the *arbustum* in Italy and the maximum expansion of the Etruscan civilization (Sereni 1961, 40-43). The argument is mostly based on onomastic grounds, but there is growing archaeological evidence for the existence of *arbustum*-like vine plantations in certain areas of *Aemilia* and *Etruria*, and around Rome, from early times onwards. For example, remains of tree trunks found together with grape seeds deep in the alluvial layers of the Po plain around Modena and Ferrara are believed to be the earliest representations of this technique in northern Italy (Malavolti 1948; Buono and Vallariello 2002). Traces of vines trenches in combination with large quadrangular tree pits – probably dating to Etruscan or Early Roman times – have been found near Magliano in Tuscany (Boissinot 2009, 121-122). Similar finds dating to the Republican period have been registered around Rome at Tor Pagnotta (Laurentina) and Centocelle (Santangeli Valenzani and Volpe 1980; 2007; 2012; Kolendo 1994). The Etruscans are also believed to have introduced the technique in *Campania*, where Roman Imperial vine plantations bordered by and/or mixed with rows of trees are known from *Pompeii* (Jashemski 1973; 1979) (Figure 5.8). Some link with this important Italic population thus indeed

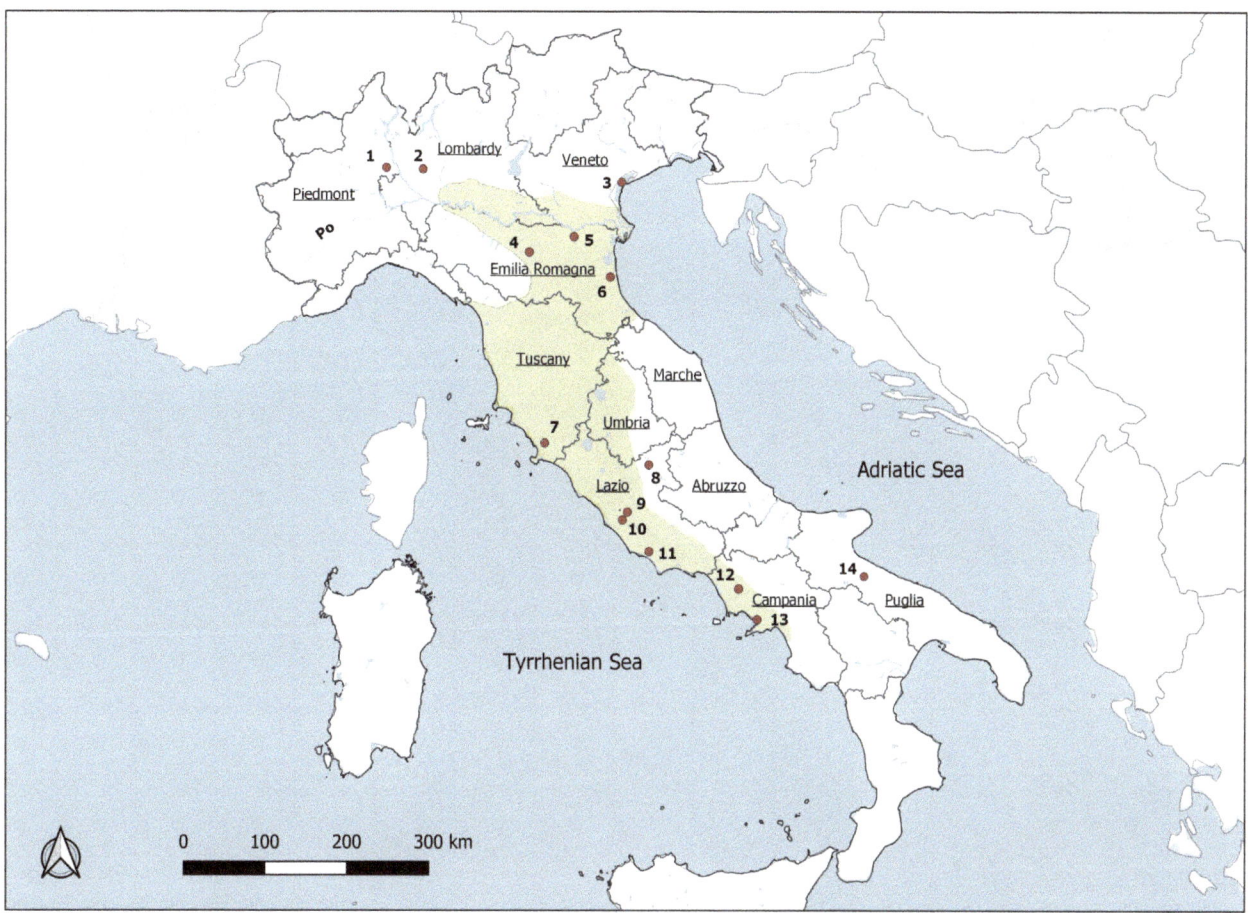

Figure 5.8: Archaeological and literary evidence regarding the historical presence of the *arbustum* in Roman Italy, with indication of the main cultivation regions, and the maximum expansion of the Etruscan civilization (green area): 1) Novara, 2) Milan, 3) Venice, 4) Modena, 5) Ferrara, 6) Ravenna, 7) Magliano, 8) Rieti, 9) Centocelle, 10) Tor Pagnotta, 11) *Ager Caecubus*, 12) Alberata Aversana, 13) *Pompeii*, 14) Canosa di Puglia (Map by author).

seems to be there when considering the overall geographic focus of the literary and archaeological evidence. On the other hand, the earliest visual representations of this technique in Italy are from both Etruscan and Greek vases in the 6th century BC (Figure 5.10), while the reference to the Greek-founded town of Canosa di Puglia also hint at a less ethnic origin.

Finally, there were also economic and practical reasons connected with the use of the *arbustum*. Its main advantage was that vines were allowed to profit from a fertile soil that had already been prepared – and was being ploughed – for cereal cultivation, while the use of host trees reduced the investment costs for wooden stakes (Plin. *HN* 17.35.203). In other words, the practicing of the *arbustum* was an economical vine growing technique, with which fair results could be obtained with less financial input and lower labour costs (Duncan-Jones 1982, 39, 59). Also important was that – besides grapes and wine – the system offered farmers the possibility to obtain a wide range of resources from the same piece of land, in the first place cereals, but also various fruits and vegetables, as well as wood and foliage for animals (Braconi 2008; Vernelli 2008). The use of olive and fig trees for supporting the vines is mentioned in the ancient sources, and one can easily imagine how other fruit trees like apples and pears were also used to this purpose.[54] Pliny's mentioning of snooping animals suggests that the space between the trees was also used as pasture land. In other words, the widespread use of the *arbustum* by peasant families in Italy was stimulated by it being a flexible and low-cost type of polyculture in which vertical viticulture could be combined with a wide range of agricultural practices.

Sustainable viticulture in central Adriatic Italy: *arbustum*, *alberata* and vine agroforestry

Unsurprisingly, the ancient agronomists – in particular Columella (*Rust.* 3.5) – did not consider the *arbustum* as the most profitable type of plantation in vine growing for the market. It follows that scholarly discussion on the *arbustum* has largely revolved around its place in subsistence farming and small-scale viticulture (Tchernia 1986, 114). Still, a closer look at some of our textual sources reveals clear productive and commercial concerns behind its use. In fact, already in the 2nd century BC, Cato recommends the *arbustum* to farmers who grow vines for the urban market. The Saserna brothers – despite

---
[54] The use of apple and pear trees is known from later periods in Italy (Cercone 2008, p. 34).

Figure 5.9: Plain of Gubbio (Italy), alberata field, 1966 (After Stefanetti and Melelli 1999). Raccolta di diapositive di Henri Desplanques, Assemblea legislativa della Regione Umbria, Perugia.

their reluctance for applying the cultivation technique – are quoted by Columella for their guidelines concerning the labour costs of an *arbustum* of 200 *iugera* (50 ha) (*Rust.* 2.12.7), and the system was highly approved by the agricultural writer Tremelius Scrofa (Col. *Rust.* 3.3.2; Pliny *HN* 17.35.199). Writing in the second half of the 1st century BC, Varro – quoting again Scrofa – discusses the practice of arranging trees within an *arbustum* according to a 'quincunxes', a regular triangular mode of planting trees that especially allow for obtaining higher wine and cereal yields (Figure 5.6, B):

"With regard to the conformation due to cultivation, I maintain that the more regard is had for appearances the greater will be the profit: as, for instance, if those who have arbusta plant them in quincunxes, with regular rows and at moderate intervals. Thus our ancestors, on the same amount of land but not so well laid out, made less wine and grain than we do, and of a poorer quality; for plants which are placed exactly where each should be take up less ground and screen each other less from the sun, the moon and the air'." (Rust. 1.7.2)

Around the same time, the method is also mentioned by Virgil (*Georg.* 2.277-278) and Cicero (*Sen.* 59.14). Later in the 1st century AD Pliny stresses the air-passing advantages (which is a strong asset in humid environments) and the pleasant sight of the system:

"In spacing out trees and plantations and planning vineyards, the diagonal arrangement of rows is commonly adopted and is essential, begin not only advantageous in allowing the passage of air, but also agreeable in appearance, as in whatever direction you look at the plantation a row of trees stretches out in a straight line." (NH 17.15.78)

In short, the literary evidence suggests that the *arbustum* had always been part of the Italian agricultural landscape, that especially from Varro's time onwards farmers were experimenting with the system for commercial reasons, and that the practice reached its most organized and perfected form in the course of the 1st century AD.

Perhaps an even greater indication for the important role of the *arbustum* in Early/High Imperial agriculture is the prominent role of similar vine agroforestry systems in commercial viticulture in Italy in later periods. Indeed, 'alberata' or 'folignata' (as these systems were now called) were particularly widespread in many central and northern Italian regions – including Marche, Abruzzo, Umbria, Tuscany, Emilia Romagna and the Po plain – up until the mid-20th century. In these areas, they were used

Growing grapes in populous landscapes

Figure 5.10: Attic black-figure amphora, ca. 540-530 BC, showing satyrs at the harvest around a vine trained to a tree and supported with props (Paris, musée du Louvre, Inv. AM 1008). Photo © RMN-Grand Palais (musée du Louvre) / Les frères Chuzeville).

in the first place for organizing vine plantations in flatter zones of the plains and the valley floors (Figure 5.9).[55] For the Marche region, an important description is offered by Andrea Bacci in what is considered one of the prime oenological studies of the Early Modern period: "*De naturali vinorum historia, de vinis Italiae et de conviviis antiquorum*". In this work, the author describes the region at the end of the 16th century as an area with fertile, flat and humid lands, where the lower and colder zones in particular suffer from excessive humidity and fog, as

such requiring vines to be trained higher on trees; that is, organized within an 'alberata' (Bacci 1596, Vol. 7, 63-90). In essence, the 'alberata' combined two long-standing agricultural traditions of central Italy: the 'piantata' and the 'coltura promiscua'. The former was characterized by the training of vines upon rows of trees (e.g. elms, maples, willows, poplars) that bordered roads or smaller paths; the latter referred to the combination of viticulture with arable or pasture land (Vernelli 2003; 2008). Several sub-types of the alberata existed, but in essence they all applied the same principle of attaching vines to trees and letting them grow between the latter on a certain distance from the ground. Illustrations from copies of the *Tacuinum Sanitatis* (Casanata Library, late 14th century) portray this practice very well and stress the strong similarity

---
[55] Based on a study by M. Ortolani in 1960, it seems that the southern limit of the alberata system along the Adriatic coast is to be situated somewhere between the Sangro and Trigno rivers, at the border between Abruzzo and Molise (Desplanques 1969).

Figure 5.11: Tree-vines in the Tacuinum Sanitatis (Casanata Library, late 14th century; after Arano 1976, fig. XLIII).

with the Roman *arbustum* in both lay-out and harvesting procedures (Figure 5.11).

The popularity of this agroforestry technique in pre-industrial Italy should not surprise us. These systems were deeply rooted in traditional polyculture farming strategies so typical for agrarian societies focused on self-sufficiency in the Mediterranean. In particular, they used little soil and allowed for a more productive use of the available land by applying vertical agriculture. Finally, as discussed above, its use in large parts of Italy was also linked to certain environmental conditions.[56] Today, these systems belong to agricultural history (Figure 5.13), but they remained in use in market viticulture until the middle of the previous century, after which they were gradually replaced by commercially more attractive (at least from a modern point of view) vineyards that now cover most of the hilly parts in Italy. One notable exception is the Caserta province in the region of Campania, where traditional winegrowers in the territory of Aversa still guide their vines high in poplar trees. The advantage of using poplars in this humid area is that these trees grow fast and straight up in the air – up until heights of more than 15m – and thus protect the grapes from the swarming fungal diseases. An additional merit is their low foliage density, which ensure ideal sun exposure conditions for the grapes. Contemporary scenes

---

[56] One example outside Italy suffices to corroborate this statement: the muggy Vinho Verde wine region in northern Portugal, where vines have been cultivated within similar vertical agroecosystems since at least the ninth century in order to protect the grapes from fungal diseases caused by high rainfall and humid summers (Stanislawski 1970; Altieri and Nicholls 2002).

Figure 5.12: Harvest in the Aversana, Caserta, near Naples, Campania (By kind permission of ©Cantine Bonaparte).

of harvesters climbing on high narrow ladders with small baskets vividly recall Pliny's description of grape harvest contracts in Campanian poplar tree vineyards, which included the costs of a funeral and a grave for the vintagers (*HN* 14.2.10) (Figure 5.12). That excellent wines could indeed be produced in the past under such circumstances is made clear by testimonies from Strabo and Martial, who both express their amazement with regard to the excellent Caecuban wine, which was made from vines grown on poplar trees in the marshy coastal grounds of the *Ager Caecubus* in *Latium Adjectum*, located in the bay area of modern-day Fondi (Figure 5.8).

Obviously, lower-trained vines arranged in rows occurred in central and northern Italy as well. In fact, historical accounts on agriculture in central Adriatic Italy make it perfectly clear that alberata coexisted with such fields in varying proportional shares through time in the Early Modern period.[57] We can easily imagine how a similar scenario must have developed in the Roman period, with the *arbustum* system being more popular in low-lying places where drainage, humidity and fog were more of a problem, while higher hill slopes and plateaus were more suitable for lower-trained vines. But these accounts also allow for identifying two main periods in which the alberata gained ground over such fields in several regions of central Italy, including Marche: a first one in the 15th-16th century, and a second one in the first half of the 19th century. Both periods were marked by significant population growth in town and country, and in both periods this caused noticeable land pressure in central

---

[57] For the Marche region in particular, see Gobbi 2003; Paci 2002; 2003; and Vernelli 2003; 2008.

Figure 5.13: Remnants of ancient alberata fields in the Marche region, central Adriatic Italy (Photographs by author).

Italy in general, and in the Marche region in particular. As a consequence, farmers looked to reorganize and expand the cultivated area. In this process, many vineyards were turned into alberata fields, while the alberata as a farming strategy was further applied onto newly cultivated land (Anselmi 1978, 76-77, 80-81; 1989; Paci 2003; Vernelli 2003; 2008). Two principle motives lay behind this strategy: first, the need to extend cereal cultivation onto the slopes and hilltops of the valley transects, and second, the desire to compensate vineyard losses by expanding tree-wedded vine systems. One could say then that in periods when local population pressure required local agriculture to produce sufficient amounts of the common foodstuffs – that is, in the first place cereals, wine and olive oil, but also fruits, vegetables and animal products – the application of the alberata proved to be a successful way to conciliate the maintenance of a respectable wine production with an efficient use of all available agricultural land in a certain territory. The fact that *quincunxes* (see supra) were again widely used in the denser alberata plantations in Marche and Umbria in the 19th and 20th century for optimizing their performance only substantiates this line of reasoning (Desplanques 1959).

So were similar processes at work in Roman Imperial times, and did such processes promote the diffusion of the *arbustum* in central Adriatic Italy, often at the expense of vineyard-type plantations? The ancient literary sources do not explicitly mention the use of the *arbustum* in central Adriatic Italy, but the wet, marshy and foggy environments that characterized many of the coastal sections along this side of the peninsula in Roman times – in particular between Rimini and Ancona, but up until Pescara to the south – suggests that the method was widely practiced in this area as well (Coltorti 1997). The longevity of tree-wedded vine plantations in this area in later times seems to confirm this picture of environmental determinism. More importantly, however, by Pliny's times the arbustum was no longer necessarily confined to damply and foggy plains:

> "On hills and dry lands the stages of the elms are spread out at a height of eight feet, and on plains and in damp locations at twelve feet." (HN 17.35.201)

Based on our literary sources then, it seems that by the mid-first century AD the *arbustum* had reached a distribution radius that surpassed its natural habit, while increasing attention was spent on its efficiency and profitability. This suggests that the system had underwent some kind of expansion and optimization from Augustan times onwards in various areas of central and northern Italy, and most likely also in Marche and Abruzzo Teramano. The available archaeological, literary and environmental data certainly leave the door open for the development of such a scenario. In fact, against the background of such transformations, one might better understand some of Columella's complaints in AD 69 about the contemporary disrepute of proper vineyard culture in certain parts of central Italy:

> *"Why then is viticulture in disrepute? Not, indeed, through its own fault, but because of human failings, says Graecinus; in the first place because no one takes pains in searching after cuttings, and for that reason **most people plant vineyards of the worst sort**; and then they do not nourish their vines, once planted, in such a way as to let them gain strength and shoot out before they wither; and if they do happen to grow, they are careless in the matter of cultivation. Even at the very start **they think that it makes no difference what kind of ground they plant; or rather they pick out the very worst section of their lands, as though such ground alone were particularly fit for this plant because incapable of producing anything else**. ... Most people, in fact, **strive for the richest possible yield** at the earliest moment; they make no provision for the time to come, but, as if living merely from day to day, they put such demands upon their vines and load them so heavily with young shoots as to show no regard for succeeding generations (my emphasis)."* (Rust. 3.3.4-6)

In the past, this much debated elegy by Columella has been seen as an act of defence against those who did no longer consider viticulture a profitable business, or as a confirmation of the idea that Italian viticulture had become increasingly unproductive, as such requiring imports from the provinces (Billiard 1913, 94-98; Martin 1971, 257-310, 370-375). Others have explicitly linked this supposed discredit of viticulture to the sector's increasing inability to find an outlet in extra-Italian markets, as such marking one of the first signs of the economic emancipation of the provinces (Rostovtzeff 1926). The same passage has also been interpreted as illustrating the resistance by a few to the widespread expansion of vineyards in the 1st century AD, or even the proliferation of a certain conservatism against an inherently speculative business (Sirago 1958, 250-274; Garnsey and Saller 2014). Some of this reasoning also lies behind the common interpretation of Domitian's famous vine edict in AD 92, which prohibited the planting of new vineyards (*vinea*) in Italy, as well as ordered the destruction of at least half the vineyards (*vinea*) in the provinces (*Suet.* 7.2). Often viewed as a protectionist measure – either to reinforce Italy's position within the Mediterranean wine market or to counteract a supposed scenario of Italian wine overproduction – this passage has often been quoted within arguments of viticultural crisis in Imperial central Italy (Tchernia 1986; Sirago 1958; Rostovtzeff 1926).

A closer reading of Suetonius's 'Life of Domitian', however, makes it abundantly clear that the edict was more inspired by a shortage for cereals in Italy, which in Domitian's eyes could be mitigated by restricting the planting of vineyards (*vinea*) empire-wide (Purcell 1985). Also, the remaining bits of Columella's discourse elucidate quite well how the agronomist is actually not deploring the downfall of viticulture *in se*, but rather the decreasing use of a particular vine cultivation system – that is, the vineyard or *vinea* – in favour of other land regimes:

*"And most people would be doubtful on this point, to such an extent that many would avoid and dread such an ordering of their land, and **would consider it preferable to own meadows and pastures, or woodland for cutting; for in the matter of arbustum there has been no little dispute even among authorities, Saserna being unfavourable to this kind of land, and Tremellius approving it most highly**. ... Meanwhile, those devoted to the study of agriculture must be informed of one thing first for all – **that the return from vineyards is a very rich one** ... and with me, that eight hundred grafted stocks of less than two years yielded seven cullei, or that first-class vineyards produced a hundred amphorae to the iugerum, when meadows, pastures, and woodland seem to do very well by the owner if they bring in a hundred sesterces for every iugerum. **For we can hardly recall a time when grain crops, throughout at least the greater part of Italy, returned a yield of four to one.**"* (my emphasis) (Rust. 3.3.1-3)

Columella's firm juxtaposition between vineyards (*vinea*) on the one side and meadows, pastures, *arbustum* plantations and grain fields on the other side suggests that the former increasingly became an inappropriate productive strategy, and thus suffered from the proliferation of more diversified cultivation strategies. This process must have been quite a thorn in the side of a purist oenophile such as Columella (Santon 1996). It is, however, as we have seen, a logical evolution in times of high food demands. The same goes for the author's criticism on the abundant use of poor lands for installing vineyards, which is a typical effect of population pressure, leading farmers to reserve the better soils for cereals and other crops, while moving vines to less fertile soils (Grigg 1980; Van Leeuwen and Seguin 2006). Columella might not always have agreed, but the widespread adoption of the *arbustum* was probably the best way in many of parts of central Italy – and certainly in the central Adriatic area – to continue cultivating the vine on an acceptable scale in times of land stress. In fact, the agronomist's doubts on the validity of the *arbustum* as an efficient vine growing strategy can be looped back to the concerns expressed by another dedicated oenologist at the beginning of the 19th century, Giovanni de Brignoli di Brunnhoff, who remarks on the poor quality of the wines from Ancona. According to the author, this situation is linked to the low quality of the vines themselves, the uncritical mixing of different varieties, the often premature harvest, and the bad custody of the vineyard; all flaws that are stimulated by the farmer's desire to produce as much wine as possible, often at the expense of its quality. Even if these complaints by Brunnhoff already show a strong resemblance with Columella's frustrations, it is especially Brunnhoff's linked discussion with the expansion of the alberata in this period that sparks our interest. Once again, the author cites the frequent mists and copious dews in the area as one of the main reasons for the system's popularity, but at the same time he doubts that the hanging of vines high in trees sufficiently compensates the damaging effects of these environmental conditions (Brignoli di Brunnhoff 1808-1809, 10, 14-15). The fact that the author expresses these doubts in a period of significant population growth (see *supra*) is – at least to me – a further incentive for considering a similar development in central Adriatic Italy in the 1st century AD.

Conclusion

In other words, turning back to the theoretical framework that was set-out earlier in this chapter, in times of high population, central Adriatic farmers resorted to a multifaceted intensification technique; that is, one that comprised '*intensification proper*' (an increase in cereal fields and a more intensive cultivation of these fields), '*specialization*' (one particular productive strategy gained ground over others), and '*diversification*' (the preferred productive strategy consisted in diversifying the output of a certain unit of land in both spatial and temporal terms). Finally, for vine growing purposes in particular, part of the land was 'intensified' vertically, as such promoting the exuberant growth pattern of these vigorous climbing plants. Further intensification strategies included the training of more vines on a single tree, or the using of higher trees, with both techniques ultimately resulting in higher wine yields per tree.[58] In an extensive way then, this strategy increased overall agricultural production of the total area under cultivation. Unsurprisingly, ideas of vertical farming are also increasingly being considered a useful strategy for coping with population growth and land stress in modern farming. Innovative variants hereby include the growing of crops over different levels in existing structures within urban environments (Despommier 2011).

Bibliography

Anselmi, S. 1978. *Mezzadri e terre nelle Marche. Studi e ricerche di storia dell'agricoltura fra Quattrocento e Novecento*, Bologna.

Anselmi, S. "Un insediamento resistente: mezzadria e reticolo urbana nell'Italia centrale." In *L'ambiente nella storia d'Italia*, edited by A. Caracciolo, G. Bonacchi, P. Pelaja et al., 39-57. Rome and Venice: Marsilio, 1989.

Altieri, M.A. and C.I. Nicholls 2002. *The simplification of traditional vineyard based agroforests in northwestern Portugal: some ecological applications*, in *Agroforestry Systems* 56, p. 185-191.

Arano, L.C. *The medieval health handbook. Tacuinum Sanitatis*. New York: George Braziller, 1976.

Attema, P. and De Haas, T. "Rural settlement and population extrapolation." In *Settlement, Urbanization and Popultion*, edited by A. Bowman and A. Wilson, 97-140. Oxford: Oxford University Press, 2011.

---

[58] Pliny is the only author who provides us with some information on the approximate number of vines that was commonly trained on each tree in Roman Italy. He recalls that farmers often grew as much as ten vines against each tree, while at the same time he condemns the practice of training less than three vines (*HN* 17.202-203).

Bacci, A. 1596. *De naturali vinorum historia, de vinis Italiae et de conviviis antiquorum, libri septem*, Rome.

Bairoch, P. "Urbanisation and the Economy in Pre-Industrial Societies: the Findings of Two Decades of Research." *The Journal of European Economic History* 18 (1989): 239-290.

Balmelle, C. and Brun, J.P. 2005. 'La vigne et le vin dans la mosaïque romaine et byzantine,' in H. Morlier (ed.), *La mosaïque greco-romaine IX.2* (Rome: École française de Rome): 899-921.

Barker, G. and Rasmussen, T. *The Etruscans*. Oxford: Wiley-Blackwell, 1998.

Beach, T., S. Luzzadder-Beach, N. Dunning, J. Jones, J. Lohse, T. Guderjan, S. Bozarth, S. Millspaugh and T. Bhattacharya 2009. *A review of human and natural changes in Maya Lowland wetlands over the Holocene*, in *QSR* 28.17-18, p. 1710-1724.

Beach, T., S. Luzzadder-Beach, R. Terry, N. Dunning, S. Houston and T. Garrison 2011. *Carbon isotopic ratios of wetland and terrace soil sequences in the Maya Lowlands of Belize and Guatemala*, in *CATENA* 85.2, p. 109-118.

Beloch, K.J. "Le città dell'Italia antica." *Atene e Roma* 1 (1898): 257-278.

Billiard, R. 1913. *La vigne dans l'antiquité*, Lyon.

Boissinot, P. 2009. *Les vignobles des environs de Megara Hyblaea et les traces de la viticulture italienne durant l'antiquité*, in *MEFRA* 121.1, p. 83-132.

Boserup, E. 1965. *The Conditions of Agricultural Growth: the Economics of Agrarian Change under Population Pressure*, London and Chicago.

Braconi, P. 2008. *Territorio e paesaggio dell'alta valle del Tevere in età romana*, in F. Coarelli and H. Patterson (eds.), *Mercator placidissimus. The Tiber Valley in Antiquity. New research in the upper and middle river valley, Atti del Convegno, British School at Rome, 27-28 febbraio 2004*, Rome, p. 73-91.

Broise, H. and V. Jolivet 1995. *Bonification agraire et viticulture antique autour du site de Musarna (Viterbe)*, in L. QUILICI and S. QUILICI GIGLI (eds.), *Interventi di Bonifica Agraria nell'Italia Romana* (Atlante Tematico di Topografia Antica 4), Rome, p. 107-116.

Brignoli di Brunnhoff, G. 1808-1809. *Istruzione sul miglioramento de vini nel Dipartimento del Metauro*, Ancona.

Brunt, J.P. *Italian Manpower. 225 B.C. – A.D. 14*. Oxford: Oxford Univesity Press, 1971.

Buono, R. and Vallariello, G. "Introduzione e Diffusione della Vite (Vitis vinifera L.) in Italia." *Delpinoa* 44 (2002): 39-51.

C. Calci, C. and Sorella, R. 1995. *Forme di Paesaggio Agrario nell'Ager Ficulensis*, in L. Quilici and S. Quilici Gigli (eds.), *Interventi di Bonifica Agraria nell'Italia Romana* (Atlante Tematico di Topografia Antica 4), Rome, p. 117-127.

Calza, G. *Scavi di Ostia: topografia generale*. Rome: Libreria dello Stato, 1926.

Carcopino, J. *Daily Life in Ancient Rome: the People and the City at the Height of the Empire*. New Haven: Yale University Press, 1940.

Cencini, C. and Varani, L. "Per una storia ambientale delle pianure costiere medio-adriatiche." *Geogr. Fis. Dinam. Quat.* 14 (1991): 33-44.

Cercone, F. 2008. *Storia della vite e del vino in Abruzzo. Dalle testimonianze romane alla diffusione del Montepulciano*, Lanciano.

Coltorti, M. 1997. *Human impact in the Holocene fluvial and coastal evolution of the Marche region, central Italy*, in *Catena* 30, p. 311-335.

Conventi, M. *Città romane di Fondazione* (Studia Archaeologica 130). Rome: L'Erma di Bretschneider, 2004.

Cristofori, A. *Non Arma Virumque: le Occupazioni nell'Epigrafia del Piceno*. Bologna: Lo scarabeo, 2004.

Crnobrnja, A.N. "Economic aspects of the use of Roman oil lamps in Moesia Superio." In *Phosporion. Studia in honorem Mariae Cicikova* (Bulgarian Academy of Sciences, National Institute of Archaeology with Museum at BAS), 408-412. Sofia.

De Graaf, P. *Late Republican-Early Imperial Regional Italian Landscapes and Demography*, Bar International Series 2330. Oxford: BAR Publishing, 2012.97

De Ligt, L. *Peasants, Citizens and Soldiers: Studies in the Demographic History of Roman Italy 225 BC – AD 100*. Cambridge: Cambridge University Press, 2012.

Delplace, C. *La Romanisation du Picenum: l'exemple d'Urbs Salvia*. Rome: L'École française de Rome, 1993.

Desplanques, H. 1959. *Il paesaggio rurale della coltura promiscua in Italia*, in *Rivista Geografica Italiana* 66.1, p. 29-64.

Desplanques, H. 1969. *Les campagnes ombriennes. Contribution à l'étude des paysages ruraux en Italie centrale*, Paris.

Despommier, D. 2011. *The vertical farm. Feeding the world in the 21st century*, New York.

Duncan-Jones, R. 1982. *The Economy of the Roman Empire. Quantitative Studies* (2nd ed), Cambridge.

Enei, F. *Progetto Ager Caeretanus. Il litorale di Alsium*. Santa Marinella: Regione Lazio/Comune di Ladispoli, 2001.

Erdkamp, P. *The Grain Market of the Roman Empire. A social, political and economic study*. Cambridge: Cambridge University Press, 2005.+

FAO. Energy and Protein Requirements: Report of a Joint FAO/WHO/UNU/Expert Consultation, 1973.

FAO. Energy and Protein Requirements: Report of a Joint FAO/WHO/UNU/Expert Consultation, 1985.

FAO. Energy and Protein Requirements: Report of a Joint FAO/WHO/UNU/Expert Consultation, 2004.

Fentress, E. et a. "Cosa in the Republic and the Early Empire." In *Cosa V. An Intermittent Town, Excavations 1991-1997*, edited by E. Fentress, 13-62. Ann Arbor: University of Michigan Press, 2003.

Fiorelli, G. *Gli scavi di Pompei dal 1862 al 1872. Relazione al Ministro della Istruzione Pubblica*. Naples: Tipografia Italiana nel Liceo V. Emanuele, 1873.

Flohr, M. "Skeletons in the cupboard? Femures and food regimes in the Roman world." In *The Routledge Handbook of Diet and Nutrition in the Roman World*, edited by P. Erdkamp and C. Holleran, 273-280. London: Routlledge, 2019.

Forni, G. 2002. *Colture, lavori, tecniche, rendimenti*, in G. FORNI and A. MARCONE (eds.), *Storia dell'Agricoltura Italiana*, Firenze, p. 63-156.

Foxhall, L. and Forbes, H. "Sitometreia: the role of grain as a staple food in classical antiquity." Chiron 12 (1982): 41-90.

Garnsey, P. and R. Saller 2014. *The Roman Empire. Economy, society and culture* (2nd edition), London/New Delhi/New York/Sydney.

Ghisleni, M. et al. "Excavating the Roman peasant." *Papers of tbe British School at Rome* 79 (2011): 95-145.

Giannecchini, M. and Moggi-Cecchi, J. "Stature in archaeological samples from Central Italy: methodological issues and diachronic changes." AJPA 135 (2008): 284-292.

Gobbi, O. 2003. *Vigne e vignaioli nel Piceno montano: secoli XV-XVI*, in *Proposte e ricerche. Economia e società nella storia dell'Italia centrale* 51, p. 23-47.

Goodchild, H. *Modelling Roman agricultural production in the middle Tiber Valley, central Italy* (Unpublished PhD thesis), 2007.

Grigg, D.B. *Population Growth and Agrarian Change. An Historical Perspective*. Cambridge: Cambridge University Press, 1980.

Hanson, J.W. *An Urban Geography of the Roman World, 100 BC to AD 300*. Oxford: Oxford Unibversity Press, 2016.

Harris, J.A. and Benedict, F.G. "A Biometric study of human basal metabolism." *Proceedings of the National Academy of Sciences* 4.12 (1918): 370-373.

Harris, J.A. and Benedict, F.G. *A Biometric study of basal metabolism in man*. Washington: Nabu Press, 1919.

Hopkins, K. *Conquerors and Slaves*. Cambridge: Cambridge University Press, 1978.

Hopkins, K. "Taxes and Trade in the Roman Empire." *Journal of Roman Studies* 70 (1980): 101-125.

Hopkins, K. "Models, ships and staples." In *Trade and Famine in Classical Antiquity*, edited by P. Garnsey and C.R. Whittaker, 84-109. Cambridge: Cambridge Philological Society; 1983.

Jackson, R.S. 2008. *Wine Science. Principles and Applications*, London.

Jashemski, W. F. 1973. *Large Vineyard Discovered in Ancient Pompeii*, in *Science* 180, p. 821-830.

Jashemski, W. F. *The Gardens of Pompeii: Herculaenum and the villas destroyed by Vesuvius*. New Rochelle NY: Caratzas Brothers, 1979.

Jashemski, W.F. "Produce Gardens." In *Gardens of the Roman Empire*, edited by W. Jashemski, K. Gleason, K. Hartswick and A. Malek, 121-151. Cambridge: Cambridge University Press, 2017.

Johnston, K.J. 2003. *The intensification of pre-industrial cereal agriculture in the tropics: Boserup, cultivation lengthening, and the Classic Maya*, in *JAA* 22, p. 126-161.

Jongman, W. "Slavery and the growth of Rome. The transformation of Italy in the second and first centuries BCE." In *Rome the Cosmopolis*, edited by C. Edwards and G. Woolf, 100-122. Cambridge: Cambridge University Press, 2003.

Kaiser, T. and Voytek, B. "*Sedentism and Economic Change in the Balkan Neolithic*." *JAA* 2 (1983): 323-353.

Kehoe, D. *The economics of agriculture on Roman Imperial estates in North Africa*. Göttingen: Vandenhoeck & Ruprecht, 1988.

Kehoe, D. "The Early Roman Empire: production." In *The Cambridge Economic History of the Graeco-Roman World*, edited by W. Scheidel, I. Morris and R.P. Saller, 543-569. Cambridge: Cambridge University Press, 2007.

Killgrove, K. "Using skeletal remains as a proxy for Roman lifestyles." In *The Routledge Handbook of Diet and Nutrition in the Roman World*, edited by P. Erdkamp and C. Holleran, 245-258. London: Routlledge, 2019.

Kolendo, J. 1994. *Praedia suburbana e loro redditività*, in *Aa.Vv* 1994, p. 59-71.

Kron, G. "Anthropometry, physical anthropology, and the reconstruction of ancient health, nutrition and living standards." Historia 54.1 (2005): 68-83.

Kron, G. "The much maligned peasant. Comparative perspectives on the productivity of the small farmer in classical antiquity." In *People, land and politics. Demographic developments and the transformation of Roman Italy, 300 BC-AD 14*, edited by L. De Ligt and S.J. Northwood, 71-119. Leiden & Boston: Brill, 2008.

Launaro, A. *Peasants and Slaves. The Rural Population of Roman Italy (200 BC to AD 100)*. Cambridge: Cambridge University Press, 2011.

Lo Cascio, E. "The Population of Roman Italy in Towns and Country." In *Reconstructing Past Population Trends in Mediterranean Europe*, edited by K. Sbonias and J.L. Bintliff, 161-171. Oxford: Oxbow Books, 1999.

Malavolti, F. 1948. *Rapporti tra alluvioni ed antichi insediamenti umani nella pianura emiliana*, in *Emilia Preromana* 1, p. 79-86.

Martin, R. 1971. *Recherches sur les agronomes latins et leurs conceptions économiques et sociales*, Paris.

Marzano, A. "Agricultural production in the hinterland of Rome." In *The Roman Agricultural Economy. Organization, investment and production*, edited by A. Bowman and A. Wilson, 85-106. Oxford: Oxford University Press, 2013.

Mattingly, D.J. "Regional variation in Roman oleoculture: some problems of comparability." In *Landuse in the Roman Empire, Atti del convegno tenutosi presso il Danish Institute di Roma nel gennaio 1993*, edited by J. Carlsen, P. Ørsted and J.E. Skydsgaard, 91-106. Rome: L'Erma di Bretschneider, 1994.

Mei, O. and Gobbi, P. "Necropoli tardoantica a Forum Sempronii: dati preliminary." In *Amore per l'antico. Dal Tirreno all'Adriatico, dalla Preistoria al Medioevo e oltre. Studi di antichità in ricordo di Giuliano De Marinis*, edited by G. Baldeli and F Lo Schiavo, 941-947. Rome: Scienze e Lettere, 2014.

Meiggs, R. *Roman Ostia*. Oxford: Clarendon Press, 1960.

Millett, M. *The Romanization of Britain*. Cambridge: Cambridge University Press, 1990.

Morley, N. *Metropolis and Hinterland: the City of Rome and the Italian Economy, 200 B.C. – A.D. 200*. Cambridge: Cambridge University Press, 1996.

Morrison, K. 1994. *The Intensification of Production: Archaeological Approaches*, in *JAMT* 1.2, p. 111-159.

Nibby, A. *Viaggio antiquario ad Ostia*. Rome: Società Tipografica, 1829.

Nissen, H. *Pompeianische studien zur Städtekunde des altertums*. Leipzig: Brietkoph und Härtel, 1877.

Osborne, F. "Demography and Survey." In *Side-by-side survey. Comparative regional studies in the Mediterranean world*, edited by S.E. Alcock and J.F. Cherry, 163-172. Oxford: Oxbow Books, 2004.

Perkins, P. *Etruscan settlement, society and material culture in central coastal Italy*. BAR International Series S788. Oxford: BAR Publishing, 1999.

Paci, R. 2002. *Vigne e vino a Jesi nel quattrocento*, in *Studia Picena* 67, p. 17-55.

Paci, R. 2003. *Dalla vigna all'arboreto: Corinaldo, secoli XVI-XVII*, in *Proposte e ricerche. Economia e società nella storia dell'Italia centrale* 51, p. 7-23.

Packer, J.E. *The Insulae of Imperial Ostia*. Rome: American Academy in Rome, 1971.

Pelgrom, J. "Population density in mid-Republican Latin colonies: a comparison between text-based population estimates and the results from survey archaeology." *ATTA* 23 (2013): 73-84.

Purcell, N. 1985. *Wine and wealth in ancient Italy*, in *JRS* 75, p. 1-19.

Quilici, L. 1992. *Una vigna nel paesaggio della Calabria*, in *Itinera. Scritti in onore di Luciano Bosio*, Padova, p. 117-129.

Rathbone, D. "Poor peasants and silent sherds." In *People, land and politics. Demographic developments and the transformation of Roman Italy, 300 BC-AD 14*, edited by L. De Ligt and S.J. Northwood, 305-332. Leiden & Boston: Brill, 2008.

Robinson, O.R. *Ancient Rome: City Planning and Administration*. New York: Routledge, 1992.

Rosenstein, N. *Rome at War: Farms, Families and Death in the Middle Republic*. Chapel Hill and London: University of North Carolina Press, 2004.

Rosentein, N. *Rome and the Mediterranean 290 to 146 BC: The Imperial Republic*. Edinburgh: Edinburgh University Press, 2012.

Rostovtzeff, M.I. *Social and economic history of the Roman Empire*, Oxford: Biblo-Moser, 1926.

Roth, J.P. *The logistics of the Roman army at war (264BC – AD 234)*. Leiden, Boston & Köln: Brill, 1999.

Salmon, E.T. *Roman Colonization under the Republic*. Oxford: Oxford University Press, 1969.

Santangeli, R. and R. Volpe 1980. *Tentativo di ricostruzione di una sistemazione agricola di età repubblicana nei dintorni di Roma*, in *ArchCl* 32, p. 206-215.

Santangeli Valenziani, R. and R. Volpe 2007. *La restituzione del paesaggio agrario della viticoltura a Roma e nel Suburbio*, in A. CIACCI et al. (eds.), *Archeologia della vite e del vino in Etruria*, Sienne, p. 48-53.

Santangeli Valenziani, R. and R. Volpe 2012. *Paesaggi agrari della viticoltura a Roma e nel suburbio*, in A. CIACCI et al. (eds.), *Archeologia della vite e del vino in Toscana e nel Lazio. Dalle tecniche dell'indagine archeologica alle prospettive della biologia molecolare*, Firenze, p. 61-70.

Santon, T.J. 1996. *Columella's attitude towards wine production*, in *Journal of Wine Research* 7.1, p. 55-59.

Sbonias, K. 1999. "Introduction to issues in demography and survey." In *Reconstructing past population trends in Mediterranean Europe (3000 BC-AD 1800)*, edited by J. Bintliff and K. Sbonias, 1-20. Oxford: Oxbow Books, 1999.

Scheidel, W. "Roman population size: the logic of the debate'. In *People, land and politics. Demographic developments and the transformation of Roman Italy, 300 BC-AD 14*, edited by L. De Ligt and S.J. Northwood, 17-70. Leiden & Boston: Brill, 2008.

Sereni, E. *Storia del Paesaggio Agrario Italiano*. Bari: Laterza, 1961.

Sirago, V.A. 1958. *L'Italia agraria sotto Traiano*, Louvain.

Spurr, M.S. *Arable cultivation in Roman Italy c. 200 BC – c. AD 100* (JRS 3). London: Society for the Promotion of Roman Studies, 1986.

Stambaugh, J.E. *The Ancient Roman City*. Baltimore: The Johns Hopkins University Press, 1988.

Stanislawski, D. 1970. *Landscapes of Bacchus: the vine in Portugal*, Austin.

Stefanetti, M. and Melelli, A. *Le campagne umbre nelle immagini di Henri Desplanques*. Perugia: Regione dell'Umbria, 1999.

Storey, G.R. "The population of ancient Rome." *Antiquity* 71 (1997): 966-978.

Taelman, D., De Dapper, M., Weekers, L. and Pincé, P. "Landscape background and Geoarchaeology in the PVS project." In *The Potenza Valley Survey (Marche, Italy). Settlement dynamics and changing material culture in an Adriatic valley between Iron Age and Late Antiquity*, edited by F. Vermeulen et al., 42-66. Rome: Fondazione Dià Cultura, 2017.

Tchernia, A. *Le vin de l'Italie romaine. Essai d'histoire économique d'après les amphores*, Rome : l'École française de Rome, 1986.

Tchernia, A. *The Romans and Trade*. Oxford. Oxford: University Press, 2016.

Van Leeuwen, C. and Seguin, G. "*The concept of terroir in viticulture*." *Journal of Wine Research* 17.1 (2006): 1-10.

Van Limbergen, D. 2011. Vinum Picenum and Oliva Picena. Wine and oil presses in central Adriatic Italy between the Late Republic and the Early Empire. Evidence and problems, *BABesch* 86, 71-94.

Van Limbergen, D. 2015. Pots, presses, people and land. The role of overseas export and local consumption demand in the development of viticulture and oleoculture in central Adriatic Italy (250 BC – AD 200) (Unpublished PhD thesis, Pisa University/Ghent University).

Van Limbergen, D. "The central Adriatic wine trade of Italy revisited." *Oxford Journal of Archaeology* 37.2, (2018a): 201-226.

Van Limbergen, D. "What Romans ate and how much they ate of it." *Revue Belge de Philologie et d'Histoire* 96 (2018b): 1049-1092.

Van Limbergen, D. "Vinum Picenum and Oliva Picena II. Further thoughts on wine and oil presses in central Adriatic Italy." *BABesch* 94 (2019): 1-30.

Van Limbergen, D., Monsieur, P. and Vermeulen, F. "The role of overseas export and local consumption demand in the development of viticulture in central Adriatic Italy (200 BC – AD 150). The case of the Ager Potentinus and the wider Potenza valley." In *The Economic Integration of Roman Italy. Rural communities in a globalizing world*, edited by T.C.A. De Haas and G. Tol, 342-366. Leiden & Boston: Brill, 2017a.

Van Limbergen, D. and Vermeulen, F. "Appendix. Topographic gazetteer of Roman towns in Picenum and eastern Umbria et Ager Gallicus." In *From the Mountains to the Sea. The Roman colonisation and urbanisation of central Adriatic Italy*, by F. Vermeulen, 165-202. Leuven: Peeters, 2017.

Van Limbergen and Vermeulen, F. "A method for estimating Roman population sizes from urban survey contexts: an application in central Adriatic Italy." In *Complexity: a new framework to interpret ancient economic proxy data*, edited by K. Verboven and J. Poblome, (in press). Palgrave Studies in Ancient Economies, 2020.

Van Limbergen, D., Vermeulen, F., Taelman, D. and Carboni, F. "Rural settlement dynamics in the Potenza corridor between 900 BC and AD 600." In *The Potenza Valley Survey: settlement dynamics and changing material culture in a central Adriatic valley between Iron Age and Late Antiquity*, edited by F. Vermeulen et al., 112-157. Rome: Fondazione Dià Cultura, 2017b.

Van Limbergen, D., Vermeulen, F., Verhoeven, G. and Verdonck, L. "Methodological approach." In *The Potenza Valley Survey: settlement dynamics and changing material culture in a central Adriatic valley between Iron Age and Late Antiquity*, edited by F. Vermeulen et al., 10-41. Rome: Fondazione Dià Cultura, 2017c.

Vera, D. "L'Italia agraria nell'étà impériale. Fra crisi e trasformazione." In *L'Italie d'Auguste à Dioclétien* (CEFR 198, 239-249. Rome: l'École française de Rome, 1994.

Vermeulen, F. *From the Mountains to the Sea. The Roman colonisation and urbanisation of central Adriatic Italy*. Leuven: Peeters, 2017.

Vernelli, C. 2003. *Vite e vino nella provincia di Ancona: fra tradizione e DOC della Lacrima di Morro d'Alba*, in *Proposte e ricerche. Economia e società nella storia dell'Italia centrale* 51, 111-133.

Vernelli, C. "La coltivazione della vite in un'area mezzadrile." *Proposte e ricerche. Economia e società nella storia dell'Italia centrale* 60 (2008): 153-174.

Wallace-Hadrill, A. "Houses and households: Sampling Pompeii and Herculaneum." In *Marriage, Divorce and Children in Ancient Rome*, edited by B. Rawson, 191-227. New York: Clarendon Press, 1991.

Wallace-Hadrill, A. *Houses and Soceity in Pompeii and Herculaenum*. Princeton: Princeton University Press, 1994.

Witcher, R. "The extended metropolis: urbs, suburbium and population." *Journal of Roman Archaeology* 18 (2005): 120-138.

Witcher, R. "Missing persons? Models of Mediterranean regional survey and ancient populations." In *Settlement, Urbanization and Popultion*, edited by A. Bowman and A. Wilson, 36-75. Oxford: Oxford University Press, 2011.

Wunderlich, C.H. "Light and economy: an essay about the economy of prehistoric and ancient lamps." *Nouveautés Lychnologiques* 1 (2003): 251-263.

U.S. Army. *Nutrition. Technical Manual 8-501.* Washington DC, 1961.

6

Population decline and wine industry: societal transformation on Late Antique Delos (Greece)

Emlyn K. Dodd

Macquarie University

Abstract: The island of Delos had a lengthy settlement history and was of great importance throughout antiquity; intermittently, it was a major religious, commercial and political centre within the Mediterranean. These periods of prosperity and growth (namely, the Hellenistic and early Roman eras) were the focus of increasingly interdisciplinary Delian studies for over two centuries; until recently, little research was completed regarding the later history and archaeology of Delos. Studies over the past 20 years confirm that Delos did experience contraction and deterioration after its Hellenistic peak, but, almost certainly, not to the extent once thought. The investigations of the present author affirm that a sizeable, albeit reduced, population remained on Delos even into Late Antiquity linked directly to a prosperous, contemporary wine industry. This illustrates the resilience and persistence of the post-Hellenistic population and elucidates the potential fate of such islands in antiquity, where, rather than collapse, they underwent some form of societal transformation more difficult to recognise through the archaeological record. Reasons for such a sequence of events are explored, linked to the discussions of Horden, Purcell, Bevan and Conolly, centred around theories of island interconnectivity and the formation of specialised, agriculturally productive niches. With these considerations in mind, Delos presents an exceptional case study to observe an influential and significant ancient city that underwent extreme change – Hellenistic prosperity to Late Antique 'decline' – but maintained notable features often associated with cities of higher-level population and prosperity.

Keywords: Delos, Wine, Presses, Viticulture, Islands

Introduction

The present volume provides a valuable opportunity to reassess the multifaceted chronology of Delos, utilising interdisciplinary research focussed on the viticultural installations and archaeological material from the latter Imperial era into Late Antiquity. While the majority of studies within this volume focus on the first three centuries of our era, this chapter represents an opportunity to observe what happened later, primarily using data from the 4th – 6th centuries AD to observe how people, their landscape and the economy were interconnected in less conspicuous times. This chapter also hopes to add to discussions on the fragility and persistence (and thus stability and resilience) of island archaeology. The recent contributions of collapse theorists, such as Curtis, Faulseit and Tainter, are considered and provide fresh perspectives on the temporal transformation of occupation on Delos (Curtis 2014; Faulseit 2016; Tainter 2016). Indeed, by combining new and reassessed archaeological material culture with the theories of Curtis, Faulseit and Tainter, among others, a refreshed line of thought concerning Late Antique Delos is revealed; one that does not revolve around terms of decline or deterioration, but rather an evolution or societal transformation.[1] It is possible, therefore, to observe the progressive resilience of the Delian inhabitants and diversity of their environment.

While the prosperity and population of later Delos was more significant than traditionally believed, it remains clear that a substantial contraction of the city and its population, along with a decline in interregional commercial importance, occurred after the events of 88 and 69 BC.[2] This chapter provides an opportunity, therefore, to examine the 'other side of the coin' from a typical Malthusian or Boserupian scenario: what are the consequences of population decline at a once affluent site and how do the associated local agricultural systems respond to this? What differentiates a region maintaining agricultural production appropriate to

---

[1] This new research on Delos supports the evolving opinion of scholars regarding the Late Antique countryside; that is, the increasing tendency to reject stereotypical concepts of decline (Chavarría and Lewit 2004: 3, n° 2).

[2] On the decline of Delos in the 1st century BC, see Laidlaw (1933: 269, 274, n° 15); Bruneau and Ducat (1966: 23-24); Brun (2004a: 81); and Zarmakoupi (2013: 6). Ancient sources also mention the decline (Strabo, *Geog.* 10.5.2; Dion. Hal., *Ant. Rom.* 1.50.1; Ant. Thess., *Anth. Pal.* 9.408, 9.421, 9.550; Paus. 8.33.2; Tert., *Apol.* 40.3, *De Pal.* 2).

a high-level population from one that reduces productive levels in line with a diminished population? Recent research by the author on viticulture on Delos supports the former scenario and is explored further below.[3] This chapter, therefore, explores a scenario in contrast to that presented elsewhere in this volume: the analysis of a case of low population, rather than population increase, along with the consequences and possibilities for agriculture.

This case study allows theories of geographic diversity to be explored, wherein the responses to population decline, as for increase, and the ability of a diminished settlement to recover almost certainly varied from site to site. Delos provides the example of an environment that allowed inhabitants to maintain high-scale agricultural productivity despite much lower population levels, reduced local demand and a shrinking economy compared to the Hellenistic commercial boom. The geographic location of Delos, a very small island close to other Cycladic islands and in a central Aegean position, encouraged interconnectivity; this was probably a leading factor in the ability of the island to absorb stress, transform and survive into Late Antiquity with such high-scale productivity.[4] Observations regarding diversity in population vs. resource balance are also noted. It appears that the reduced Late Antique population effectively maintained much greater, possibly surplus, wine production using the larger ratio of available land to population size; if true, a relative economic expansion and increase in per capita income. Such considerations reveal a perceived degree of resilience within the population on Delos – a population that wishes to prioritise and maintain a prosperous wine industry, despite a number of exogenous and endogenous stresses (including, violent past events, an evolving political situation, and a much reduced population).

The chapter starts with an outline of what is known of Delos in Late Antiquity, with an emphasis on the material culture, in particular the structural evidence destroyed during early excavations, in order to provide the reader with a firm grounding in the contemporary academic discourse and archaeological material.[5] This is followed by an overview of the evidence for viticulture, including discussions on the topography, geography, ethnographic comparison, ancient agricultural evidence and comparisons to the Hellenistic epigraphic record. The core of the study is an analysis of relevant vinicultural installations recently surveyed by the author, including quantified calculations of productivity and processing capacities, crop estimation, land use, and potential vintage yields. This dataset is supplemented by a discussion of the population decline and agricultural response of Late Antique Delos along with its inclusion in the debate of persistence and fragility within Mediterranean island archaeology.

Late Antique Delos: settlement characteristics and the evidence to date

Throughout the 1st–2nd century AD, the extant ancient literature describes a sense of isolation and abandonment on Delos, with the grand structures of the 3rd–2nd century BC still standing, but beginning to ruin, and the remaining inhabitants surrounded by increasingly abandoned neighbourhoods and an encroaching countryside.[6] The archaeological evidence, however, clarifies that the island was not entirely abandoned with continued occupancy on a somewhat prosperous scale throughout the Imperial period, more significant than traditionally acknowledged.[7] A range of small finds were recovered from this era, including inscriptions, coins, glass, ceramics, vases, lamps, and, near the Bay of Skardhana, a large quantity of crushed *murex* over the abandonment layer.[8] A spatial assessment reveals that, while certain neighbourhoods were indeed abandoned, others were undoubtedly densely inhabited.[9] Furthermore, in the 3rd-4th centuries AD, the construction of an aqueduct supplying running water to the city for the first time and two large bathing complexes – significant ventures of capital expenditure and construction – prompts a reconsideration regarding the character and activities of the settlement at this time (Moretti and Fincker 2011; Zarmakoupi 2014-2015: 127; Bouet and Le Quéré 2016; Le Quéré 2018b: 276-277; 2018c: 348f).

It is clear, therefore, that there was a higher degree of settlement continuity across the Hellenistic-Imperial-Late Antique eras than traditionally acknowledged, and it no longer appears likely that the early Christian community sprang from nothing in the 4th century.[10] While the settlement remained small – a shadow of its Hellenistic self – the archaeological evidence suggests continuity in population and spatial use post-Imperial era. This is not

---

[3] Much of the data used in this paper derive from surveys completed between 2014 and 2016 on Delos by the author, presented in more detail in Dodd (2020).

[4] The great success of many islands in the commercial, trade and agricultural spheres, along with their ability to form lucrative 'niches' for specialised production, is largely attributed to their generally superior interconnectivity compared to mainland or land-locked sites. A number of examples are given in Horden and Purcell (2000: 225-240); Bevan and Conolly (2013: 6-10, n° 95).

[5] The 'Hellenistic obsession' of the 19th-early 20th century excavations led to the frequent destruction of Imperial to Late Antique layers and structural remains (Bruneau 1968: 694; Kourtzian 2000: 49, n° 6; Zarmakoupi 2013: 49; Hasenohr 2014: 291). See particularly, Dodd 2020: 97, n° 725.

[6] Bruneau (1968: 693, n° 4) suggests that the ancient literature, while not completely realistic, portrays a metaphoric, striking comparison between the huge city of the past and the semi-abandoned community of their time.

[7] This argument is sustained in the publication of Hasenohr (2014), on the Imperial occupation of the lower theatre quarter, and strengthened in recent work by Le Quéré (2018a; 2018b; 2018c; 2018d).

[8] Cf. the inventories of Bruneau (1968: 694-700).

[9] Hasenohr (2014) proposes that finds in the upper theatre quarter suggest abandonment through the Imperial period, while others in the Skardhana quarter, Synagogue, Agora of the Delians, port area, and Hypostyle Hall demonstrate continual occupancy or use (Bruneau 1968: 694-700). Le Quéré (2018a; 2018b) and Bouet and Le Quéré (2016) clearly outline the extent of the Imperial and later settlement around the Sanctuary and later grand bathing complexes (cf. the useful illustration in Le Quéré 2018d: 113, fig. 2).

[10] Suggested by Kourtzian (2000: 49). Though islands in antiquity were susceptible to and often experienced large-scale population fluctuation, even from abandonment back to prosperity; this is explored in Bevan and Conolly (2013: 8-9).

only observed in the artefacts and structural remains of the period, but also now by the agricultural facilities, which provide a quantified indication of consumable products produced by the inhabitants. Significant habitation and social stratification is also indicated through the relatively high quality workmanship evident on many of the early Christian structural remains; notably, the iconography and construction of the basilica of St. Cyrique, the architectural items with relief found over the *Ekklesiasterion*, and, to a lesser extent, the Rue 5 'large' counterweight. Such quality provides evidence for the presence of individuals with significant expertise in construction and artistry and suggests social aspirations of individual artisans.

The known extent of the Late Antique community is appreciated graphically by mapping the evidence described in Dürrbach and Jardé (1905: 254-257), Orlandos (1936), Gallet de Santerre and Tréheux (1947-1948: 412-418), Bruneau (1968: 696-709), Hasenohr (2002: 106-107; 2014) and Le Quéré (2013: 267ff; 2018a: 118-47; 2018b: 273-77; 2018c; 2018d) (Figure 6.1). Those structures destroyed by early excavations should be noted, including:

1. At least three religious structures built over the *Ekklesiasterion* (*Thesmophorion*) near the southeastern corner of the Hypostyle Hall (barely visible at present and not indicated in the most recent published surveys) (Cabrol and Leclercq 1920: 568-69; Orlandos 1936: 86-88, fig. 17; Moretti et al. 2015: map 22; Le Quéré 2018a: 120), and a possible baptistery that reused an apse of the *Ekklesiasterion* (Le Quéré 2018a: 188; 2018b: 273);
2. Monastic remains built over houses of the 2nd century AD (over the older Hypostyle Hall), including a private cross-shaped bathroom, hypocaust heating and baptistery of the early 5th century AD (Orlandos 1936: 94-94, fig. 29; Le Quéré 2018d: 114);
3. A church constructed across two galleries of the Portico of Philip V – demolished during excavation (Vallois 1923: 3; Orlandos 1936: 96-7; Le Quéré 2018a: 143; 2018b: 275). It is not dated, although from the associated contextual iconography (cf. Vallois 1923: 3, 167, figs. 3, 231; Orlandos 1936: 95-98, figs. 30-35) it appears close in date to the other early Christian churches and basilicas on the island, possibly also the Rue 5 counterweight.
4. Domestic quarters: over the *Prytaneion* and *Bouleuterion* in the 4th-5th centuries AD (Le Quéré 2018a: 147); nearby the Temple of Apollo in the 3rd-5th centuries (Le Quéré 2018a: 143); and a neighbourhood stretching from the *Oikos of the Naxians* to the church over the Portico of Philip V (Le Quéré 2018b: 275).
5. A large commercial area (>500m^2) to the south of the Hypostyle Hall (Le Quéré 2018d: 114).

This spatial distribution of artefacts and structures from the 3rd century AD onwards illustrates occupancy from the Bay of Skardhana, through the central area of the ancient city and just past the port; it does not appear to extend past magasin γ, although the area to the south and southeast is largely unexcavated.[11] Indeed, the wall of Triarus reaches the sea just past magasin γ; the archaeology and literature suggest that the Late Antique community was based largely within this boundary. The settlement continued up into the theatre quarter as well as towards and around the Mount Kynthos sanctuaries. Evidence was also found at the farther removed locations of the Synagogue and the Maison de Fourni.[12] Finally, the discovery of a 4th or 5th century early Christian chapel in the northern region of the adjacent island, greater *Rheumatiaris*, suggests that the community extended there, either as a settlement or a place of worship (Gallet de Santerre and Treheux 1947-1948: 414-418; Bruneau 1968: 706). The fact that this chapel, and several other contemporary structures on the southeast coast of greater *Rheumatiaris*, reuse large quantities of earlier material demonstrates a relationship between Delos and the surrounding islands even at this late date; the later inhabitants of greater *Rheumatiaris* sourced disused building material from Delos with which to characterise new structures on the neighbouring island.[13]

From the 1st to 7th centuries AD the archaeological evidence provides an abundance of data at times, and almost none at others. One might speculate that, despite probable continuous occupation, there was some degree of population and economic fluctuation; taking into account the usual caveats regarding archaeological bias and the inconsistent survival rate of material culture. Considering the dates prescribed to this cumulative data, a relative peak of Late Antique prosperity around the 5th century AD is observed; the contemporary epigraphic evidence supports such a notion, with the majority dated to the 5th and 6th centuries AD (Orlandos 1936: 69; Brunet 1999: 9, n° 20; Kiourtzian 2000: 49-60). It also appears that many of the larger structures from early Christian Delos fit into this century, including the basilica of St. Cyrique, the basilica (of St. John?) and other religious structures near the *Ekklesiasterion* and the monastery over the Hypostyle Hall – the numismatic evidence supports this (Leroux 1909: 57-58; Orlandos 1936: Bruneau 1968: 703).[14] The

---

[11] Exceptions do exist (e.g. the Synagogue, sanctuaries of Kynthos and the Maison de Fourni), and further exploration to the W of the theatre and S of the theatre quarter will undoubtedly clarify the full extent of the Late Antique settlement.

[12] Orlandos (1936: 69, 84-86) mentions the discovery of a basilica in the vicinity of the Asklepieion; it is now possible to equate this with that at the Maison de Fourni.

[13] Including architectural fragments, basins, entablature, architraves, masonry, funerary stele and Doric capitals (Gallet de Santerre and Tréheux 1947-1948: 416-418). It is not unusual to observe early Christian finds and settlement remains on many small Aegean islands, e.g. Telendos, Saria (G. Deligiannakis, pers. comm. Nov. 2016); recent research suggests that notable communities were present on many of the small islands at this time – a fact often overlooked within earlier scholarship.

[14] The structures near the *Ekklesiasterion* possess similar characteristics to the 5th century Basilica of St Cyrique; a tripartite plan with an eastward facing apse and architecture and decoration suggestive of construction in the late 4th-5th century AD (Le Quéré 2018a: 120). One of the basilicas built near the *Ekklesiasterion* is tentatively ascribed a dedication to St John, after the discovery of a stamp mould for Eucharistic bread nearby the *Porinos Oikos*, with the inscription: Εὐλογία Ἁ(γίῳ) Ἰωάν(ν)η (Cabrol and Leclercq 1920: 568-69; Déonna 1938: 232-33; Dürrbach and Jardé 1905: 256 with fig. 5; Le Quéré 2018a: 120). The monastery is

space delimited by the Portico of the Naxians in the west, Temple of Apollo in the north, *Prytaneion* in the east and Agora of the Delians in the south was composed of residential dwellings and probably also dates to this time, including those poorly constructed near the Agora of the Competaliasts (Hasenohr 2002: 106-107; Le Quéré 2018b: 274-75). The existence of a contemporary chapel and other structures on greater *Rheumatiaris* attest to a large spatial settlement area (Dürrbach and Jardé 1905: 255-257).

The Maison de Fourni basilica appears slightly earlier, around the early 4th century, and the church remains in the Portico of Philip V are undated, though are probably contemporary with 5th century dwellings nearby the Portico of the Naxians (Orlandos 1936; Bruneau 1968: 703; Le Quéré 2018a: 139; 2018b: 275). Although not contemporary, perhaps the Maison de Fourni remains illustrate intensification in early Christian construction leading up to the 5th century AD. Epigraphy and iconography associated with the viticultural industry dates to the 5th/6th centuries AD; such a date is supported by the often significantly higher stratigraphic position of the viticultural remains.[15] This evidence endorses intensification in habitation and industry at this time.

This increase and eventual peak in population, architecture and economic prosperity is also seen through a simple quantitative approach regarding structures and artefacts from this era. In the 3rd-4th centuries AD, the city constructed two large public bath complexes, one near the Agora of the Delians (460 m^2) and a second over the Agora of the Competaliasts (237 m^2), and, for the first time, a supply of running water through an aqueduct (above); in the following century it possessed at least five basilicas and churches.[16] This suggests that a relatively large population of faithful people lived permanently on the island and demanded basic, Roman facilities. The evidence of artisanal activity[17] along with the high quality lithic sculpture and iconography, bronze work, and industrial viticultural activity also suggests a more complex type of inhabitation than previously recognised.

This occupation and relative success was not long lived, and the evidence begins to dwindle through the 6th century.

It is known that the Delian bishopric, if it existed, was abandoned before the 8th century, probably replaced by Syros in the 7th century AD, and the absence of coins and sculpture after the mid-6th century suggests that perhaps the island was finally abandoned in the 7th century AD.[18] This interdisciplinary evidence provides a useful *terminus ante quem* for viticultural activity on the island.

The evidence does not suggest that the Late Antique occupation of Delos was timid in any way. The continuity of occupancy through and past the Imperial period, presence of resurgent viticultural industry, significant civic, domestic and ecclesiastical building activity, and spatially widespread settlement indicates a relatively successful and prosperous community; one not equal to that of the booming Hellenistic era, but one that should not be labelled as 'collapsed' or 'abandoned'. A detailed examination of the viticultural evidence, which follows, provides further indications of settlement continuity, prosperous agricultural enterprise and economic endeavour.

Viticulture on ancient Delos

The viticulture of ancient Delos is a problematic topic that has been debated periodically over the past century, culminating in the years 1980 – 2000 when a number of studies were completed in an attempt to further comprehend agricultural life on the island in antiquity (Bruneau and Fraisse 1981: 142-143; Bruneau and Fraisse 1984: 721; Bruneau 1987; Brunet 1990: 674-676; Brun 1993; Brun and Brunet 1998: 605-607; Brun 1999: 136-137, n° 89; Brun 2000: 284-285, n° 45-47). The problem, and our relative lack of understanding, stems from three main areas: biased epigraphic records, almost non-existent literary references, and the, often, highly organic nature of viticultural material. Ancient textual sources, where one typically encounters comment on regional ancient wine types, such as Pliny, Theophrastus or Columella, bear no mention of Delian wine or the viticulture associated with the island. Archaeological evidence recognised over the past century fills some *lacunae*, particularly in relation to the vinicultural processes, and this is the emphasis of the present chapter. The fortunate nature and history of the island post-antiquity also aids preservation of the rural areas to the north and south of the city; here, well preserved walls, terraces and landscapes remain as they did in antiquity (Brunet 1993: 202). Nonetheless, a great deal of the agricultural (viticultural) side of the industry remains unknown.

---

tentatively labelled as such; Deligiannakis (2016) summarises the issues involved in identifying monasteries of this era (cf. Sweetman 2016: 1).

[15] On the epigraphy, see Bruneau and Fraisse (1984: 729); Brun (1999 : 153-154); and Kourtzian (2000: 49-60). On the iconography, see Orlandos (1936); Bruneau and Fraisse (1984). Dodd (2020) provides a discussion on the stratigraphic position of the viticultural architecture.

[16] Bouet and Le Quéré 2016; Bruneau 1968: 707-8; Hasenohr 2014; Le Quéré 2018b: 276-77; 2018c: 348; 2018d: 114. Earlier publications (Orlandos 1936; Déonna 1948) state that six paleochristian churches were present on the island; however, this can now be reduced to five, with the possible addition of a baptistery. The iconographic evidence from the Rue 5 'large' press and the counterweight near the Establishment of the *Poseidoniastae* belong to agricultural installations rather than ecclesiastical structures.

[17] Architectural and foundry remains reveal that bronze work might have continued in the area of the Granite Palestrae in an artisanal fashion during Late Antiquity (Delorme 1961: 37, 152-153). It is possible that ceramic production also occurred here (Le Quéré 2018c: 359; 2018d: 117; Dodd 2020: 115).

[18] Orlandos (1936: 69-70); Hasenohr (2002: 108). Some believe it was abandoned as early as the late 6th century AD (Zarmakoupi 2013: 62, n° 110). There is some doubt whether Delos was actually elevated to the level of bishopric, with bishop Sabinos present at the fourth ecumenical council at Chalcedon in 451 AD (Kiourtzian 2000: 50). Nonetheless, that Delos was named in such a list indicates some level of religious importance placed upon the site at this time, particularly in its regional context.

Population decline and wine industry: societal transformation on Late Antique Delos (Greece)

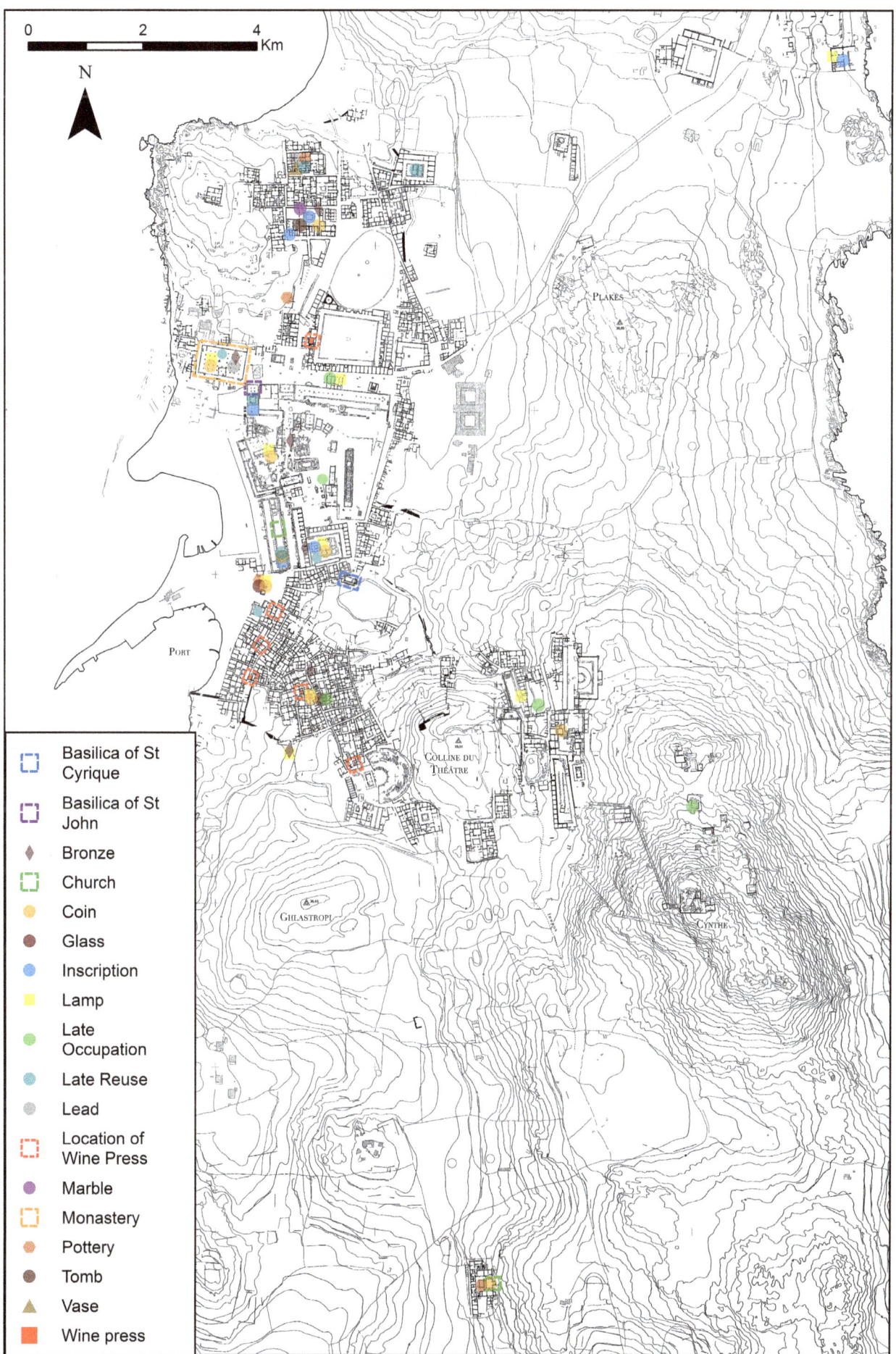

Figure 6.1: Archaeological evidence from Late Antiquity on Delos (post-3rd century AD) (Map by author).

The climatic, topographical or, otherwise, general agricultural suitability for the island of Delos to support thriving ancient grapevine growth has not been studied scientifically, as it has in other eastern Mediterranean regions.[19] Although not always entirely accurate, it is possible to gain some idea of the viticultural potential of Delos by observing the prosperous modern industry on Santorini and the contemporary suitability of Mykonos for the vine in an ethnographic manner; both islands with topographic, climatic and geological similarities to Delos. Indeed, previous studies suggested that the agricultural and viticultural methods of cultivation on Delos in antiquity might be similar to those currently practiced on Santorini (Brunet 1993: 202).

The nature of the land on Delos, with its granite substrata and shallow topsoil, along with strong winds and other difficult climatic conditions, encourage these increasingly plausible comparisons (Brunet 1993: 202). These features were no different in antiquity – Pindar (*Isthm.* I.1) identifies the island as "rocky" and Strabo (*Geog.* 10.5.2; 17.793) describes the fierce winds. It seems likely that viticulture was practiced here in a different manner to that on the contemporary Italic peninsular and, thus, comparisons might be better found in the ancient eastern Mediterranean or in modern Cycladic practices.[20] The island does possess a similarly temperate, arid, and occasionally challenging Mediterranean climate to that found in coastal Rough Cilicia and Hippocrates confirms this in antiquity (Hippoc., *Prognosticon* 25).[21]

Brunet and Poupet surveyed the remaining terraces and fields to the north and south of the ancient city in the 1990s (Brunet 1993: 202; Brunet and Poupet 1997: 776-782; Brunet 1999); however, this study focussed on the Hellenistic era and largely neglected any possibility of continued Late Antique use. It is conceivable that the cultivation of vines was concentrated around the south end of the island (explored further below), where the south and southeastern facing terraces afforded protection from the strong northern *meltemi* winds (Greek: ἐτησίαι ['*etesian*']) (Bruneau and Ducat 1966: 14; Brunet 1993: 202). These strong, sometimes gale-force, winds blow during the summer months, beginning as early as May and lasting until September; modern climatic studies suggest that they were even stronger in antiquity (Neumann and Metaxas 1979: 185f; Murray 1987: 140, n° 8, 10; Morton 2001: 48, n° 18-20).[22] They are, thus, an influential, potentially damaging, climatic feature against which protection was required during the vital summer period of grape growth. The surviving archaeological evidence and ethnographic comparanda shows an awareness and concerted effort of protection against these natural features.

A low, creeping vine with wide spaces between plants was probably grown on ancient Delos (Brunet 1993: 202);[23] one not dissimilar to that found in modern times on Santorini. Here, in 2014, the author observed vines grown low to the ground, with their branches trained and woven in circles to create a basket shape (a traditional process called '*koulara*'); the aim of which is to protect grape bunches on the inside from the harsh winds and sun of the Cyclades. This traditional method of cultivation is seen across the island on small plots of land, with mixed grape varieties grown together haphazardly. The roots of the vines on Santorini are kept close to the surface in order to absorb more water in the arid climate – this increases their yield. Keeping vines low to the ground also provides more warmth and, consequently, higher sugar content in the berries. This is a good method to make sweet wines from sun-dried grapes, especially on poor, stony soils (lime, granite, gneiss etc.) like those of Delos, as they still do today in parts of the Mediterranean.[24] In 2016 the author again observed that these cultivation techniques are still commonly used on other Cycladic islands, including Paros and Mykonos – both close by Delos. Such concepts were known to the ancients and might be accurately applied in locations like Delos in order to combat the equally hot and arid conditions, harsh winds, as well as the shallow granite substrata. Low-lying, or 'woven' vine cultivation, does not require the many trellises, stakes, supporting trees or trenches that leave noticeable traces in the archaeological record and are observed at other Roman viticultural sites.[25] These somewhat non-intrusive cultivation techniques might serve to explain the relative lack of ancient agricultural evidence surviving in the Delian, and broader Aegean, landscape.

Significant research has already been completed regarding the epigraphic records from the temple of Apollo and how these contribute to our understanding of the Delian wine industry. Unfortunately these records only apply to the period from c. 434 to 156/7 BC and, thus, not to the present study. While the temple of Apollo no longer held control over land in Late Antiquity, it is possible that the Church took this place and a similar arrangement existed in later eras. Nonetheless, it remains difficult to determine

---

[19] A variety of modern scientific methods, with relative success, were used within Rough Cilicia (Rauh *et al.* 2006; Rauh *et al.* 2009).

[20] Brunet (1993: 202) also discourages a comparison to the Tauric Chersonese.

[21] The similarity is explored further in Dodd (2020).

[22] Strabo (*Geog.* 10.5.2; 17.793) describes all manner of strong winds affecting Delos and how, in Alexandria, the *etesians* began in the summer. Ap. Rhod. (*Argon.* 2.498-99; 2.524-27) describes them as "gusting" and "...sent by Zeus (to) cool the land for forty days...". Wachsmuth (1837: 427-430) provides an extensive summary of the *etesian* winds in the ancient literature. Murray (1987) provides a useful general account of comparing modern and ancient winds along with the associated, supporting methodologies.

[23] From the 4th century BC, the Greeks favoured growing their vines without props, trellises or other supports (Forbes(1956: 132). Perhaps this method continued throughout history and was favoured on both traditional habit and regional suitability.

[24] As part of a wider study using material from Delos and Rough Cilicia (Turkey), the present author suggested that early Christian Delians might have produced a sweet wine, similar to the *passum* of Rough Cilicia, based on similarities in topography, production style, chronology among other socio-cultural and economic factors (Dodd 2020: 113-131).

[25] For example, the vine trenches excavated in the *suburbium* of Rome or the evidence for trellised cultivation and root systems at Pompeii (Jashemski 1968; 1973a; 1973b; 1979; 1993; Volpe 2004; 2009).

the extent and character of land use continuity from the Hellenistic period into Late Antiquity. Many other aspects of the island saw great change (including population, increasing rurality and the contraction and selective concentration of urban areas); therefore, it is conceivable that such extensive vine cultivation across the island did not endure. The archaeological evidence supplemented by the present study, however, suggests the contrary: that a wine industry of significant scale, potentially even larger than that of Hellenistic Delos, was present on the island in Late Antiquity. Characteristically, small islands are able to form extremely productive niches with their "all-round connectivity" and close proximity to the sea – "the prime medium of communication and redistribution" – of which wine holds pride of place among the 'island monocultures' (Horden and Purcell 2000: 224-227). It is not surprising, therefore, that the inhabitants of Late Antique Delos were able to establish (or re-establish) such an intense viticultural industry.

The current state of knowledge regarding Delian wine-related agricultural practices is, therefore, sparse, particularly for Late Antiquity. The extensive archaeological remains utilised in the following section serve as the backbone to our understanding of wine production in this era and, combined with other interdisciplinary data, act as an impetus for understanding how this relates to the contemporary Delian population.

Quantified analyses

Data from six viticultural installations is used in this study; all located within the bounds of the main Late Antique settlement, which had shrunk to concentrate itself around the Hellenistic Sanctuary of Apollo and extend no further than the wall of Triarius (Figure 6.1). The present author has previously proposed that they are all roughly contemporary, at least within Late Antiquity, and were potentially used in one contiguous phase of early Christian habitation between the 4th and 6th centuries AD (Dodd 2020: 75ff, 108-112).[26] Detailed datasets of the six installations are highlighted in Dodd (2020: 82-103).

In order to estimate a maximum annual production quantity for known installations on Delos in the applicable time period and, therefore, elucidate new data on characteristics of the Late Antique settlement, the capacity of each collection vat was recorded (Table 6.1). Following the theory that installations of this type in this region utilised in-vat fermentation and were filled 7-9 times per vintage, the maximum vintage production of 113,750 – 146,250

Table 6.1: Total capacity of known collection vats.

Installation	Capacity (L)
Agora of the Italians	2200
Rue 5 'large'	6260
House of Cleopatra	4480
Agora of the Competaliasts	2270
Rue 5 'small'	1040
Theatre	unknown
Total	**16,250**

L is reached.[27] This calculation is contingent upon one assumption: that the listed installations are all roughly contemporary. A lack of decisive chronological evidence prevents absolute certainty; however, it is possible to suggest that they were in use within the same 200-year timespan.[28] While the phrase 'maximum annual production quantity' is used, it must also be recognised that these calculations do not take into consideration the contribution of the Theatre press to an overall production figure, as the capacity of the collection vat at that installation is unknown at present, along with any undiscovered installations, which may further increase productivity.

Crop estimation and suitability

It is more difficult to quantify agricultural production in the fields than at the installations; however, using a combination of modern agricultural knowledge, ethnographic data, archaeological traces and descriptions in the ancient literature it is possible to estimate potential yields in antiquity. Van Limbergen provides an excellent introduction to the concept in his study of Late Antique north Syrian oil and wine production and it is not necessary to repeat here what is already known in that respect (Van Limbergen 2015: 176-179). Unlike in northern Syria, at Delos it is almost impossible to calculate the proportion of arable fields dedicated to vine cultivation in Late Antiquity.[29] The present author proposes a range of estimations and possibilities using the archaeological data (in Dodd 2020: 117-118) to check for validity and accuracy. The known processing capacities of the installations (Table 6.1) also assist in determining a minimum cultivated area that was necessary for their economically viable operation.

---

[26] The House of Cleopatra, Agora of the Italians, and Rue 5 'large' installations were previously ascribed tentative dates, from the 4th-6th centuries AD, based on paleography, iconography, stratigraphy and construction – the stratigraphy of the Rue 5 'small' and Agora of the Competaliasts installations indicates a similar date (Bruneau and Fraisse 1984; Brun 1999). Lengthy discussion, comparison and conclusions regarding their dating and probable contemporaneity are found in Dodd (2020: 108-112).

[27] Brun 2003: 63; Decker 2009: 144; Van Limbergen 2015: 175.
[28] Cf. Dodd (2020: 108-113) and the comparison between vintage production and agricultural yield (below).
[29] This problem is compounded by the inability to recognise and quantify contemporary Late Antique domestic areas or farmsteads, and, thus, relate population size or familial groups to plots of land and the crops produced. Van Limbergen (2015) shows how these details are useful to estimate what proportion of land was dedicated to certain crops. To the best knowledge of the present author no contemporary olive presses exist; thus, if the linear relationship of Van Limbergen (2015: 179) is followed, it is suggested a vast proportion of the agricultural terracing was probably devoted to viticulture. This hypothesis is supported by the quantification of vinicultural installations (Table 6.1) and a comparison of viticultural output at the fields and grapes required to operate the presses.

Figure 6.2: Terraced regions on the island of Delos (in red) (Map by author).

The island of Delos covers a total area of approximately 360 ha.[30] Satellite imagery and ground survey, along with existing publications that discuss the geography and agriculture of Delos in the Hellenistic period, reveal regions of the island dedicated to agricultural cultivation unchanged from antiquity.[31] Based on the surviving archaeological evidence in the form of terraces and man-made agricultural features, it is possible to quantify these areas for the first time; at least 47.91 ha, or 13.3%, of the island was dedicated to terraced agricultural cultivation (indicated in red on Figure 6.2). Previous studies suggest that much of this was dedicated to cereal cultivation and livestock grazing during the Hellenistic boom, with a smaller area set aside for olive and vine growth.[32] There is now sufficient ethnographic comparative data to suggest that, in times of population increase and pressure, the cultivated area was expanded with the best soils given to food crops and grain; vines were typically allowed only the poorer soils (Grigg 1980: 5, 32).

Table 6.2: Maximum annual vintage production of known installations.

Max. vintage production (L)	113,750 – 146,250
Quantity of LR1 amphorae (~20 L) (Decker 2009: 144, n° 70)	5688 – 7313
Grapes required (kg)[a]	167,279 – 215,074

[a] Using the conservative estimation of 0.68 L/kg (Khalil and Al-Nammari 2002: 48; Şenol and Waltz 2015: 36, n° 6).

On Late Antique Delos, characterised by a comparatively lower population to earlier eras, it is possible that this situation was reversed; cultivation of vines increased and, consequently, they were provided with more favourable soils.[33] The absence of population pressure allowed these smaller communities to grow more profitable crops, like the vine for wine production, rather than subsistence crops necessary to sustain a large population.[34] It is possible, therefore, that a greater proportion of the known 47.91 ha was dedicated to vine cultivation in Late Antiquity than the Hellenistic era. This theory is supported by comparing the quantified vat capacities of the installations (Table 6.1) and the amount of produce required for their annual operation (below). The notoriously problematic ability to accurately date ancient terraces allows for their possible use in later eras, despite the fact that previous studies suggest they date

---

[30] Historically, measurements exist ranging from 343 – 360 ha. For the purpose of this study, the figure of 360 ha is used, as given by the ÉfA (http://www.efa.gr/index.php/fr/recherche/sites-de-fouilles/cyclades/delos/delos-presentation-geographique). This places Delos on the extreme end of the "small island" category given by Bevan and Conolly (2013: 6). Almost all of its land is in direct contact with the coastline, thus greatly pronouncing the impact of connectivity upon its character

[31] The author used Google Earth Pro to measure and quantify recognised terraced areas across the island based on ground survey, published data and satellite imagery captured from 2004-2014.

[32] The evidence for cereal and livestock comes largely from the Hellenistic farms found within the fields, including threshing floors for grain production (Brunet and Poupet 1997; Brunet 1999; Bruneau and Ducat 2005). At the time of writing, such features are not found for Late Antiquity, although limited survey and excavation is completed

[33] Into the future it may be possible, if not already, to suggest this as a trend for Late Antiquity and the Byzantine eras. As an earlier example, it is, perhaps, no coincidence that the period of great wine export from Roman Italy directly precedes Italy's population boom (D. Van Limbergen, pers. comm. May 2016).

[34] Viticulture in general, however, still requires a substantial population base to provide labour for intensive processes of pruning, picking and harvesting.

Table 6.3: Total vine, grape and must yield (S & SE terracing).

	S+SE terraced area (ha)	N° of vines[a]	Yield	
			Grapes (kg)	Must (L)[b]
Poor year (1.5 kg/vine)	45.16	90,320	135,480	92,126
Average year (3kg/vine)	**45.16**	**90,320**	**270,960**	**184,253**
Bumper crop (4.5 kg/vine)	45.16	90,320	406,440	276,379

[a] Based on a yield of 2000 L/ha.
[b] Based on a ratio of 0.68 L/kg of grapes.

only to the Hellenistic period.[35] It is now possible to refute the existing argument that terraces on Delos date solely to the Hellenistic and Classical periods because there were no other historical periods that justified "an intensification of agricultural production and landscaping with terraces".[36] The current chapter clearly illustrates viticultural industry on Late Antique Delos, which required large quantities of terraced land with a suitable exposure.

Based on arguments of topography and other features, it is assumed for the purpose of estimation that the entire south and southeast terraced area of the island was dedicated to vine cultivation in Late Antiquity: an area of 45.16 ha.[37] A conservative average density of 2000 vines/ha and yield of 3 kg grapes/vine are used in the following calculations (Van Limbergen 2015: 177-178).[38] It should be noted, however, that a much wider range of planting densities are given in the ancient literature, ranging from 1300 to 8100 vines/ha; epigraphic and archaeological traces support a similarly diverse range.[39] The use of 2000 vines/ha is further supported by the nature of vine cultivation and growth on Cycladic islands that occurred in antiquity, and continues today, where vines are grown close to the ground without staking or trellising (above). This somewhat haphazard growth pattern leads to lower vine densities than the intensively trellised methods more common to the central and western Roman Mediterranean.

Table 6.3 illustrates the results for the south and southeast terraced areas of Delos. The figures for an average year suggest that 6000 kg of grapes (or 4080 L) were produced per ha.[40]

It is possible to further test the validity of these calculations by comparing the total quantified capacities of the pressing installations (Table 6.1; 6.2) to the quantification of the agricultural terraces (Table 6.3). The known installations were capable of producing between 113,750 to 146,250 L of must annually. Crop estimations indicate that, in an average year, the 45.16 ha of south and southeast terracing on the island produced approximately 184,253 L of must. While the figures do not equate exactly, the latter assumes that the entire terraced area was devoted to the cultivation of vines; this was probably not the case. If it is assumed that only three quarters of the terraced area was dedicated to vines (33.87 ha), then a more reasonable figure of 138,190 L is reached. This fits within the estimation for the annual production capacity of the installations. Agricultural terraced areas might also be dedicated to cereal, livestock, fruit and vegetable production. It is possible that one quarter of the south and southeast terracing was used in this manner, along with terraces in the north region of Delos and areas closer to the – now contracted and ruralised – Late Antique city. The possibility for poor or bumper vintages should also be taken into consideration, and may account for the slight differences between the productive capacity of the installations and the agricultural fields.

Generally, it appears that the maximum productive capacity of the wine installations on Delos fits closely within the scale of possible poor, average, and bumper vintages for the terraced south and southeast agricultural areas of the island. This not only provides further support for the argument that these installations were used in conjunction with the terraced areas visible today, but also permits the

---

[35] On the problematic dating of ancient terraces: Foxhall (1996: 44-45); Price and Nixon (2005); Bevan and Conolly (2011); Dimakopoulos (2016: 1, 9); and Kinnaird et al. (2017: 67-69). The Hellenistic attribution of the Delian terraces is based on ceramic sherds recovered during brief excavations in the 1990s (Brunet and Poupet 1997: 778-780, figs. 3-5; Brunet 1999). This, in itself, is problematic; farming practices and other taphonomic processes often lead to significant post-depositional disturbances that are notoriously hard to detect, and it is equally as difficult to determine whether ceramics were discovered in their primary context or whether they have been re-deposited (Kinnaird et al. 2017: 68).
[36] Cf. Price and Nixon 2005: 670; Dimakopoulos 2016: 9. A number of other criteria presented by Price and Nixon (2005) in relation to the Delian terraces are also questionable. The present author does not deny the use of these terraces in earlier periods, but simply reinforces the likelihood of their later use in connection with contemporary wine presses.
[37] S and SE exposures were best suited to vine growth and protection from harmful elements and climatic conditions (e.g. the *meltemi* winds).
[38] Alternate estimations are also given: 1.5 kg grapes/vine for a poor year and 4.5 kg grapes/vine for a bumper crop. In most cases of modern wine production, these are still relatively conservative figures (although yields in modern production are highly variable) (Greven 2007: 10, table 2; Reynolds 2015).
[39] An inscription from Rhodes results in estimations of 4000 and 7000 vines/ha, and archaeological traces in Sicily and France result in anything from 1100 to 9800 vines/ha (Van Limbergen 2015: 177, n° 41-42). The excavation of a vineyard at Pompeii (*insula* II.5) revealed root positions and exact planting densities from 79 AD, which resulted in the ability to estimate productivity from the c. 4000 vines planted in c. 0.75 ha (Jashemski 1973a: 824, 828, fig. 3; Jashemski 1973b: 36). This leads to an estimated density of 5333 vines/ha, roughly at the midpoint of those described by the ancient sources. The same study also estimated that this climatic region produced 3.63 kg of grapes/vine and 0.63 L/kg; comparable figures to those used both by Van Limbergen (2015) and within the present study.
[40] Notably, this figure fits within the range provided by Columella (*Rust.* 3.3.3; 3.3.10) supporting its validity for use within future analysis. For a modern comparison, the nearby Muğla region of Turkey yielded 7500 kg of grapes (or 5100 L) per ha in 1984 (Höhfeld 1998: 249, chart 2; Şenol and Waltz 2015: 35).

suggestion that the entire catalogue of installations were contemporary and in operation at roughly the same time. In a similar manner, the combined quantified productivity of the installations demands a suitable amount of arable land on the island, one that can only be found among the visible terraces in the south-east of the island. This reinforces the argument that these terraces were used in not only in the Hellenistic and Classical periods, but also in Late Antiquity.

Conclusions from the dataset

The quantified analysis of viticultural installations reveals relatively high-level productivity; one that necessitated the use and upkeep of large terraced agricultural fields. This does not take into account any undiscovered or destroyed installations that might raise the vintage output further; counterweights that may belong to a contemporary period and are no longer *in situ* reinforce this possibility, such as that near the Establishment of the *Posedoniastae*. The numerous ecclesiastical structural remains, domestic quarters, small finds and the spatial spread of the early Christian settlement support a notable population size that demanded basic Roman necessities, such as wine.

A quantified analysis indicates that the combined productivity of the six vinicultural installations fits within the estimated agricultural output of south and southeastern ancient terraces. This not only suggests that the installations were used directly in conjunction with these fields, but also that the installations are relatively contemporaneous in order to cater for this quantity of raw agricultural output. Space is left at the northern end of the island, along with a smaller percentage of the south and southeast terracing, for other agricultural activities. An apparent favouritism for viticultural enterprise might be explained by the reduction in population pressure, whereby the settlement no longer required large quantities of subsistence produce and could instead focus on a 'cash crop' along with the continued demand of wine for regular ritual and domestic purposes (White 1970: 268-269).[41]

Population decline and agricultural response on Late Antique Delos

Post-Hellenistic and Imperial Delos was historically viewed as a city in a steady state of decline, barely surviving until its eventual abandonment somewhere between the 6th and 8th centuries AD. Recent studies by Le Quéré (2018a; 2018b; 2018c; 2018d) among others, along with the evidence presented above, suggests an alternate reality; one of a notable settlement evolved from a prosperous earlier self, with clear markers of continual socio-economic prosperity and, later, characterised by a renewed viticultural industry. Indeed, these recent archaeological reassessments, combined with the newly quantified viticultural data, suggests that perhaps the terms 'decline' and 'collapse'

Table 6.4: Population size on Delos.

Period	Estimated Population Size
Independence (314-167 BC)	1500 – 2000
Second Athenian Domination (167-69 BC)	>15,000
Peak	20,000 – 30,000
First century BC	6000 (incl. 1200 citizens)
Late Antiquity	Unknown

are not as applicable as once thought. Use of the term 'decline' subconsciously implies the negative trajectory of a community, in socio-cultural and -economic terms, often inevitably leading to collapse. There is undoubtedly some degree of population and economic loss present on Delos from the Hellenistic to Late Antique eras, but not necessarily with the blanket negative connotations frequently prescribed to the label 'decline.'. The thriving Late Antique viticultural industry on Delos illustrates change, continuance and success of one kind, while the construction of ecclesiastical structures of relatively high quality and substantial domestic quarters demonstrate another. We might more aptly apply the recently popularised terms of resilience, societal transformation, evolution and diversity. As outlined by Curtis (2014: 18), resilience is for the most part dictated by endogenous societal responses, including: the capacity of a society to protect itself and, following trauma, to create sufficient recovery over the medium and long term. It is a slow moving phenomena and, over a period of many centuries, is clearly present at Delos. Here, in particular, it is is defined by the ability of the population to absorb stresses and traumatic events, evolve socio-economic structures, and redefine the character of their community to suit an ever-changing Mediterranean socio-political context. When these high-level requirements are met, the population is able to survive over many centuries with a degree of prosperity and at least some regional importance.[42]

After the Hellenistic boom, the importance of Delos as a strategic midway port between Italy and the eastern Mediterranean was removed, along with a multitude of other factors that contributed towards a rapid change in character and role for the island. Delos faced comparative economic stagnation, invasion and a sharp reduction in interregional importance – both chronic and acute stresses (Faulseit 2016: 6). A settlement that finds itself at this impasse faces a number of possibilities. Two of the most common responses include: submission to the stresses (often leading to decline and eventual collapse); or evolution and transformation in character and purpose (often a longterm process that presents an opportunity for continued settlement prosperity under a new guise). In the case of Delos, rather than succumbing to these stresses, socio-cultural evolution and a degree of resilience within the population allowed the settlement to continue with

---

[41] Col. (*Rust.* 3.3.8-13) praises viticulture as one of the most profitable agricultural endeavours.

[42] Illustrated by the potential elevation of Delos to Cycladic bishopric (see above).

some semblance of prosperity (see the Imperial evidence above). Although the archaeological record portrays the city as a shadow of its former self (indeed, at face value it was, with a much reduced urban fabric and fewer monuments constructed), a change of role lead to continued prosperity in Late Antiquity of a type much more difficult to distinguish within the material culture. Ancient viticultural activity is notoriously problematic to detect in the archaeological record; the majority is organic in nature and the more permanent structures are often difficult to differentiate from other agricultural or artisinal installations. It is fortunate, in the case of Delos, that we have well attributed agricultural press structures and ancient terracing still visible; a relationship between them is supported by the quantified analysis above.[43]

When the markers of prosperity typically associated with Hellenistic Delos – that is, those of an important port, commerial and religious site – largely disappear in the Roman era, the island is traditionally thought to be in a state of collapse; this is, perhaps, the most notable oversight of earlier analyses. Our inherent bias to value complexity as modern researchers, particularly in its most obvious forms, and our often stereotypical and narrow association of Delos with its Hellenistic character contribute towards this. When the character of Delos changes from a commercial port city to an agriculturally productive settlement it is more appropriate to consider it a societal transformation rather than the removal of complexity and foundation of collapse; indeed, the evidence of viticultural industry, civic, domestic and religious construction and continued habitation argue against any labels of wholly negative decline, particularly when aspects of high quality artisanal production and construction are added (Faulseit 2016: 4ff). Through this lens, we can suggest that Delos did not entirely exhibit traits of decline and collapse in the Imperial and Late Antique eras; rather, the society displayed a resilience to adapt and survive, with forms of monumental architecture and high-level agricultural productivity, despite the evolution of the surrounding Mediterranean and a reduction in importance as a commercial and cult centre. Such fluctuations in population and prosperity are typical of islands, where changes in success or productive development through time are linked to their connectivity within wider networks or socio-economic systems; in this case, probably either the Church or regional cabotage (see below) (Horden and Purcell 2000: 224-230).

Even more interesting is the ability of Late Antique Delos to satisfy only some of the characteristics provided by Renfrew in his archaeological definition of collapse (Renfrew 1984: 367-369; Faulseit 2016: 5). While there was certainly an "abandonment of elite residences" and population decline, there remained some continuation of public construction, notably the multiple large bathhouses and ecclesiastical structures, and a connection with a central administration, particularly if Delos was the seat of a Cycladic bishopric (Dodd 2020: 76, n° 570). The island is listed within the province of Hellas (Achaea) in the *Synekdemos* of Hierokles (c. AD 535) and it is possible that wine was exported from Delos and paid to the State as a form of taxation.[44] Relatively contemporary references exist to the Aegean islands supporting Athens with corn as a form of tax; based primarily on historical reasons, it is possible that Delos provided tax to Athens in a similar manner using wine (Eunap., *VS* 509-513; Julian, *Or.* 1.21-23). Equally, around the 5th century AD, the province of Hellas produced large quantities of olive oil for the Late Antique *annona*; perhaps now Cycladic wine from Delos can also be added to those other products listed in the *Expositio Totius Mundi et Gentium* (lii 4-16). Such theories of surplus export direction, however, are difficult to confirm at this stage; some argue that the Aegean possessed more of a 'free' commercial economy than that of, for example, post-Vandal North Africa, where it appears there was a more restricted 'State' circulation of goods.[45] It is equally likely at this stage that the Delian viticultural industry supports a model of estate production, where surplus was exploited commercially. Regardless, such interconnectivity stemming from a viticultural industry supports the notion that we should consider islands within the pattern of contiguous landscapes or connective corridors rather than as isolated features (Bevan and Conolly 2013: 6). Similarly, the survival of Delos in such a productive state might largely be attributed to the interconnected position of the island within a wider socio-economic framework at this time, which caused continual demand for the wine produced.

The existence of wine production in a centralised and organised manner further rejects any notion of collapse; indeed, the listed viticultural installations are situated in locations that were highly urbanised in the Hellenistic period, and probably continued to be urban in Late Antiquity (Figure 6.3).[46] Traditionally, press facilities were often preferred close by the fields for economic and practical reasons and this is reflected at many sites (Curtis 2001: 298, n° 77).[47] Agricultural installations in the eastern Mediterranean from Late Antiquity and the

---

[43] The attribution of the structural evidence to viticulture (rather than oleiculture or another commodity) is discussed at length in Dodd (2020: 104-108).

[44] Mass produced amphorae from neighbouring Cycladic islands might be filled with Delian wine, redistributed from large ports (e.g. Naousa, Paros) in a second phase and sent to the State as a tax (Diamanti 2016; Dodd 2020: 128, n° 989).

[45] Reynolds 2004; W. Bowden (pers. comm. Oct. 2016). It is also possible that the Church used distribution mechanisms independent to the State and free economies (Reynolds 2004: 246, n° 58). Although this is difficult to recognise in the archaeological record, it might be equally applicable to Delos.

[46] The theatre press is marginal in this respect, as the upper theatre quarter is considered largely abandoned at this time with minimal structural evidence and small finds indicating occupation. The remaining five installations are all situated in areas where large quantities of 'late' evidence and dense structural remains exist.

[47] For archaeological evidence of this practice, see Diler 1995: 88; Frankel 1997: 74; Baratta 1999: 139; Lefort 2002: 256; Aydınoğlu and Alkaç 2008: 280; and Baldiran 2010: 303. A lack of surviving archaeological evidence indicates that portable (wooden) presses might be constructed or brought into the fields during the Hellenistic period on Delos (Brunet 1993: 203-205).

Byzantine, however, are found regularly situated within urban environments, as they are at Delos.[48] It should be stressed that this process of re-ruralisation also appears in the west throughout Late Antiquity and probably illustrates a changing economic mentality, in which town and country melt together into a continuous productive landscape (D. Van Limbergen, pers comm. Dec. 2016). Curtis recognises that, in times of instability (e.g. warfare, economic collapse, invasion, disease), the rural environment is most susceptible to collapse and, hence, a contraction often occurs (Curtis 2014: 6-7). This might illuminate why press facilities were often constructed within the city walls in Late Antiquity – a similar trend is noticeable regarding later burials.[49]

It is also possible that the Delian installations were owned/operated by a collective, perhaps the Church, and were situated within the urban environment to form a centralised pressing operation.[50] Most of the installations are found nearby the remains of ecclesiastical structures and might conceivably be directly associated with their economy: the Agora of the Italians installation is near the three Church structures of the *Ekklesiasterion* and monastery over the Hypostyle Hall; the Agora of the Competaliasts installation is close to the basilica of St. Cyrique; and the Rue 5 installations are within what might be conceived as a domestic quarter of the Late Antique settlement (compare Figures 6.3 and 6.1). The position of agricultural presses within the urban fabric appears to continue a trend that began around the second Athenian domination (167/6–69 BC), where the city of Delos was characterised by "mixed-use neighbourhoods", which blended domestic, commercial and productive spheres (Zarmakoupi 2013: 24).

A possible relationship between the Church and viticultural enterprise is tantalising, albeit almost impossible to confirm at this stage. It is important to consider such possibilities, however, in an attempt to discover who implemented the changes necessary to commence (or continue) such large-scale viticultural production and, ultimately, fabricate a new role for the island – a necessity for the survival of Delos into Late Antiquity. Production of this scale, incorporating the architecture and technology indicated through the archaeological record, required considerable initial (and long-term) capital investment.[51] It is equally possible that wealthy, smaller operators owned these installations. Regarding contemporary production in the Dodecanese, it is believed that "large, although fragmented, monetised estates constituted the dominant form of rural land exploitation in this period" (Bowden 2016: 1711; Deligiannakis 2016). If the installations on Delos were part of a larger estate, perhaps the key notion is that they were, indeed, highly fragmentary. The vintage was processed at installations spatially distant to each other, compared to the earlier Roman villas where multiple presses were typically situated either in the same or adjacent rooms. What this means for ownership and operation is difficult to determine. Were these spatially distant installations owned by one person or group but managed by various individuals? Or did they combine their produce as one 'estate' but were owned separately?

It is entirely possible to suggest a similar scenario here to that which was previously suggested for Byzantine oleiculture:

> *"Early Byzantine pressworks rarely made use of more than two (presses), and generally in privately owned structures (in a few cases, monasteries also had their own presses). The oil production was thus no longer concentrated; it was in the hands of small-scale operators, whatever their status might otherwise be, and whatever the distribution pattern of the commodity – whether in-kind payment of a portion of the harvest to the village's landowner or the direct sale to oil merchants"* (Morrison and Sodini 2002: 198).

The lack of multiple presses, private and small-scale ownership all recall distinct similarities to the Delian installations, if we are to believe that they are not part of a large, fragmented estate owned by the Church or a wealthy individual.

The presence of early Christian epigraphy and iconography on a number of press counterweights does not confirm either theory; it is equally conceivable that private individuals commissioned their prayers and wishes to be inscribed upon daily pieces of equipment or that the Church crafted the relatively high quality iconography seen at the Rue 5 'large' press counterweight (Figure 6.4).[52] Regardless of the exact solution, we can conclude that a group of individuals or a collective institution was

---

[48] Probably linked to the ruralisation of many sites in Late Antiquity and the change from large cities to smaller dispersed settlements. Many cases exist, particularly in the east, where Late Antique agricultural installations reuse pre-existing urban monumental structures. The author observed this at: Athens, Aegina, Antiochia ad Cragum, Asar Tepe, Kestros, and Elaiussa Sebaste. The transportation of grapes from field to centralised 'urban' press might be a common feature of Late Antiquity and the Byzantine (Kazhdan 1991: 2200; Dodd 2020).

[49] The majority of pagan burials were extra-mural, yet around the advent of Christianity there was a change in burial practice and consequent swing in the 3rd-4th centuries AD to mostly intra-mural burial. Contraction of cultural features like this are a common feature of Late Antique settlements (Morrisson and Sodini 2002).

[50] Similar inward-facing transport of produce is possible regarding oileries in the Hellenistic city, where the excavated presses are all situated directly in the congested urban fabric with no space nearby for olive tree cultivation (Brun and Brunet 1997; Brun 1999; Brun 2000).

[51] The cost of constructing, maintaining and operating wine presses, particularly those with mechanical equipment, is made clear in Amouretti *et al.* (1984: 418-420) and Robinson (2006: 545). Comparable to an olive press (Drachmann 1932:46-49). New research (Burton and Lewit 2019: 577) suggests that mechanical lever presses, like those at Delos, were much more expensive to construct than direct-pressure screw presses.

[52] See Dodd (2020: 110, n°821-824) for examples of shrines and pagan prayer inciptions at wineries. Although the inscription on the counterweight near the Establishment of the *Poseidoniastae* is rudimentary, it is on a marble block, a very hard stone that required skilled craftsmen and the appropriate tools to inscribe upon, and there was some attempt to orientate and centre the text correctly. While the rudimentary nature suggests that it was probably completed locally, it was not a simple matter to inscribe into marble with clarity and the above features indicate that it was probably commissioned. Much the same can be said for the iconography present on the marble Rue 5 'large' counterweight (Figure 6.4).

Population decline and wine industry: societal transformation on Late Antique Delos (Greece)

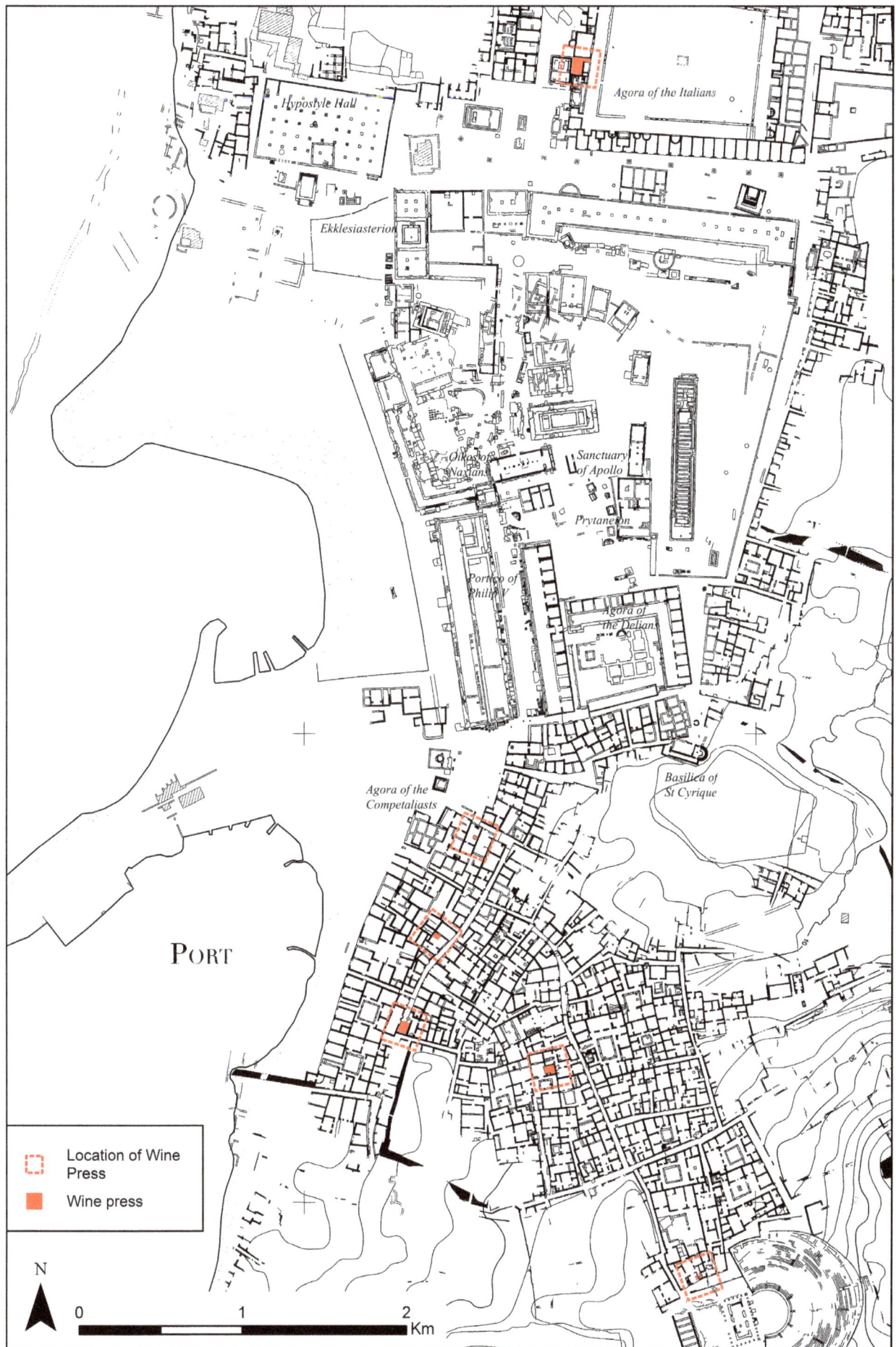

Figure 6.3: Location of Paleochristian vinicultural features on Delos (Map by author).

Figure 6.4: Type 14 counterweight with Paleochristian iconography at the Rue 5 'large' installation, c. 5th century CE (photo: E. Dodd, 2015).

present and possessed enough wealth to construct, operate and maintain this equipment.[53] Beyond this, unfortunately it is impossible, at this stage, to determine exactly who owned these installations.

Conclusion

Delos exemplifies the case where a once monumental and affluent city, situated in the centre of a pan-Mediterranean trade network, experienced severe change in its role and character caused by a range of endogenous and exogenous factors. The settlement did not, however, succumb to the stress of negative change and, instead, utilised the removal of population pressure and its inherent island characteristics to create new economic opportunities; in this case, a flourishing viticultural industry. We must use this case study to illuminate the possibility that prosperity does not necessarily follow population on a path of decline. The people of antiquity were knowledgeable, resilient and diverse enough to enact societal transformation and change the character of their settlement when enough factors permit survival and prosperity (geographic, environmental, socio-cultural, economic etc.). In the case of Delos, this meant a change in role for the island and its occupants. The present study suggests that the viticultural industry evident through the archaeological record might have played a large part in this change.

A combined analysis of the archaeological dataset, socio-cultural and socio-economic history, along with the studies of Curtis, Faulseit and Tainter, reveals the resilience of the Delian inhabitants and diversity in their ability to adapt and utilise their surrounding environment over the longue dureé. Although Curtis' (2014: 19) "three main markers" for measuring resilience are not necessarily completely met – Delos certainly experienced shrinking population levels and destruction/abandonment of housing – the remaining inhabitants reinvigorated and maintained the majority of agricultural land on the island and constructed new civic and religious structures of considerable quality. It is clear that a degree of resilience was present over the first 500 years of our era.

This study also illuminates a case in contrast to the typical Malthusian or Boserupian scenario and shows how a diminished population might respond agriculturally. Nonetheless, the scenario described in this chapter fits

---

[53] The fact that such relatively significant wealth was present on the island at this time further evidences against labels of 'decline' or 'collapse.' It also suggests the existence of a larger lower-class population necessary to support such social stratification and wealth.

well with an alternative Malthusian perspective: that population decline can birth niche economic expansion and subsequent increase in per capita income – a step in the other direction from 'Malthusian catastrophe'.

A range of new research questions are revealed upon consideration of this case study, foremost of which: why Delos? What allowed this small island to recover from such a tumultuous history, dramatic socio-cultural and economic transformation, absorb heavy exogenous and endogenous stress and still avoid rapid decline, complete collapse and abandonment? Many ancient settlements that experienced such events were not so fortunate. Indeed, Delos possessed significant habitation for another 700 years after the dramatic events of the 1st century BC. What internal or external factors were present that allowed this to occur?

At least part of the answer must lie in the geographic location of the island. While its importance as a commercial Graeco-Roman port between the Near East and Italy was lost, it still lay almost midway between the eastern and central Mediterranean and North Africa and the Black Sea.[54] Such a strategic location, along with longstanding ritual significance, surely encouraged continued habitation on Delos. Similarly, while long distance direct trade in larger vessels now circumvented the island, its central location in the Aegean and close proximity to many other Cycladic islands probably encouraged continual use as a small-scale trade or regional cabotage port. This was only further bolstered by the increased connectivity afforded by a small island like Delos with its all-round close proximity to the sea and typical island characteristics. Long years of peace in the later Roman Empire meant no immediate threat of invasion or violence and this, too, encouraged prosperous occupation in the Imperial era (as illustrated in Le Quéré 2013; 2018a; 2018b; 2018c; 2018d).

The existence of Delos as the seat of a Cycladic bishopric, if we are to believe this, surely confirms some support and connection to the Church and State with access to a wider economic network. The possibility for Delian wine to be exported as tax or army supplies in relation to the *quaestura exercitus* or *annona* also reinforces potential connectivity on a previously unrecognised scale. The fertile terraces and historically established agricultural systems of the Hellenistic period, noted through the epigraphic record of temple estates and surviving archaeology, encouraged later intensification of land use for vine growth; particularly if mature wild or domesticated vines survived on the island. The pre-existing terraces, irrigation systems, paths and farmhouses expedited the renewal or continuation of vine cultivation and allowed more rapid profits from this industry.[55] It was probably a combination of factors, both external and internal, along with a change in socio-cultural and economic focus that allowed this high-level viticultural industry to flourish despite a reduced population.

Further research into the possibility of Cycladic Late Roman amphorae produced on Delos or neighbouring islands (e.g. Paros or Tenos) for Delian wine is necessary to illuminate suggestions on local use vs surplus export. Tentative conclusions based on preliminary research suggest that it is possible wine was either exported in a multi-phase manner using imported amphorae or in local Delian ceramics.

This discussion also offers new opportunities to understand ancient socio-cultural and settlement transformation. The typical markers of prosperity often highly visible in the archaeological record – monumental architecture, elite housing, civic construction, large cohesive settlements, and rich burials – do not provide the only indications of a complex and prosperous society (where prosperity is the absence of decline and collapse and presence of positive economic motion). Prosperity also occurred through less perceptible, sometimes invisible, means within the archaeological record. We must, therefore, not label a site with 'decline' or 'collapse' until its archaeological record is investigated deeply and thoroughly; such hasty conclusions gloss over less distinctive forms of evidence that might suggest an alternate history.

Bibliography

Amouretti, M.C., Comet, G., Ney, C., and Paillet, J.-L. "À propos du pressoir à huile : de l'archéologie industrielle à l'histoire." *Mélanges de l'école française de Rome* 96, vol. 1 (1984): 379-421.

Aydinoğlu, Ü., and Alkaç, E. "Rock-cut wine presses in Rough Cilicia." *Olba* 16 (2008): 277-290.

Baldiran, A. "Lykaonia bölgesi sarap işlikleri (beyşehir-seydişehir civarı)." In *Olive oil and wine production in Anatolia during antiquity: symposium proceedings 06-08 November 2008, Mersin, Turkey*, edited by Ü. Aydinoğlu, and A.K. Şenol, 303-317. Istanbul: Ege Yayinlari, 2010.

Baratta, G. "Gli impianti di produzione." In *Elaiussa Sebaste I: campagne di scavo 1995-1997*, Bibliotheca archaeologica 24, edited by E. Equini Schneider, 129-141. Rome: L'Erma di Bretschneider, 1999.

Bevan, A., and Conolly, J. *Mediterranean islands, fragile communities and persistent landscapes: Antikythera in long-term perspective.* Cambridge: Cambridge University Press, 2013.

Bouet, A., and Le Quéré, E. "Les thermes impériaux de Délos: l'infrastructure publique d'une ville ἄδηλος?"

---

[54] The island of Antikythera is in a similarly advantageous location, described as "sometimes strategic, sometimes marginal" characterised by an "often discontinuous history of human presence, as well as a range of unusual activities" (Bevan and Conolly 2013: 2).

[55] The combination of early niche-construction (terracing, field clearing, cultivation), population fluctuation and later reuse characterises the cultural island landscape of Delos as one of persistence and fragility; best understood in tandem rather than isolation (Bevan and Conolly 2013: 9-10).

Bulletin de Correspondance Hellénique 139-140.1 (2016): 417-462.

Bowden, W. "Georgios Deligiannakis. The Dodecanese and the eastern Aegean islands in Late Antiquity." *Antiquity* 90 (2016): 1710-1711.

Brun, J.-P. "La discrimination entre les installations oléicoles et vinicoles." In *La production du vin et de l'huile en Mediterranée*, Bulletin du correspondance hellénique supplément XXVI, edited by M.-C. Amouretti, and J.-P. Brun, 511-537. Athens: École française d'Athènes, 1993.

Brun, J.P., and Brunet, M. "Une huilerie du premier siècle avant J.-C. dans le quartier du théâtre à Délos." *Bulletin de Correspondance Hellénique* 121, vol. 2 (1997): 573-615.

Brun, J.-P. "Laudatissimum fuit antiquitus in Delo insula: la maison IB du quartier du stade et la production des parfums à Délos." *Bulletin de Correspondance Hellénique* 123, vol. 1 (1999): 87-155.

Brun, J.-P. "The Production of Perfumes in Antiquity: The Cases of Delos and Paestum." *American Journal of Archaeology* 104, vol. 2 (2000): 277-308.

Brun, J.P. *Le vin et l'huile dans la Méditerranée antique: viticulture, oléiculture et procédés de fabrication*, Paris: Editions Errance, 2003.

Brun, J.-P. *Archéologie du vin et de l'huile dans l'empire romain*, Paris: Editions Errance, 2004.

Bruneau, P., and Ducat, J. *Guide de Délos*, 2nd edition. Athens: Editions de Boccard Limoges, 1966.

Bruneau, P., and Ducat, J. *Guide de Délos*, 4th edition. Athens: Editions de Boccard Limoges, 2005.

Bruneau, P. "Contribution à l'histoire urbaine de Délos à l'époque hellénistique et à l'époque Imperial." *Bulletin de Correspondance Hellénique* 92, vol. 2 (1968): 633-709.

Bruneau, P., and Fraisse, P. "Un pressoir à vin à Délos." *Bulletin de Correspondance Hellénique* 105, vol. 1 (1981): 127-153.

Bruneau, P., and Fraisse, P. "Pressoirs déliens." *Bulletin de Correspondance Hellénique* 108, vol. 2 (1984): 713-730.

Bruneau, P. "Deliaca." *Bulletin de Correspondance Hellénique* 111, vol. 1 (1987): 313-342.

Brunet, M. Contribution à l'histoire rurale de Délos aux époques classique et hellénistique. *Bulletin de Correspondance Hellénique* 114 (1990): 669-682.

Brunet, M. "Vin local et vin de cru, les exemples de Délos et de Thasos." In *La production du Vin et de l'Huile en Mediterranée*, Bulletin du correspondance hellénique supplément XXVI, edited by M.-C. Amouretti and J.-P. Brun, 202-212. Athens: École française d'Athènes, 1993.

Brunet, M. "Le paysage agraire de Délos dans l'antiquité." *Journal des savants* 1, vol. 1 (1999): 1-50

Brunet, M., and Poupet, P. "Délos." *Bulletin de Correspondance Hellénique* 121, vol. 2 (1997): 776-789.

Burton, P., and Lewit, T. "Pliny's presses: the true story of the first century wine press." *Klio* 101.2 (2019): 543-598.

Cabrol, F., and Leclercq, H. *Dictionnaire d'Archéologie Chrétienne et de Liturgie*, vol. 4. Paris, 1920.

Chavarría, A., and Lewit, T. "Archaeological Research on the Late Antique Countryside: A Bibliographic Essay." In *Recent Research on the Late Antique Countryside*, Late antique archaeology 2, edited by W. Bowden, L. Lavan, and C. Machado, 3-54. Leiden - Boston: Brill, 2004.

Curtis, D.R. *Coping with crisis: the resilience and vulnerability of pre-industrial settlements*. Ashgate: Routledge, 2014.

Curtis, R.I. *Ancient food technology*, Technology and change in history 5. Leiden - Boston: Brill, 2001.

Decker, M. *Tilling the hateful earth: agricultural production and trade in the Late Antique east*, Oxford studies in Byzantium. Oxford: Oxford University Press, 2009.

Deligiannakis, G. *The Dodecanese and the eastern Aegean islands in Late Antiquity: AD 300-700*, Oxford monographs an classical archaeology. Oxford: Oxford University Press, 2016.

Delorme, J. *Les palestres*, Exploration archéologique de Délos 25. Paris: École française d'Athènes, 1961.

Déonna, W. *Le mobilier Délien*, Exploration archéologique de Délos 18. Paris: École française d'Athènes, 1938.

Déonna, W. *La vie privée des Déliens*. Paris: Éditions de Boccard, 1948.

De Simone, G.F., and Martucci, C.S. "Thirst for wine? An amphora assemblage from Vesuvius and the problem of self-sufficiency in Late Antique Campania." *Rei Cretariae Romanae Fautorum Acta* 44 (2016): 127-135.

Diamanti, C. "The Late Roman amphora workshops of Paros island in the Aegean sea: recent results." *Rei Cretariae Romanae Fautorum Acta* 44 (2016): 691-697.

Diler, A. "The most common wine-press type found in the vicinity of Cilicia and Lycia." *Lykia* II (1995): 83-98.

Dimakopoulos, S. "Agricultural terraces in Classical and Hellenistic Greece." *LAC 2014 proceedings* (2016) (URL: http://lac2014proceedings.nl/article/view/61/37).

Dodd, E. *Roman and Late Antique wine production in the eastern Mediterranean: a comparative archaeological study at Antiochia ad Cragum (Turkey) and Delos (Greece)*. Oxford: Archaeopress, 2020.

Dürrbach, F., and Jardé, A. "Fouilles de Délos." *Bulletin de Correspondance Hellénique* 29 (1905): 169-257.

Faulseit, R.K. "Collapse, resilience, and transformation in complex societies: modeling trends and understanding diversity." In *Beyond collapse: archaeological perspectives on resilience, revitalization, and transformation in complex societies*, Visiting scholar conference volumes, Occasional paper (Center for Archaeological Investigations) 42, edited by R.K. Faulseit, 3-26. Carbondale: Southern Illinois University, 2016.

Foxhall, L. "Feeling the earth move: cultivation techniques on steep slopes in Classical antiquity." In *Human landscapes in Classical antiquity: environment and culture*, Leicester-Nottingham studies in ancient society 6, edited by G. Shipley, and J. Salmon, 44-67. London: Routledge, 1996.

Frankel, R. "Presses for oil and wine in the southern Levant in the Byzantine period." *Dumbarton Oak Papers* 51 (1997): 73-84.

Gallet de Santerre, H., and Tréheux, J. "Chronique des fouilles et découvertes archéologiques en Grèce en 1946." *Bulletin de Correspondance Hellénique* 71-72 (1947-1948): 407-419.

Greven, M. *Marlborough: manual for yield assessments. Sustainable farming fund: final report June 2007*, unpublished report. Marlborough, 2007.

Grigg, D.B. *Population growth and agrarian change: an historical perspective*, Cambridge geographical studies 13. Cambridge: Cambridge University Press, 1980.

Hasenohr, C. "L'Agora des Compétaliastes et ses abords à Délos: topographie et histoire d'un secteur occupé de l'époque archaïque aux temps byzantins." *Revue des Études Anciennes* 104, vol. 1-2 (2002): 85-110.

Hasenohr, C. "Le bas quartier du théâtre à Délos à l'époque Impériale." *Topoi* 19 (2014): 291-308.

Höhfeld, V. "Antike terrassenkomplexe am nordwesthang des Kolaklar Tepesi (Yavu-Bergland, Lykien)." In *Feldforschungen auf dem Gebiet von Kyaneai (Yavu-Bergland)*, Lykische Studien 4, Asia Minor Studien 29, edited by F. Kolb, 243-250. Universität Münster, 1998.

Horden, P., and Purcell, N. *The corrupting sea: a study of Mediterranean history*. Oxford: Wiley-Blackwell, 2000.

Jashemski, W.F. "Large vineyard discovered in ancient Pompeii." *Science* 180, vol. 4088 (1973a): 821-830.

Jashemski, W.F. "The discovery of a large vineyard at Pompeii: University of Maryland excavations, 1970." *American Journal of Archaeology* 77, vol. 1 (1973b): 27-41.

Jongman, W. "Slavery and the growth of Rome: the transformation of Italy in the second and first centuries bce." In *Rome the cosmopolis*, edited by C. Edwards, and G. Woolf, Cambridge: 100-122. Cambridge University Press, 2003.

Kazhdan, A.P. *The Oxford dictionary of Byzantium*. Oxford: Oxford University Press, 1991.

Khalil, L.A., and Al-Nammari, F.M. "Two large wine presses at Khirbet Yajuz, Jordan." *Bulletin of the American Schools of Oriental Research* 318 (2000): 41-57.

Kinnaird, T., Bolòs, J., Turner, A., and Turner, S. "Optically-stimulated luminescence profiling and dating of historic agricultural terraces in Catalonia (Spain)." *Journal of Archaeological Science* 78 (2017): 66-77.

Kiourtzian, G. *Recueil des inscriptions grecques chrétiennes des cyclades de la fin du IIe au VIIe siècle après J.-C.* Paris: Editions de Boccard, 2000.

Laidlaw, W.A. *A history of Delos*. Oxford: Basil Blackwell, 1933.

Le Quéré, E. *Les Cyclades sous l'Empire Romain (Ier s. av. J.-C. – IIIe s. ap. J.-C.): formes et limites d'une renaissance éconmique et sociale*. Unpublished PhD Dissertation, Université Panthéon-Sorbonne – Paris I, 2013.

Le Quéré, E. *Les Cyclades sous l'Empire Romain: histoire d'une renaissance*. Rennes: PUR, 2015.

Le Quéré, E. "Édifices et constructions des époques Impériale et Protobyzantine dans le Sanctuaire d'Apollon." In *Le Sanctuaire d'Apollon à Délos. Tome I. Architecture, topographie, histoire*. Exploration archéologique de Délos 44, edited by R. Étienne, 117-151. Athens, 2018a.

Le Quéré, E. "Un paysage sacré en mutation, du Ier s. av. J.-C. à la fin de l'Époque Impériale." In *Le Sanctuaire d'Apollon à Délos. Tome I. Architecture, topographie, histoire*. Exploration archéologique de Délos 44, edited by R. Étienne, 249-288. Athens, 2018b.

Le Quéré, E. "Un lot de céramiques d'époque Imperial Romaine provenant du puits du Prytanée de Délos." *Bulletin de Correspondance Hellénique* 142.1 (2018c): 317-402.

Le Quéré, E. "Was Delos really Adelos? A reappraisal of the Delian solitude in Roman Imperial times." In *What's new in Roman Greece? Recent work on the Greek mainland and the islands in the Roman period*. Mélétimata 80, edited by V. Di Napoli, F. Camia, V. Evangelidis, D. Grigoropoulos, D. Rogers and S. Vlizos, 111-121. Athens, 2018d.

Lefort, J. "The rural economy: seventh-twelfth centuries." In *The economic history of Byzantium: from the seventh through the fifteenth century, vol. 1*, Dumbarton Oaks studies 39, edited by A.E. Laiou, 231-310. Washington

D.C: Dumbarton Oaks Research Library and Collection, 2002.

Leroux, G. *La salle hypostyle*, Exploration archéologique de Délos 2. Paris: École française d'Athènes, 1909.

Moretti, J.C., Fadin, L., Fincker, M., and Picard, V. *Atlas*, Exploration archéologique de Délos 43. Athens: École française d'Athènes, 2015.

Morrisson, C. and Sodini, J.-P. 2002. "The sixth-century economy." In *The economic history of Byzantium: from the seventh through the fifteenth century, vol. 1*, Dumbarton Oaks studies 39, edited by A.E. Laiou, 171-220. Washington D.C: Dumbarton Oaks Research Library and Collection, 2002.

Morton, J. *The role of the physical environment in ancient Greek seafaring*, Mnemosyne Supplementum 213. Leiden-Boston: Brill, 2001.

Murray, W.M. "Do modern winds equal ancient winds?" *Mediterranean Historical Review* 2, vol. 2 (1987): 139-167.

Neumann, J., and Metaxas, D.A. "The Battle between the Athenians and Peloponnesian fleets, 429 B.C., and Thucydides' "wind from the gulf (of Corinth)"." *Meteorologische Rundschau* 32 (1979): 182-188.

Orlandos, A.K. "Délos chrétienne." *Bulletin de Correspondance Hellénique* 60 (1936): 68-100.

Price, S., and Nixon, L. "Ancient Greek agricultural terraces : evidence from texts and archaeological survey." *American Journal of Archaeology* 109, vol. 4 (2005): 665-694.

Purcell, N. "Wine and wealth in ancient Italy." *Journal of Roman Studies* 75 (1985): 1-19.

Rauh, N.K., Dillon, M.J., Dore, C., Rothaus, R., and Korsholm, M. "Viticulture, oleoculture, and economic development in Roman Rough Cilicia." *Münstersche Beiträge zur Antiken Handelsgeschichte* XXV, vol. 1 (2006): 49-98.

Rauh, N.L., Townsend, R.F., Hoff, M.C., Dillon, M., Doyle, M.W., Ward, C.A., Rothaus, R.M., Caner, H., Akkemik, Ü., Wandsnider, L., Ozaner, F.S., and Dore, C.D.. "Life in the truck lane: urban development in western Rough Cilicia". *Jahreshefte des Österreichischen Archäologischen Institutes in Wien* 78 (2009): 253-312.

Renfrew, A. C. *Approaches to social archaeology*. Edinburgh: Edinburgh university press, 1984.

Reynolds, A. 2015. *Grapevine breeding programs for the wine industry*, Woodhead publishing series in food science 268. Cambridge: Elsevier.

Reynolds, P. "The Roman pottery from the triconch palace." In *Byzantine Butrint: excavations and surveys 1994-99*, Butrint archaeological monograph series 1, edited by R. Hodges, W. Bowden, and K. Lako 224-246. Oxford: Oxbow Books, 2004.

Robinson, J. *The Oxford companion to wine*, 3rd edition. Oxford: Oxford University Press, 2006.

Şenol, A.K., and Waltz, S. "Vine cultivation areas on the Carian Chersonesos in antiquity: estimating the wine production potential of Bybassos and its territory." In *Olive oil and wine production in eastern Mediterranean during antiquity: international symposium proceedings 17-19 November 2011 Urla – Turkey*, Ege Üniversitesi Edebiyat Fakültesi yayınları 189, edited by A. Diler, K. Şenol, and Ü. Aydınoğlu, 31-40. Izmir: Ege Üniversitesi Edebiyat Fakültesi Yayınları, 2015.

Sweetman, R. "A Survey of the Aegean Islands in Late Antiquity." *The Classical Review* 67.1 (2016): 1-2.

Tainter, J.A. "Why collapse is so difficult to understand." In *Beyond collapse: archaeological perspectives on resilience, revitalization, and transformation in complex societies*, Visiting scholar conference volumes, Occasional paper (Center for Archaeological Investigations) 42, edited by R.K. Faulseit, 27-39. Carbondale: Southern Illinois University, 2016.:

Tchernia, A. *Le vin de l'Italie romaine. Essai d'histoire économique d'après les amphores*, Bibliothèque des Écoles françaises d'Athènes et de Rome 261. Rome: École française de Rome, 1986.

Tchernia, A. *Les Romains et le commerce*, Études 8, BiAMA hors collection. Naples: Centre Jean Bérard and Centre Camille Jullian, 2011.

Vallois, R. *Les portiques au sud du héron. Le portique de Philippe*, Exploration archéologique de Délos 7. Paris: École française d'Athènes, 1923.

Van Limbergen, D. "Figuring out the balance between intra-regional consumption and extra-regional export of wine and olive oil in Late Antique northern Syria." In *Olive oil and wine production in eastern Mediterranean during antiquity: international symposium proceedings 17-19 November 2011 Urla – Turkey*, Ege Üniversitesi Edebiyat Fakültesi yayınları 189, edited by A. Diler, K. Şenol, and Ü. Aydınoğlu, 169-190. Izmir: Ege Üniversitesi Edebiyat Fakültesi Yayınları, 2015.

Volpe, R. "Lo sfruttamento agricolo e le costruzioni sul pianoro di centocelle in età Repubblicana." In *Centocelle I: Roma S.D.O le indagini archeologiche*, Studi e materiali dei musei e monumenti comunali di Roma, edited by P. Gioia, and R. Volpe, 447-462. Rome: Rubbettino, 2004.

Volpe, R. "Vini, vigneti ed anfore in Roma repubblicana. In *Suburbium II. Il suburbio di Roma dalla fine dell'età monarchica alla nascita del sistema delle ville*, Collection de l'École française de Rome 419, edited by R. Volpe, 369-381. Rome: École française de Rome, 2009.

Wachsmuth, E.W.G. *The historical antiquities of the Greeks with reference to their political institutions*, English trans. by E. Woolrych. Oxford, 1837.

White, K.D. *Roman farming*, Aspects of Greek and Roman life (Cornell University). Ithaca, New York: Cornell University Press, 1970.

Zarmakoupi, M. *The quartier du stade on Late Hellenistic Delos: a case of rapid urbanization (fieldwork seasons 2009-2010)*. ISAW Papers 6. New York : Institute for the Study of the Ancient World, 2013.

Zarmakoupi, M. "Hellenistic and Roman Delos: the city and its emporion." *Archaeology in Greece* 61 (2014-2015): 115-132.

7

Cities and sustenance in Roman Asia Minor

Rinse Willet

University of Leuven

Abstract: According to Philostratus and Josephus, Asia Minor was once dotted with some 500 cities in the Roman period and indeed both older works, such as the Cities of the Eastern Roman Provinces, and newer projects, such as the Barrington Atlas or Pleiades database, show a vast number of towns and cities to have existed during the Roman Empire. Particularly the western part of Asia Minor was probably one of the most urbanized regions of the ancient world and the ruins of many (large) cities attest to this, such as Pergamon, Ephesos, Miletos, Alexandreia Troas or Kyzikos. Furthermore, there seems to have been an increase in the number of cities from the Hellenistic period to the Roman imperial period, while at the same time archaeological research shows a multitude of cities expanding in size in this period and this seems to happen not to be exclusive to large cities on the west coast, but also has been observed for smaller cities to the east as well, such as Sagalassos, Herakleia Pontika, Laodikeia on the Lykos, Sardis, Aspendos, Selinous, Perge and possibly Smyrna and Tarsus. Although some cities disappear in this period as well and there is still much archaeological work to be done, a pattern of growth seems to emerge from this data.

This chapter critically reviews this pattern and considers its demographic implications. Is this increase in number of dots on the map indeed indicative of an expansion of the population and if so, what were the effects on agricultural exploitation of this area? To answer this, the cities of Sagalassos, Kyaneai, Ephesos and Pergamon will be taken into closer consideration, since all of these have a longer (ongoing) history of archaeological research, which will serve to explain the changes observed in other, sometimes only superficially researched cities. In the end, the question is raised whether the developments in this part of the Roman Empire should be understood from either a Malthusian or a Boserupian perspective.

Keywords: Asia Minor, Kyaneai, Sagalassos, Ephesos, Pergamon, Roman urbanism, Roman demography, Agriculture, Population reconstructions

Introduction

Although Asia Minor has long been known to be dotted with cities in Antiquity, relatively little attention has been spent on how the inhabitants of these places were fed. Examples of agricultural carrying capacity reconstructions for Roman Imperial cities of Asia Minor are mostly appearing in the last two decades (e.g. Blanton 2000; Höhfeld 2006; Koparal 2014). Still, the richness of archaeological, epigraphic and historical material allows us to perceive changes in the urban pattern from the Hellenistic to Roman periods in some detail. The overall observations seems to include an expansion of the urban pattern, with more cities appearing from the 2nd century BC to the 3rd century AD, which triggers questions a) is this increase in number of cities entailing demographic growth, b) how is an increased urban population sustained by agriculture and c) does this result in an increased agricultural exploitation during the later Hellenistic and Roman Imperial period? This paper will try and start answering these difficult questions by taking a closer look at the apparent increase in number of cities from the Hellenistic to Roman periods and whether this constitutes a demographic change on the one hand, while on the other making an attempt at qualifying this change on the basis of reconstructed population sizes and carrying capacities by using four case-studies, namely Sagalassos, Kyaneai, Pergamon and Ephesos.[1]

---

[1] This chapter was part of my study on Roman cities in Asia Minor in the framework of the ERC funded project 'An Empire of 2000 cities: urban networks and economic integration in the Roman Empire', directed by Luuk De Ligt and John Bintliff (De Ligt, Houten and Willet 2014). The research presented in this paper is made possible by the European Research Council, ANAMED Research Center for Anatolian Civilizations of Koç University, and the Sagalassos Archaeological Research Project. The draft of this chapter was first drawn up in 2016 and slightly updated in 2017. A section based on this chapter was published as part of chapter 5 in Willet 2020: 181-191. For this, a few changes in calculations and new figures were added, hence the descrepancies between Willet 2020 and this chapter. The conclusions are, in broad terms, similar.

Increased urbanisation or more dots on the map?

The twentieth century has produced several major works on cities of the Hellenistic and Roman Imperial periods, relying in no small part on ancient geographical descriptions by Strabo, Pliny the Elder, Ptolemy and others, numismatic evidence, a vast body of epigraphic evidence, and, to a lesser degree, archaeological evidence. These works place the development of the urban pattern in the history of the region. Archaeological research on cities in Anatolia has focused mostly on heavily monumentalized centres. A large part of the knowledge on smaller cities and towns is derived from extensive surveys or reconnaissance projects in pursuit of epigraphic material, which leaves many places relatively obscure. For example the polis of Alia in Phrygia is not yet located, while the poleis of Midaion in Phrygia or Soatra in Lykaonia are only superficially described.[2]

Still, on the basis of this problematic evidence and using earlier work by Louis Robert and A.H.M. Jones, Thomas Broughton suggested a growth in the number of cities in Asia Minor, from the Hellenistic to Roman Imperial period. Broughton focused on self-governing cities, i.e. settlements with the official status and institutions of a city in Roman Asia Minor. For Asia Minor, these consists of colonies, but mostly peregrine cities or poleis. He provided a list of cities according to a (rough) periodization of Late Republican, Julio-Claudian, Flavian-Severan era and the 3rd century AD, which show a steady increase in the number of cities (Broughton 1938: 700-702, 705-706, 737-739). At the beginning of Augustus' reign, Broughton records 265 self-governing cities with 6 probable cities and this number would grow to 368 at the end of the Severan period, with 4 more cities being founded during the third century, totalling a growth of 107 cities, or 40 % in under three centuries. (Table 7.1; Figures 7.1 & 7.2). The Pleiades website shows a similar diachronic development (https://pleiades.stoa.org/). This website lists apart from cities, all settlements known to have existed. The dates provided by Pleiades are rather rough with periodization based in many cases on the earliest historical record for a place, adjusted with archaeological data (if available), and categorized in broadly Classical (or even earlier), Hellenistic, Roman Imperial, Late Roman or Late Antiquity and later. Still, the pattern of increased settlements for the Roman Imperial period as observed in the data provided by Broughton is confirmed, from c. 314 settlements until 330 BC to 1,207 in AD 300 or an increase of 384 %.

Broughton's list shows relatively few new cities appearing in Bithynia et Pontus during the Imperial period, where an urban pattern was already set up by Pompey in 63 BC during the provincial organization after the defeat of Mithridates VI (Marek 1993: 59, 101; Plin. *HN* V.43). In contrast, the Pleiades data shows an increase in number of settlements for this region, which can mostly be explained by the greater increase in known secondary settlements or settlements that are not officially cities (E.g. Matthews and Gatz 2009: 241-242). Other areas saw active policies of (re)founding cities, such as by Emperor Hadrian in the region Mysia, an area relatively empty of cities (Boatwright 2000). Yet for many others the origins are obscure, such as Panemoteichos in Pisidia, which existed as a settlement dependent on neighbouring Ariassos already in the Archaic and Classical periods, but the scant numismatic evidence dating from Julia Domna and Gallienus, suggests that this place was raised to a city by the late 2nd to early 3rd century AD, although the size and monumentality remained rather limited (Aydal et al. 1997: 160; Aulock 1977: 43).

My study of Roman cities in Asia Minor attempts to map the pattern of self-governing cities during the 2nd and 3rd centuries AD and is indebted to previous work already mentioned, but incorporates recent archaeological studies and other sources (e.g. the Roman Provincial Coinage database). This collection of cities, shows similar pattern with Asia having most cities, but with Pisidia, Lycia et Pamphylia having a higher density of cities (Figures 7.1 & 7.2, Table 7.1). When the densities are observed further, it is obvious that Galatia et Cappadocia and Bithynia et Pontus are virtually empty of cities in comparison with the other provinces. Clearly the focus of the urban pattern lies on the Western, South-Western and Southern coastal regions of Asia Minor.

Despite an apparent growth in number of cities, it is necessary to critically review such data. First of all these data consist of either officially recognized cities or of settlements that are not necessarily archaeologically well understood.[3] The size of these places is often obscure and

Table 7.1: Number of located self-governing cities (n=437) using the provincial division of 117 CE. Note that 20 of the 437 total are left out since these could not be located exactly. The percentages therefore are based on all the located cities. The provincial boundaries are a simplification for the second century AD, derived from the Barrington Atlas.

Province	Area (in 1000 km²)	No. of cities (% of total)	Urban sites per 1000 km²
Total research area	543	417 (95.4%)	7.7
Asia	135	199 (45.5%)	14.7
Pisidia, Lycia et Pamphylia	39	85 (19.4%)	21.8
Cilicia	38	48 (10.9%)	12.6
Galatia et Cappadocia	264	54 (12.4%)	2.1
Bithynia et Pontus	68	20 (4.6%)	2.9
Outside provinces within research area	\	11 (2.5%)	\

---

[2] Alia (Drew-Bear 1980: 951); Midaion (Humann and Puchstein 1890: 23; Aulock 1987: 33); Soatra (Aulock 1976: 48; Belke 1984: 223).

[3] Thonemann 2013: 31 notes that the smallest poleis are often particularly obscure, with a few cities only known through provincial coin emissions, such as Sala and Akkilaion in Phrygia.

Cities and sustenance in Roman Asia Minor

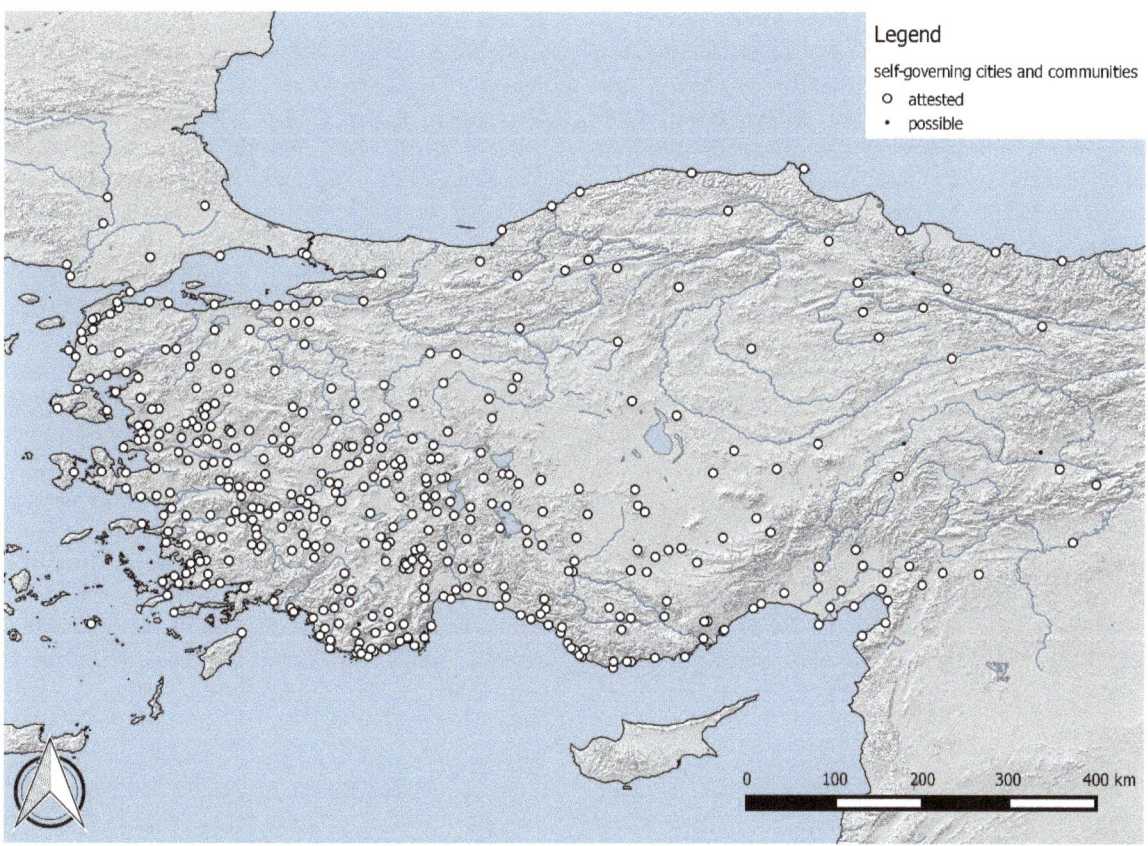

Figure 7.1: Map of the self-governing communities and cities in Roman Asia Minor at the beginning of the third century CE (n=446).

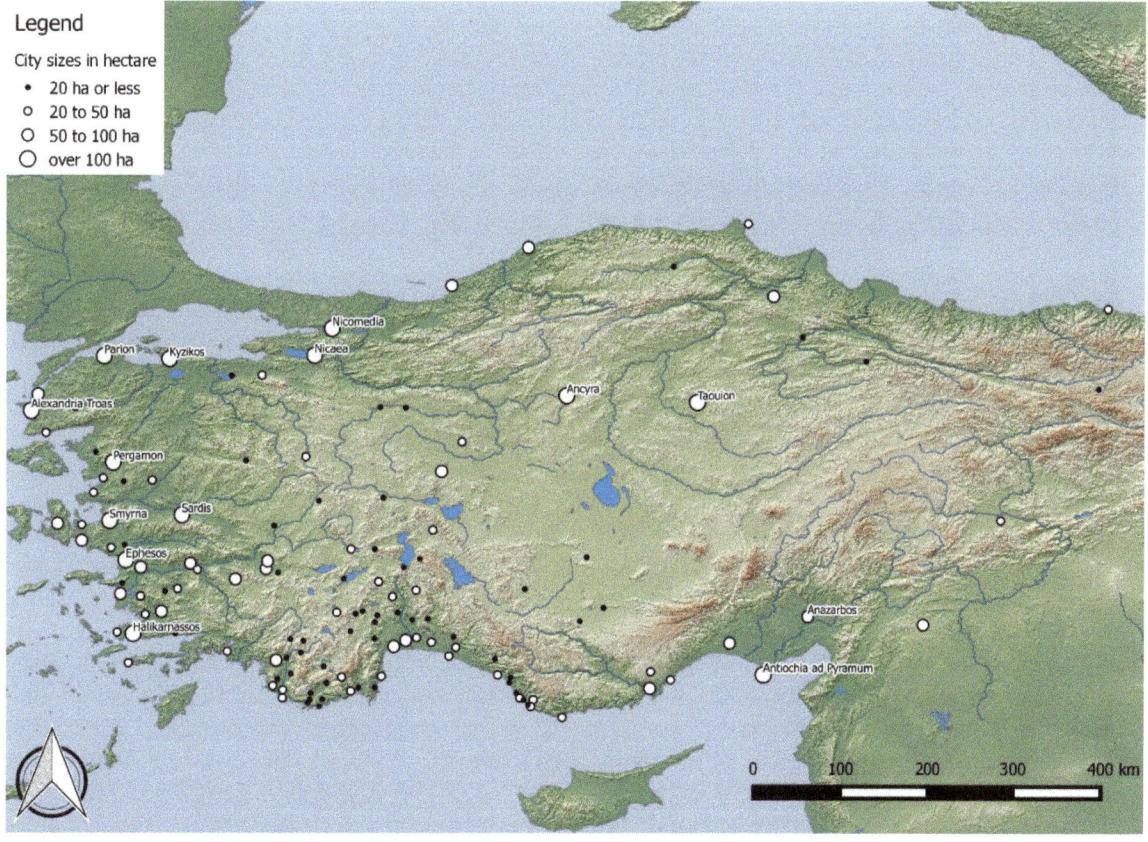

Figure 7.2: Map of cities according to physical size. Sizes are obtained through archaeological publications and/or by measuring the physical remains using satellite imagery (Google Earth) with published maps. Wherever possible, the figure for built-up city is used. The following arbitrary categories are used to differentiate between small (less than 20 ha, n=57), small to medium (20 to 50 ha, n=43), medium to large (50 to 100 ha, n=21) and large cities (over 100 ha, n=13).

the number of people inhabiting these settlements can in the vast majority of cases only be guessed at. For the data from Broughton, the increase in number of cities is first and foremost an expansion in urban administrative centres and not per se an expansion or growth of the physical attributes of cities, being size (and possibly population) and public buildings for example. Furthermore, there are indications of cities disappearing during the Hellenistic and Roman periods. Several examples of synoikosmos (multiple settlements/cities being combined into one) or sympoliteia (multiple cities merged together) are known for the Hellenistic period.[4] For the densely settled region of Lycia, Strabo mentioned that its League of cities was 23 strong, however Pliny the Elder mentions specifically that the league dwindled from 70 to 36 cities (Strabo 14.3.3; Plin. *HN* V.101). A few of the self-governing 'cities' will in reality probably not have had a proper physical urban centre and will have been *de facto* tribal societies that are officially recognized, such as the communities of the Tmolos Mountain (Jones 1971: 38, 78-81, 397, n. 84; Habicht 1975: 76). Lastly, many other cities must have been physically very small (See Figure 7.2) (Thonemann 2013: 32.).

Is this increase phenomenon then purely an administrative phenomenon of the Roman Empire, or is there actual urbanisation happening? An inscription from Orkistos dating to the time of Constantine is indicative as it contains a (successful) plea by the inhabitants of the village of Orkistos to the Emperor to become officially recognized (*MAMA* VII, 305). Amongst the various arguments why Orkistos should become raised to a city (had been a polis in the past, connection to four roads, availability of fresh water, watermills, presence of an agora etc.) is the fact that Orkistos had a sizable permanent population, which easily filled the marketplace. Frank Kolb therefore argued positively that the official recognition of a city must have implied the presence of physical urban amenities and a (sizable) population. Peter Thonemann interpreted this inscription more negatively, suggesting that the specific stipulation of the presence of a sizable population is indicative for the fact that many self-governing cities in fourth century Phrygia did not or no longer had a great number of inhabitants (Kolb 1993; Tacitus, *Agricola*, 21; Thonemann 2013: 32). Unfortunately, Orkistos yields little archaeological information to confirm either of these interpretations. Thonemann argued from another inscription, which describes the raising of Tymandos to a city, probably under Diocletian, that the ability of a community to reliably provide enough wealthy men, who could serve as *decuriones* of the city, was the main consideration for city status (*MAMA* IV, 236). However, to me these two inscriptions and interpretations do not necessarily seem mutually exclusive, as a community which was able to reliably sustain a wealthy elite must have had an urban and/or rural population to sustain the elite. Thonemann is in essence arguing for the relatively small scale of many of the Phrygian (urban) populations. Here, I would agree with him not only for Phrygia but even for most of the cities of Asia Minor. Still, both inscriptions seem to at least hint that the raising of places to self-governing cities is related to the presence of a sustained (put perhaps small) urban population and wealthy elite.

If the archaeology, however incomplete, is taken into consideration it is obvious that many places saw extensive monumentalisation during the Hellenistic to Roman periods. For example, of the 446 cities studied in Asia Minor, many have remains of spectacle buildings (129 cities have a theatre, 30 an odeion, 29 with stadium and 6 with an amphitheatre), baths (present in 65 cities), gymnasia (48), agorai (52), aqueducts (87), temples (79) or fortifications (present in 56 cities, although most of these are Hellenistic or earlier in date).[5] For 190 cities, the presence of any of these buildings can be attested for the later Hellenistic or Roman Imperial periods, leaving the vast majority of the cities without known public buildings. Here, the lack of knowledge 'on the ground' for many of these places must be to blame, as it is striking that even very small places can have monumental buildings. A diachronic perspective on these buildings, with the admission that not all buildings have been studied in great detail, let alone dated, further provides an image of expansion similar to the increase of cities and settlements discussed above, albeit on a much smaller scale. The number of settlements with theatres in Asia Minor is 61 in early Imperial times, while this number has risen to 87 by the 3rd century AD with several cities receiving a second theatre or an expansion of the *cavea* to receive more spectators. Some 35 settlements have bath buildings during the first century, which would rise to 69 baths in the third century. Increased monumentalisation in and of itself does not prove that that the increase of dots on the maps reflect a massive increase in urbanisation and urban population, but at the very least seems to suggest an increased accumulation of wealth or capital in these cities, which facilitated the construction of these buildings. The accumulation must have relied on the taxation/exploitation of a non-elite urban and rural population.

A last indicator are the archaeologically and historically attested instances of expanding cities. Heraclea Pontica, a city on the Black Sea coast, was enclosed by classical city walls, an area of c. 42 ha, but expanded beyond the city walls, probably during the 2nd century AD, to the north and grew to c. 80 ha as evidenced by a new city wall and monuments dating to the reign of Trajan (Hoepfner 1966: 21, 28-30, 40). Sagalassos, located in Pisidia, expanded eastward beyond its Hellenistic enceinte in the early Imperial period to reach a size of some 37.5 ha (Martens, Richard and Waelkens 2008, 1337; Willet and Poblome 2015: 135; Cleymans forth.). Tarsus had a walled enclosure

---

[4] For synoikismos, e.g. Alexandreia Troas (Tscherikower 1927: 16; Cohen 1995: 145; Ricl 1997: 1); Antioch on the Maiandros (Cohen 1995: 250-251; Magie 1950: 128; Labuff 2016 for sympoliteia of Karia).

[5] These data are derived from numerous publications, notably some specialist catalogues, e.g. Sear 2006; Nielsen 1993 and Farrington 1995.

of some 73 ha and Dio Chrysostom alludes to the growth of the city in the 1st century AD.[6] Richard Blanton saw an expansion for Selinus in Rough Cilicia, based on the area where ceramic finds were scattered observed with extensive survey, from the Hellenistic (2.4 ha) to Early Roman (10.85 ha) to its maximum size of c. 24 ha in the Late Roman period.[7] Aspendos expanded in Hellenistic to Roman times with some buildings built outside the main city to an area of c. 35 ha (Kessener and Piras 1998: 151; Smith 2007: 217; Grainger 2009: 234). Perge in Pamphylia expanded probably already in Hellenistic times (based on the 2nd-century BC city walls) beyond its main hill into the lower city, extending to c. 50 ha (Schütte-Maischatz and Winter 2004: 8; Hellenkemper and Hild 2004: 193; Grainger 2009: 29, 234; Brandt 1992: 47; McNicoll 1997: 127). Laodicea ad Lycum, situated on a hill (154 ha) in the plain of the Maeander River seems to have grown significantly from its smaller Hellenistic phase, up to covering c. 93 ha of the hill.[8] Sardis was already sizable in pre-Roman times and possibly extended to 250 ha in Late Roman-Early Byzantine times. G.M.A. Hanfmann argued that Sardis reached its largest population-size in AD 200-395. More recent research shows that Sardis would reach a more modest maximum size of 130 ha during Imperial times.[9] Some of the largest cities in Roman Asia Minor seem to have enjoyed a period of expansion in the Hellenistic to Roman Imperial periods. Smyrna is overbuilt by the modern city of Izmir, although the reconstructed city-wall area suggests a size of 186 ha (Akurgal 1993: 50; Mcevedy 2011: 354; Broughton 1938: 718). Ephesos, with a walled area of c. 386 ha, expanded during the Hellenistic to Imperial periods to a size of 185 ha (Groh 2006: 101). Lastly Pergamon, quite possibly the largest city of western Asia Minor during the Roman Empire, expanded from beyond its Hellenistic city walls (90 ha) to a total area of c. 190 ha.[10]

Although these examples do not denote an overall physical expansion of cities all over Asia Minor, they do indicate, together with increased monumentalisation and increasing numbers of cities and settlements, that a change is happening in Asia Minor during the Hellenistic period to Roman Imperial periods and in all likelihood, this will have been accompanied with increased urban and/or rural population. Whether this new situation entails an increased urban population and whether this is sustainable by the territories of these cities will be explored in the next section using the examples of Kyaneai, Sagalassos, Ephesos and Pergamon.

Population and sustainability

As population statistics are virtually non-existent for antiquity, demographic studies of this epoch have to rely on educated guesses and reconstructions. Many methods have been employed, from using the seating capacity of theatres to historically mentioned figures on army-sizes, using urban area together with an assumed population density, using the (reconstructed) carrying capacity of the territory or even using the water supply as indicators of population size.[11] Of all these, the usage of settlement surface area as indicator for population size is very common and for many of the cities of Hellenistic and Roman Imperial Asia Minor it is the only option to reconstruct population levels (Chamberlain 2006: 126-133).

The cities taken into closer consideration – Kyaneai, Sagalassos, Ephesos and Pergamon – are of interest as all of these are relatively well known in terms of urban size, and population levels have been reconstructed. Furthermore, for both Kyaneai and Sagalassos, attempts have been made to reconstruct their carrying capacities, while for Pergamon the territory has been reconstructed, which can be used to provide a very preliminary insight at the carrying capacity of the territory. Furthermore, these cities can be taken as representative for different sizes of cities in different regions of Asia Minor, with Kyaneai being very small, Sagalassos mid-tiered and Ephesos and Pergamon as very large cities. What follows is a critical summary of both population reconstruction levels and carrying capacities per city.

Kyaneai

Kyaneai, located on a steep hill at Yavu in the south of Lycia, has been researched using extensive survey for the surface remains, consisted of a monumentalized main settlement enclosed by a Hellenistic city wall, with a necropolis and theatre located in the southwest outside the perimeter. The perimeter itself measures some 4 ha (5 ha including the theatre).[12] Kolb suggested on the basis of c. 500 graves (of which 380 sarcophagi) an urban population of c. 1000 inhabitants (Kolb 2008: 284). Werner Tietz suggested on the basis of house-remains (c. 22 of Classical-Hellenistic date), that in the 3rd-2nd century BC some 500+ inhabitants were present and 700 in the 1st century BC, while in Imperial times 24 more residential buildings are attested (Tietz 2010: 20-23). For the Imperial period,

---

[6] Hellenkemper and Hild 2004: 191; Dio Chrys. *Or.* 40.11, also alludes here to the growth of Smyrna and Ephesos as well.
[7] Blanton 2000 : 33; Rauh et al. 2009: 288 provide the figure of 41.4 ha, which must be too large as when measured, the plan on p. 287 gives a largest size of c. 24 ha.
[8] Using a plan in Şimşek 2011: 336 a measurement of 93 ha was made using GIS and Google Earth imagery; Traversari 2000: 19-20 notes that activities extended beyond this area, since the northern necropolis was also the location of a painting workshop.
[9] 250 ha (Hanfmann and Waldbaum 1975: 31-32; 130 ha (Rautman 2011: 11); the author measured the city in Satellite imagery, including the acropolis and slopes of the mountain, to some 125 ha.
[10] Wulf 1993 notes 230 ha, however her reconstruction has recently been revised and during RAC/TRAC 2016, Felix Pirson mentioned the figure of 190 ha.

[11] Hansen 2006 suggested that a combination of methods can provide a range of possible population size. See Willet 2012 for a review of methods for Roman Corinth.
[12] Kolb 2008: 169; Tietz 2010: 2-24 states that the enclosure even shrunk slightly only to have the city wall extended from 3.34 ha to 4.39 ha in Byzantine times.

Kolb states that not more than c. 150 houses were present at Kyaneai, which constitutes an increase.[13]

For the rural population a similar pattern can be observed. The hinterland of c. 100 km² was studied with extensive survey and the number of settlements rose from 7 (with an estimated 35 farms) plus 150 single farms (outside settlements) in the Archaic-Classical period to 55 settlements (containing an estimated 275 farms) and 250 single farms in the Hellenistic-Imperial period.[14] Using the figure of 7 people per farm, Volker Höhfeld reconstructed an increase of rural population from 1295 people to 3675 people for the Hellenistic-Imperial period, which would increase even further to its height during the Late Roman period. Continuing to carrying capacity, the arable land in the territory of Kyaneai (meaning flat lands, arable terraces and fields) is estimated at c. 5000 ha, of which at the moment of research in the 1990s and early 2000s, only 1430 ha was used for agriculture (Höhfeld 2006: 195-197). To reconstruct ancient carrying capacity, Höhfeld used multiple scenarios of intensive and extensive cultivation of cereals, vines and olive-culture, which fundamentally meant a division of all the arable land into plots for farms.[15] Using a figure of 5 people per farm, a total of 3,300 – 6,500 people could be sustained using extensive or intensive farming of cereals. From this and in congruence with Kolb's estimates (1000 people living in the city and 5000 in the countryside), a total sustained population of at least 6000 people is reconstructed. Despite the thoroughness of this reconstruction, it is reliant on (reasonably) reconstructed farm-sizes and different scenarios of agriculture. Using other figures for land per farm present in the literature, slightly higher figures emerge at c. 6,250 - 8,300 people sustained.[16] If a yield in barley or wheat is considered (cf. Engels) a lower sustained population results at some 4,200 people.[17] Still considering that the settlements are rather small and theoretically half of the territory is arable, the other areas being available for pastoralism, it seems hard to imagine Kyaneai could not sustain its own population even when growth took place.

Sagalassos

As already mentioned Sagalassos expanded during the late Hellenistic to Roman Imperial period eastward to a size of 37.5 ha. Recently a thorough reconstruction for the population of Sagalassos was made, on the basis of population densities for smaller cities and a reconstruction on the number of houses.[18] A cautious estimate of 1,500 to 5,000 inhabitants for the city was used, with indications from the housing suggesting a low figure in this spectrum.[19]

The primary catchment area of Sagalassos consisted of the c. 29 km² area of the Ağlasun and Yeşilbaşköy valleys. As for Kyaneai, reconstructions are used on the basis of yield in barley equivalent and calorific need per person or reconstructions on the basis of land needed for a farming family of 5, resulting in a sustained population of c. 2,000-4,500 people. This figure does not include the potential influx of agricultural products from the wider territory of Sagalassos (c. 1,200 km²). Sam Cleymans suggested that the valleys in the territory, measuring some 450 km², are the primary sources for arable land. Here, basing on Hartwin Brandt's estimates on arable land in the Pisidian Valleys, some 70 % of the 450 km² was probably arable (c. 315 km²), which would constitute 25 % of its territory. Using a safe margin of 20-35 % of the total territory as arable land, or 240-420 km², and using similar yield/calorific need methods or land needed for a farming family of 5, a capacity of 17,500 - 52,500 (Engels 1990), 29,000 - 66,900 (De Angelis 2000) or 32,400 - 55,700 (Bintliff 2002) is reconstructed. Based on their research on geomorphology, erosional processes, grain yields and soil fertility, Gert Verstraeten and Maarten Van Loo reconstructed that the territory of Sagalassos could sustain 40,000 people (Willet and Poblome 2015: 139).

This estimate of carrying capacity is in agreement with estimates for the rural population. Historically, the region of Pisidia is known to have been very fertile and filled with villages and extensive survey in the territory revealed that it was settled intensely during the Roman Imperial period (Mitchell 2000: 145; Livy, *Ab urbe condita libri* 38.15; Justinian, *Novella* 24.1). An increase in the number of sites in the territory from the Hellenistic to Roman Imperial period is observed with the presence of 41 larger local centres or villages (10,000+ m²) and 27 farms for the Imperial period (late 1st century BC until mid-5th century AD) with a further increase particularly in Late Antiquity (Vanhaverbeke and Waelkens 2003: 241-246; Poblome 2015). Jeroen Poblome and the author estimated the rural population cautiously at 6,000 and 25,000 people on the basis of assumed ratio's between central urban: rural population and reconstructed population densities. A population in the tens of thousands for the entire territory has furthermore been recently corroborated by research on the mitochondrial DNA of 40 human skeletons excavated

---

[13] Kolb 2008: 284-285; Price 2011: 22 calculated 110 houses maximum for Kyaneai, resulting in a population density of 80 people/ha.
[14] Kolb 2006: 3; Höhfeld 2006: 202 uses 5 farms per settlement as estimate.
[15] The following figures are averages from Höhfeld 2006: 198-200: extensive cereal farm – 7.5 ha, intensive cereal farm – 3.5 ha, wine farm – 2 ha and olive farm – 3.5 ha.; Koparal 2014 for the carrying capacity of Klazomenai, uses this division of agriculture too, but instead focuses on the yield of viticulture and olive orchards per hectare.
[16] 3-4 ha for a farm of 5 (De Angelis 2000); 3.6 ha for a farm of 5 (Bintliff 2002).
[17] Engels 1990: production (barley) = area arable * yield /year. Yield is set at 12 hl/ha per 2 year (fallowing). 1 hl weighs 61.8 kg. Yearly calorific need in barley per person is 438 kg, resulting for 5000 ha in 1,854,000 kg or c. 4,233 people sustained. Garnsey 2004 uses similar methods for Attica, but builds in more assumptions.

[18] Willet and Poblome 2015; in 8.7 ha some 92 houses were counted, extrapolated to 328 houses for the entire city, excluding the eastern suburbium. Using a ratio for elite houses (n=10-12 people per household) and lower class houses (4-5 people per household; Bagnall and Frier 1994: 68) as observed at Cosa (Fentress and Badel 2003: 24; De Ligt 2012: 220), where a population of 1,492 - 1,850 is estimated. This is consistent with a calculation using the urban area of 37.5 ha and a population density for unplanned towns (Price 2011: 23) at 40-60 people per ha, hich resulted in 1,500-2,250 people.
[19] Recent work by Cleymans (forthcoming) corroborates this.

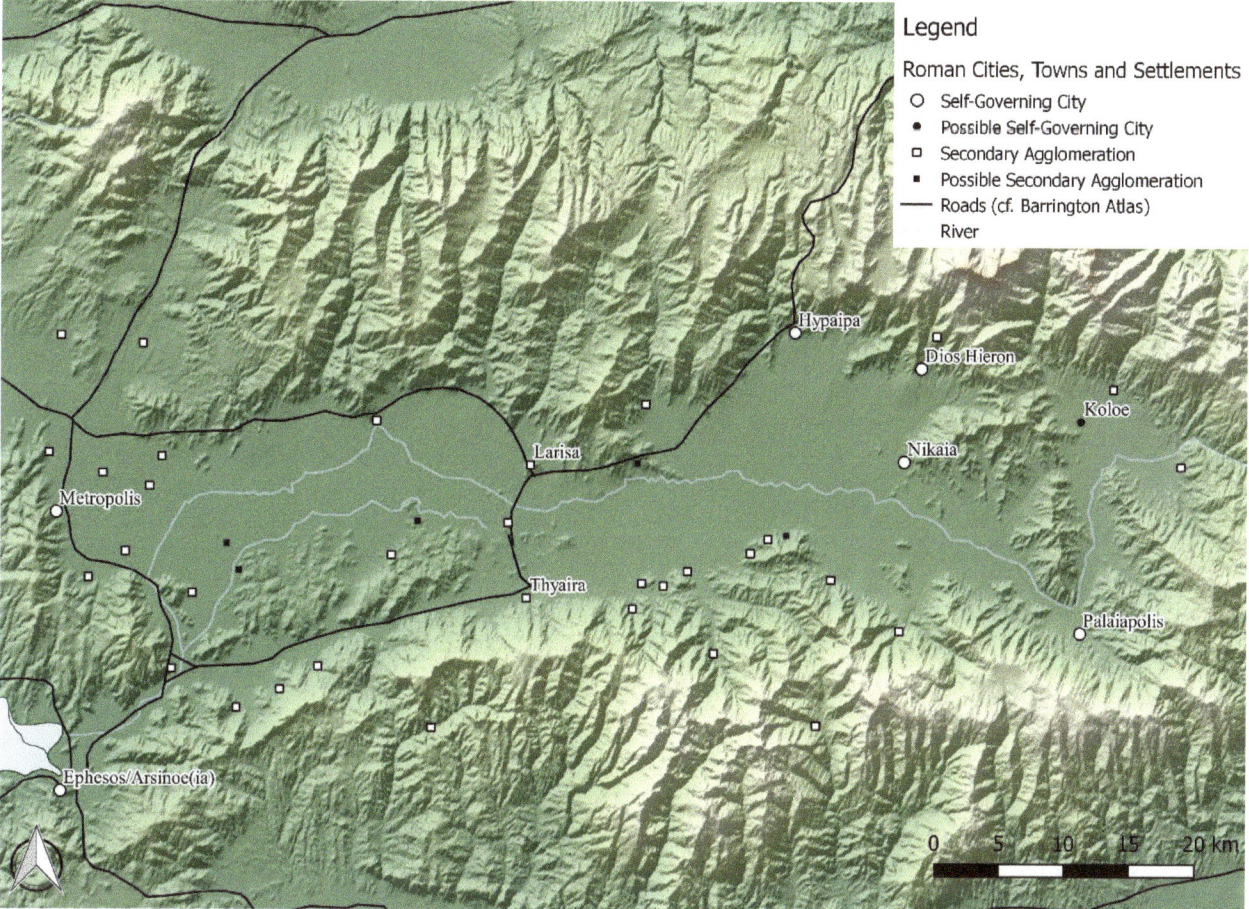

Figure 7.3: Kaystros River valley with the settlements present in Roman time (after Schuler 1998, Meriç 2009, Altınoluk 2013 and the Barrington Atlas).

at Sagalassos. This study suggests a population of c. 26,000 and 32,000, based on modelling the variation and changes in the gene pool (Ottoni et al. 2016). These figures are within the range of the carrying capacity.

Ephesos

Ephesos was one of the biggest cities in Asia Minor and Seneca actually counted it among the largest cities in the east, next to Alexandria (Sen. *Ep* 17.2.21; Scherrer, 'Ephesos', DNP). As already mentioned, the city plan was reconstructed with archaeological geophysical prospection and excavation and expanded from the late Hellenistic to early Imperial period to 185 ha, which contained c. 108 ha residential area, 34 ha harbour and residential area and 43 ha for public/civic buildings and spaces (Groh 2006: 101). Back in the 19th century, Karl Beloch estimated that this city must have had a vast population, at some 225,000 inhabitants.[20] Stefan Groh reconstructed for the residential area of 90-120 ha, (or 200-250 insulae out of 300) estimated 25,000 inhabitants (equal to a density of c. 200-280 people/ha) and a total of 30,000-70,000 inhabitants in the wider city area (equal to 160-380 people/ha for 185 ha) (Groh 2006: 112-113). Groh presents comparative estimates and notes that the theatre, which could hold c. 30,000 spectators, suggests a large populous.[21] It seems logical to assume a slightly higher density for the residential area than for the wider city area and table 6.5 contains a suggestion to calculate the urban population of Ephesos, taking into account the residential area with a higher population density and the remaining urban area with a lower density. The resulting range of 23,000-39,750 inhabitants is lower than some of the other estimates, yet at the same time higher than results from other population reconstruction methods by John Bintliff and Mogens Hansen. A good vista for future reconstructions would involve counting the number of houses in (a part of) Ephesos and try to extrapolate population levels, as data for this are at the moment lacking.

Ephesos controlled a large area in the Kaystros valley, although measurements are lacking (Figure 7.3). Larisa, probably situated in the Kaystros valley near Çatal, approximately 42 km distant, is described by Strabo to

---

[20] Beloch 1886: 230 reasons that Ephesos was larger than Pergamon with 40,000 citizens, and suggests Ephesos had 50,000 citizens. From this, he extrapolates a total free populous of 150,000 with a further 75,000 slaves. This high figure was partially corroborated by an overestimation of the physical size of Ephesos at 415 ha (set against 500,000 for Alexandria at 920 ha).

[21] Another estimate by Hanson (2011: 254) estimates 33,600-56,000 inhabitants for Ephesos (150-250 p/ha) but erroneously uses a size of 224 ha. His work contains many of such deviations, including Perge (26 ha, most publications reach around 50 ha), Sagalassos (19 ha) and Sardis (356 ha).

Table 7.2: Population estimates for Ephesos.

Estimates for the number of inhabitants of Ephesos		
source	parameters	urban population
Residential area	90-120 ha residential area, 150-250 people/ha	c. 13,500 - 30,000
Urban area	185 ha urban area, 100-150 people/ha	c. 18,500 - 27,750
Residential plus urban area	90-120 ha residential, 150-250 people/ha plus 65-95 remaining urban area, 100-150 people/ha	c. 23,000 – 39,750
Hansen 2006, p. 61	185 ha urban area, 50 percent residential area, 150 people/ha	c. 13,900
Bintliff 1997, p. 235	185 ha urban area, 56 percent residential area, 225 people/ha	c. 23,300
Estimates for the carrying capacity of the Kaystros Valley		
source	parameters	sustained population
Engels 1990, p. 203	production (barley) = area arable * yield/year; yield = 12 hl / ha per 2 years (fallowing); 1 hl barley = 61.8 kg; pop. sustained = production / necessity pp; necessity (barley) pp = 438 kg / year; area arable = 96600 ha	c. 81,780
De Angelis 2000, p. 118-125	Area of farming plot for family of 5, 3 - 4 ha; area arable = 96600 ha	c. 120,750 - 161,000
Estimates for the number of inhabitants for Ephesos and the territory		
source	parameters	total population
Bintliff 1997, p. 235.	23,000-39,750 urban population; urban:rural population 4:1	c. 28,750 – 49,690
Hansen 2006, p. 22, 43, 65.	23,000-39,750 urban population; urban:rural population 2:1	c. 34,500 – 59,625
De Graaf 2012, p. 91-92 and new suggested urbanisation rates by de Ligt, de Graaf, Bintliff	23,000-39,750 urban population; urbanisation rate: 10-25 % or urban:rural population 1:3 – 1:9	c. 92,000 (69,000 rural) – 397,500 (357,750 rural)

have been a village in the territory of Ephesos, and it is suggested that the entire valley of the Kaystros came under control of Ephesos (Strabo 13.3.2; Meriç 2009; Davies 2011: 185). Other self-governing cities existed in this valley, with their own territories, such as Metropolis, Hypaipa, Dios Hieron, Nikaia, and the territories of the Kilbianoi (probably centred around the cities of Nikaia, Palaiopolis and Koloe) all of which attested by their own coinages during the imperial period (RPC Online). It thus becomes difficult to accurately reconstruct the amount of land in this area under Ephesian control, while this will have been the main region of arable land for Ephesos. Furthermore, some of the land belonged to the Temple of Artemis at Ephesos and in the territory of Metropolis, there was the sanctuary of Pegaseum Stagnum (Meriç 2009: 64). Still, the archaeological evidence for these cities is fairly limited, with only Metropolis studied archaeologically in more detail (Aybek et al. 2009). This was a small city with public buildings and measured some 10 ha. Hypaipa was probably a city of greater size, with a theatre of some 65 m diameter, it probably measured some 32 ha.[22] The other cities in the Kaystros valley probably were much smaller, as archaeological remains are scant. In terms of food requirements, we can assume that Metropolis had a slightly higher need than Kyaneai and Hypaipa slightly less than Sagalassos. Assuming that these settlements were not in need of vast quantities of arable land to sustain their populations, the Kaystros valley, tentatively measured at some 96,900 ha, can be considered as the primary supplier of food to Ephesos.[23] Using similar methods for Kyaneai using either calorific yield or size for a farm of 5, a sustained population of c. 80,000 - 160,000 people emerges. To reconstruct the territorial population at this point in time, only methods relying on an assumed ratio between urban and rural population can be applied (Table 7.2). The methods as applied by Hansen and Bintliff give results that are not conflicting with the (very preliminary) estimates on carrying capacity, however more modern estimates which assume a much lower ratio of urbanisation by de Ligt, de Graaf and Bintliff, result in enormous populations. Although Ephesos was one of the largest cities in the east, it must be remembered that the Kaystros valley contained more cities (and Asia is relatively highly urbanized, see above), the higher

---

[22] Using the plan and the location of monuments produced by Altinoluk 2013: 252.

[23] This figure is derived using plan 2 in Meriç (2009) by measuring the flat area (excluding hills/mountains and modern built-up area of larger towns) east of the ancient coastline around Efes and the Kaystros as far inland as Palaiapolis (Beydağ), Koloe (Kiraz) and Kalekoy (Gevele). In essence this figure constitutes a rough measurement of satellite imagery using modern agricultural fields. This figure is not intended to be definitive but rather as a first tentative exploration.

urbanisation ratio seem more reasonable in the light of the carrying capacity figures. A total population of 50,000 - 100,000 seems not unlikely with 25,000 - 40,000 living in the city. The urban population of the cities in the hinterland, such as Hypaipa (say 5,000 people as a maximum) and Metropolis (2,000 inhabitants as a maximum), should be added. But with the agricultural reconstructions, there is some leeway to sustain these populations as well. Still, it must be remembered that these figures rely still a very preliminary assessment and if a high figure for population is to be assumed, the implication is a very high intensification of agriculture in the Kaystros valley. Recep Meriç in his study on the hinterland of Ephesos lists 38 sites that were (possible) secondary agglomerations (small dependent towns, villages and hamlets) during the Roman Imperial period. On average, the sites are spaced 5.7 km from each other, with a minimum distance of 1.4 km and a maximum of 14.7 km. When buffers of 5 km or 1 hour walking distance are drawn around these sites, most of the flat area of the Kaystros River valley is covered.[24] This may indicate that the valley was intensively occupied and exploited for agriculture, although future intensive field survey in the countryside should be able to shed more light on this.

Pergamon

The impressive site of Pergamon today is partially overbuilt by the modern city of Bergama. It played a major role in the Hellenistic period as the capital of the Attalid dynasty and was during the Roman Imperial period, as Ephesos, not only the centre of a large assize but also held the Neokoros or wardenship over the Imperial Cult two times (Habicht 1975; Friesen 1993; Price 1984; Knitter et al. 2013). The city expanded beyond its Hellenistic area from the main city mountain, which towers some 300 m high, outward to the lower plain southward in the Roman Imperial period as attested by several large public buildings of Roman date (Wulf 1994).[25] The substantial amount of public buildings denote the significance of this place in Antiquity. Although not lauded by Strabo or Seneca in the same way as Ephesos, the physician Galenos, who was born in Pergamon during the 2nd century AD, famously remarked that Pergamon had a population of four *Myriads* (= c. 40,000) of citizens, which he extrapolates to a total population of at least twelve *Myriads* (Galen, *De Propriorum Animi Dignotione et Curatione* 9 (Kühn 5.49 = Corpus Medicorum Graecorum 5.4.1.1.33); Radt 1999: 57-59). This has been reconstructed to by Beloch, who used the 40,000 male citizens as not to include women, children and the slaves, which counted half of the citizenry, were certainly not included, resulting in 120,000 for the city and territory, although he considers 180,000 more likely (Beloch 1886 : 236). Ulrike Wulf reconstructed population on the basis of the urban area which she reconstructed to 230 ha, of which 200 ha residential (87 %), and the size of a house of 6 residents (300-350 m2). The resulting estimate of 34,300 to 40,000 inhabitants is somewhat high compared with other methods that could be applied on the data. Furthermore, the demographic expansion from the Hellenistic to Roman Imperial period was substantial. The reconstruction of the Roman city has recently been questioned somewhat as some of the roads do not align with her reconstruction (Pirson 2008: 120; 2011 : 138-140 & n. 122). Recent geophysical survey on the city mountain, plus excavations in the lower city confirm that the size estimation by Wulf was somewhat too large, probably laying closer to 190 ha. Still, a cautious population estimate of 30,000 - 40,000 seems reasonable when compared to Ephesos and Sagalassos, assuming a higher population density for these large cities (Table 7.3) (Pirson 2014: 105-141).

The territory of Pergamon has been reconstructed on the basis of natural boundaries and historical evidence by Kai Sommerey. Here, a clear expansion of the territory is suggested for the Hellenistic/Imperial territories of Pergamon over the preceding Classical phase (Sommerey 2008: 135-170, compare Abb. 2 & 3). The port of Elaia, a polis in proximity of Pergamon, was itself a relatively large place of some 46 ha and already in Hellenistic times tied to Pergamon (Strabo 13.3.5; Pirson 2004: 208-209).[26] The territory as reconstructed by Sommerey in the Imperial period measured c. 135,000 ha maximum, comparable in size to the territory of Sagalassos. A large part of this terrain is rather hilly to mountainous, so the primary catchment area must have been the Kaikos river plain, a very fertile area and which Strabo calls about the best in Mysia (Strabo 13.4.2). Although no calculations on carrying capacity have been attempted for Pergamon yet, by measuring the river plain within the reconstructed territory a tentative reconstruction can be made. This results in a conservative measurement of 44,762 ha arable land (33 % of the total), which is enough to sustain a population of 37,000-75,000 (Figure 7.4). Reconstructions of the rural population are just as tentative if not more as for Ephesos without further study of the settlement pattern, although ongoing survey and research, notably by Daniel Knitter, will undoubtedly shed more light on this in the coming years. But the results from tentative reconstructions using a low urbanisation rate results in total populations which are very high, while higher urbanisation rates fit the preliminary carrying capacity estimates better. Furthermore, the terrain outside the river plain was potentially useful for agriculture and pastoralism as well. But even if the higher population estimates are (somewhat) accurate and the lower carrying capacity figures as well, the central situation and role of Pergamon with to the west fertile areas to the coast around Elaia, Pitane and the former city of Atarneus, and to the

---

[24] Meriç 2009: 130-131 lists less sites, but the 38 include isolated grave finds and sarcophagi that may be indicative of settlement in close proximity.
[25] A period of extensive construction of public buildings started already around the middle of the first century BC.

[26] Politically, Elaia must have been a separate entity from Pergamon as evidenced by coinage which continued into the Roman Imperial period; the old polis of Atarneus was abandoned in the first century BC (Pirson and Zimmerman 2012: 64).

Table 7.3: Population estimates for Pergamon.

Estimates for the number of inhabitants for Pergamon		
Author	parameters	urban population
Wulf 1994, p. 166	200 ha residential, houses at 300-350 m2 of 6 residents each	c. 34,300 – 40,000
Urban area Hellenistic	90 ha urban area, 100-150 people/ha	c. 9,000 – 13,500
Urban area Imperial	190 ha urban area, 100-150 people/ha	c. 19,000 – 28,500
Hansen 2006, p. 61	190 ha urban area, 50 percent residential area, 150 people/ha	c. 14,250
Bintliff 1997, p. 235	190 ha urban area, 56 percent residential area, 225 people/ha	c. 23,940
Estimates for the carrying capacity of the territory of Pergamon (Sommerey 2008)		
author	parameters	sustained population
Engels 1990, p. 203	*production (barley) = area arable * yield/year* *yield = 12 hl / ha per 2 years (fallowing)* *1 hl barley = 61.8 kg* *pop. sustained = production / necessity pp* *necessity (barley) pp = 438 kg / year* *area arable = 44762 ha*	c. 37,900
De Angelis 2000, p. 118-125	*Area of farming plot for family of 5, 3 - 4 ha* *area arable = 44762 ha*	c. 56,000 – 74,600
Estimates for the number of inhabitants for Ephesos and the territory		
author	parameters	total population
Bintliff 1997, p. 235.	30,000 - 40,000 urban population; urban:rural population 4:1	c. 37,500 – 50,500
Hansen 2006, p. 22, 43, 65.	30,000 - 40,000 urban population; urban:rural population 2:1	c. 45,000 – 60,000
De Graaf 2012, p. 91-92 and new suggested urbanisation rates by de Ligt, de Graaf, Bintliff	30,000 - 40,000 urban population; urbanisation rate: 10-25 % or urban:rural population 1:3 – 1:9	c. 120,000 (90,000 rural) – 400,000 (360,000 rural)

Figure 7.4: Image of the measured area of the Kaikos river plain as located in the territory of Pergamon (Sommerey 2008). Note that a wide margin is taken from the hills and that in reality, more field are currently under cultivation. This constitutes a conservative measurement of the current agricultural situation.

east, plains of the river Kaikos further upstream towards Stratonicaea, meant the city could have accessed sources of food relatively close-by. It seems fair to say that the Kaikos valley could sustain the population of Pergamon and its territory to a large extent. Still for both Ephesos and Pergamon this topic deserves more critical attention and more in-depth discussion than the tentative models presented here.

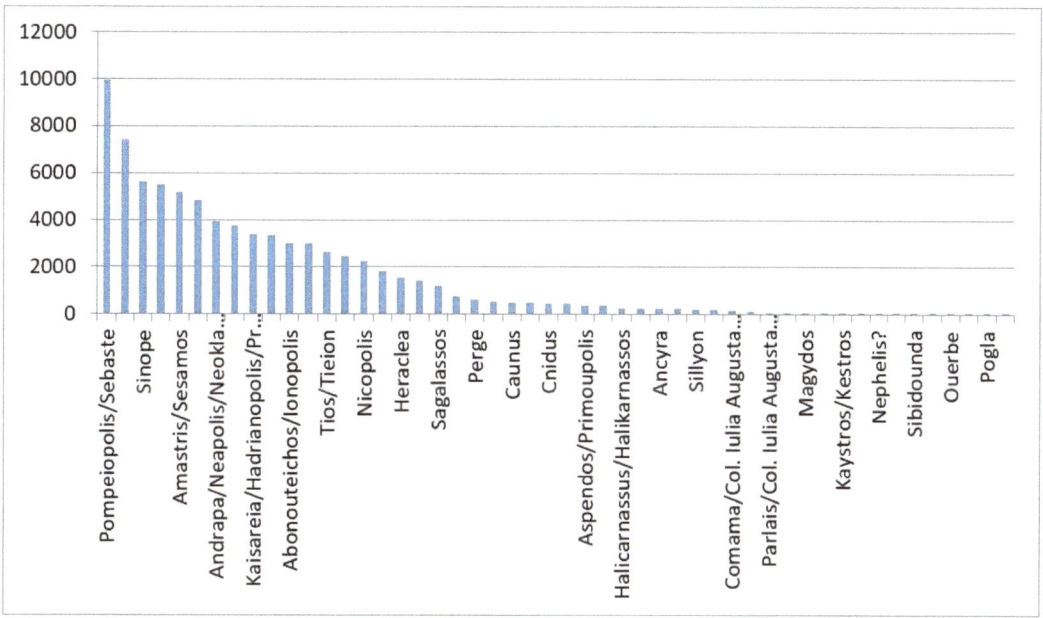

Figure 7.5: Published reconstructed territories (n=50; source from numerous publications, e.g. Marek 1993, Grainger 2009, Levick 1967 and Blanton 2000). Size in km². See Willet 2020 appendix II for all references.

Discussion: Malthus or Boserup?

The growth in cities in Asia Minor from the Hellenistic to Imperial periods has been assessed through four growing cities, and the expansion of the small and mid-sized cities of Kyaneai and Sagalassos seems to have been easily sustained by their respective hinterlands. The growth of the urban centre was accompanied by an increase in the number of sites in the hinterland. Ephesos and particularly Pergamon grew, which must have affected the intensity of cultivation in their direct hinterlands. If the carrying capacities for Ephesos and Pergamon are even remotely correct and the higher population estimates are assumed, then the logical conclusion is that the direct (assumed) arable hinterland was not enough to sustain their respective populations, which is at odds with the other two smaller cities. For the hinterland of Ephesos this seems to have been the case with the sites from reconnaissance survey suggesting that the Kaystros River valley could have been intensively exploited. For Pergamon, we eagerly await further study of the hinterland in the near future.

It is possible to provide a cautious characterization of the increase in number of cities in general in Asia Minor and for these four examples in particular. First, the vast majority of the cities that could be measured was small (100 of 134 smaller than 50 ha) and it can be assumed that the overwhelming majority of the cities belonged to the smaller category. The larger cities probably will have had more monuments, higher chances of remains being visible and therefore, a higher chance of attracted interest by archaeologists. Many of these places remain rather obscure and this is partially a result of their size. Second, the territories of many of the cities in Asia Minor, when any attempt at reconstruction has been made, were substantial, particularly in Northern Anatolia (Figure 7.5). For example, a measurement of the reconstructed territory of Pompeiopolis by Christian Marek, a city of some 16 ha, resulted in 9,957 km² (Marek 1993: Beilage 3 & 4, 33-46; Berghausen et al. 2009: 44).

Third, modern figures on agricultural productivity suggest that the four cities more closely studied, all lay in areas with good prospects for grain production, with the possible exception of Kyaneai. Calculations for agricultural yields in cereal production of 1927 show that the region around Bergama particularly, but also the area around Efes and Sagalassos had substantial yields in cereal production, while near Kyaneai only marginal yield are recorded. Such figures hint at differences in fertility of the territories under study (Table 7.4).

Fourth, a look at Ottoman Anatolia provides a perspective on the demographic and agricultural possibilities of this area. Suraiya Faroqhi reconstructed the population-levels of the larger cities in the period AD 1550-1600 using tax-records of the number of registered tax-payers. Bergama (Pergamon) ranked among the larger cities of Ottoman Anatolia, although it was much smaller than in antiquity (7,000-9,000 inhabitants on the basis of 2,444 taxpayers).[27] For the region of Ephesos, the focus of population had clearly shifted more inland, with Ayasoluğ (Ephesos) and Birgi (ancient Dios Hieron) being listed amongst the smallest Ottoman cities.[28] Tire (ancient Thyaira, a small town in the territory of Ephesos), however, is similarly large as Bergama (2,374 taxpayers) (Faroqhi 1984: 303).

---

[27] Faroqhi 1984: 303; I was unable to obtain Ottoman figures for the other places, although these, together with Ottoman agricultural censuses must exist (De Planhol 1959, Pers. Comm. Oktay Özel).

[28] For Ayasoluğ and Birgi, Faroqhi (1984: 12-13) gives no estimates, but they reside in the smallest category of taxpayers (between 400-999). In the line of Faroqhi's reconstructions, this would result in population-sizes below 3,000 people and can be categorized as semi-rural market settlements.

Table 7.4: Cereal production of 1927 in the regions in or close to the territories of the ancient cities discussed. The figures represent yield in tons per 100 km². Figures derived from Stratil-Sauer 1933.

Vilayets in the vicinity of ancient city.	*Wheat*	*Barley*	*Oats*	*Rye*
Kyaneai				
Finike	26-50.9	30-59	1-4.9	1-4.9
Kaş	26-50.9	30-59	0-0.9	5-14.9
Sagalassos				
Bucak	201-400	120-249	5-14.9	0-0.9
Burdur	401-700	120-249	1-4.9	1-4.9
Ephesos				
Bayindir	201-400	120-249	0-0.9	15-29
Kuşadasi	101-200	30-59	5-14.9	1-4.9
Ödemiş	51-100	120-249	0-0.9	1-4.9
Tire	201-400	250-399	0-0.9	15-29
Torbali	101-200	30-59	5-14.9	0-0.9
Pergamon				
Bergama	401-700	120-249	1-4.9	15-29
Soma	401-700	250-399	5-14.9	15-29

For the Ottoman successors of Sagalassos and the region of Kyaneai, Faroqhi provides no figures.

For two Ottoman cities there are clear indications that an agricultural system could be pushed beyond its sustainable limits by an increasing population and by demands of the central authorities. For Akşehir (Roman Philomelion, also an assize centre) and Konya (the Roman colony of Ikonion), both located in Lykaonia north of the Taurus Mountains, Faroqhi clearly shows cases where there was a dramatic increase in population during the 16th century (Faroqhi 1984: 194-218.). Akşehir saw an increase in population of 89 %, while the value of the wheat and barley harvests rose by 15.8 % and 22.3 % respectively. It is the production of gardens and vineyards that grew more dramatically (67.2 %). Still, after the 16th century, 1/3 of the settlements in the city's district disappeared and the tax revenue dropped significantly. For Konya, the results are as dramatic: 82 % population increase; 31.2 % barley and wheat harvest value increase and 63 % vineyard/garden harvest value increase. The data is unclear on how many settlements disappeared in the district of Konya. Central authorities tried to settle nomadic people to stimulate agriculture, but this had only limited success (Faroqhi 1984: 199-203). Tax pressure from the central authorities and unsuccessful agricultural intensification led to rising grain prices as well, with prices at Konya increasing 1,600 percent between 1566 and 1651. These examples are situated in areas which had grain yields above those of Kyaneai, but markedly lower as the other cities. The Ottoman socio-economic context obviously differs from the Roman Empire. Yet these attested historical examples demonstrate that a Malthusian equilibrium between natural resources and urban population did not necessarily have to emerge and that historical context could indeed push the production limits of agriculture. That this ultimately was not sustainable is perhaps an argument in favour of the forces of natural constraints. However changes in (geo-) political and socio-economic context of the Ottoman Empire as a whole, reaching its 'Golden Age' in the 16th century after which it would decline, must have contributed to these changes as well.

Yet most cities of Asia Minor during the Empire probably did not outgrow their agricultural potential, as evidenced by Kyaneai and Sagalassos, making the Malthusian axiom less usable (Chamberlain 2006: 4-5). The big cities may be a different story, with Ephesos and Pergamon possibly coming close to or outgrowing the agricultural potential of their direct hinterland. Yet these cities fulfilled important roles in their respective regions, e.g. as assize centres or centres for the imperial cult, and may have been able to rely on regular influx of goods or foodstuffs (Dalla Rosa 2012; Dio Chrys. Or. 35.15-17). However, without good intensive survey data, it is as of yet not possible to accurately confirm whether agricultural intensification took place, although it seems very likely. Both for Kyaneai and Sagalassos we can see an increase in number of sites and farms, and a few other surveys, like the Paphlagonia project in Northern Anatolia, show this increase as well (Matthews and Glatz 2009). The presence of large estates in Central Anatolia, where few cities were located during Imperial times, further indicate a greater agricultural exploitation which are not the result of a (dramatic) local increase in demand for food, but rather for economic exploitation (Mitchell 1993; Thonemann 2013). In that sense, the agricultural layout of Anatolia was more complex than the simple city-territory division. It seems fair to characterize the livelihood of most small cities as not reaching a Malthusian ceiling, although population growth and the expansion of the area under agricultural exploitation did occur: The growth in number of cities focused on southwestern Anatolia from the Hellenistic to Imperial times, but it is also the case that new cities appeared in less densely settled areas, such as in central Anatolia. This suggest that a greater area of potentially arable land came into use. Furthermore, the growth of the size of cities (some of them becoming very large) suggests that the arable land already under use in the territories of these cities must have been exploited more intensely. This would fit a Boserupian model of population growth much better, where population growth is sustained through agricultural intensification and the increase of lands under cultivation or extensification.

Taking a step back, we still are confronted with a) the growth of individual cities in size and b) the growth in number and spread of cities and settlements throughout Asia Minor from the Hellenistic period onwards, accompanied with growth in monumentalisation. Part of the explanation can be sought in the legacy of the Classical polis, which resulted in a fragmented political landscape of many mostly small towns (Hansen and Nielsen 2004). During the Hellenistic phase, this modular model of the city was adopted and further spread, particularly through new Seleucid and Attalid foundations (Cohen 1995; Ma 1999). In turn, the Romans would use this model to exercise control and levy taxes further inland in Asia Minor.

Yet it seems that another important piece of the puzzle must be found in overall demographic growth in Asia Minor. The expansion of self-governing cities implies the expansion of urban elites, who not only required food that was produced by others, but also required items of conspicuous consumption and status display. This in turn required specialist artisans and / or stimulated trade. The elite not only levied rents from their agricultural holdings, but also controlled to some extent the amount of taxes levied from the territory in their role as members of the city-councils. This must have required an increase in agricultural productivity as well. Last, the requirement of the central government for taxes will have had a similar effect and quite possibly stimulated the exchange of goods as well (Hopkins 1980). One might even suspect that there is a scalar property to this aspect of urbanism, as being argued for modern cities, in that the larger cities were not only limited by their size of population and need for food, but could rely due to their socio-cultural and –economic importance on a larger 'catchment' area as they attracted lines of trade from/forced control over neighbouring territories and their agriculture (Bettencourt 2013).

Ultimately, a Boserupian intensification of agriculture must be evidenced not necessarily through the archaeology of cities, but through the archaeology of villages and hamlets as sites of residence for farmers, and the archaeology of rural sites in Asia Minor of the Roman period in general. Unfortunately, these are poorly studied in Asia Minor. However, it is possible to collect the evidence of historically attested secondary agglomerations (i.e. all settlements without civic autonomy, mostly rural settlements like villages; Willet 2020). In my recent work on urbanism of Asia Minor, a collection of these settlements resulted in a total number of 743 secondary agglomerations, of which 479 associated with a located archaeological site. Although this number must be too low (intensive survey projects reveal many more sites that are not historically attested; e.g. Willet & Poblome 2019), the spatial patterning shows that the highest number of secondary agglomerations are present in the areas with the densest distribution of autonomous cities (i.e. southwestern Anatolia; Willet 2020: 125). As these areas are also characterized by having the highest agricultural potential of Asia Minor, the dense pattern of villages suggests a relatively high agricultural exploitation compared to other regions (e.g. central or north-eastern Anatolia) during the Roman Empire. The fact that this zone of dense distribution of secondary agglomerations, cities, high agricultural potential and high levels of monumentality in these cities again suggest that agricultural exploitation (as an economic foundation for cities) was high and that income from agriculture and other economic activities was enough to invest in urban monumentalisation.

Paul Erdkamp stated (following others) that urbanisation as a whole, can be regarded as an indicator of economic growth and performance as well, as a denser population requires higher levels of agricultural productivity. Urbanisation results in (relatively) fewer people involved in working the land and greater economic specialization and the presence of more cities reflects the ability of sustaining of the non-agricultural sectors (Erdkamp 2015). More urban markets stimulate greater agricultural exploitation and intensification, as they create stable markets that allowed specialisation and sustained increased production (Erdkamp 2016: 9-10). Erdkamp further posits that agricultural production was more elastic than the Malthusian model assumes, since increased production and improved cultivation as a results of specialization and increased scale of production. Weak markets and risk-avoiding strategies resulted on average that a large part of arable land was not used to an optimum. Greater market integration could improve this. Erdkamp further proposed to focus on the inefficiency of economic performance and in that sense, the fact that many if not most cities in Roman Asia Minor did not grew to an hypothetical Malthusian maximum must be sought in this inefficiency, that the (relatively weak) market integration of the urban system did not allow for greater agricultural productivity.

Although large private and imperial estates have been attested in Asia Minor (alongside domains in control by sanctuaries), the majority of agricultural cultivation took the form of small holdings. The fragmented local power, legal authority and fiscal competence in the hands of local rulers and institutions in Asia Minor may have hampered the development of the economy as well. Yet the fact remains, that the number of cities, the level of monumentalisation reached a new high during the Empire. Furthermore specialized production (e.g. tableware production at Sagalassos and in the region of both Ephesos and Pergamon to Phokaia) also roughly coincided with the Roman occupation from Late Hellenistic times onward (Willet and Poblome 2015; Poblome and Zelle 2002). It is perhaps therefore better to speak of a system, not stuck in a low-equilibrium trap, but in a gradually improving equilibrium. More studies should be committed to rural sites to pinpoint agricultural intensification or improvements / changes in agricultural practices. The fact that Asia Minor was to some extent already urbanized, with a few sizable cities, propelled the growth and spread of cities as well. This last fact may also explain that the system was resilient enough to continue to exist and even grow, long after the Roman West had fallen. Yet, the resilience of hinterlands and the social-political and economic framework of individual cities may have been decreased by population growth. The Ottoman figures for Pergamon and the region of Ephesos show that these cities were markedly smaller in Early Modern times than the relatively conservative estimates for the Roman period.

That being said, this paper has hopefully shown mostly that much more work is needed in Asia Minor on solving these issues. The very preliminary reconstructions for Ephesos and Pergamon are in that sense by no means intended to be definitive, but rather to present first explorations to such questions and invite/persuade more work on these topics from their respective impressive archaeological projects.

Bibliography

Akurgal, E. *Eski Çağda Ege ve İzmir*. Izmir: Tükelmat, 1993.

Altinoluk, S. *Hypaipa. A Lydian city during the Roman Imperial period*. Istanbul: Ege Yayinlari, 2013.

De Angelis, F. "Estimating the Agricultural Base of Greek Sicily." *Papers of the British School at Rome* 68 (2000): 111–148.

Van Aulock, H. *Münzen Und Städte Lykaoniens*, Istanbuler Mitteilungen. Beiheft 16. Tübingen: E. Wasmuth, 1976.

Van Aulock, H. *Münzen Und Städte Pisidiens. Teil I*, Istanbuler Mitteilungen. Beiheft 19. Tübingen: E. Wasmuth, 1977.

Van Aulock, H. *Münzen Und Städte Phrygiens. Teil II*, Istanbuler Mitteilungen. Beiheft 27. Tübingen: E. Wasmuth, 1987.

Aybek, S., Meriç, A.E., and Öz, A.K. *Metropolis. A Mother Godess City in Ionia*. Istanbul: Ege Yayinlari, 2009.

Aydal, S., Mitchell, S., Robinson, T. and Vandeput, L. "The Pisidian Survey 1995: Panemoteichos and Ören Tepe." *Anatolian Studies* 47 (1997): 141–172.

Bagnall, R.S., and Frier, B.W. *The Demography of Roman Egypt*, Cambridge studies in population, economy and society in past time 23. Cambridge: Cambridge University Press, 1994.

Belke, K. *Galatien Und Lykaonien*, Österreichische Akademie der Wissenschaften. Philosophisch-Historische Klasse: Denkschriften 172, Tabula imperii Byzantini 4. Vienna: Austrian Academy of Sciences Press, 1984.

Beloch, K.J. *Die Bevölkerung Der Griechisch-Römischen Welt*. Leipzig: Verlag Von Duncker & Humblot, 1886.

Berghousen, E., Fassbinder, J.W.E., Gorka, T., Kühne, L., Summerer, L. and Von Kienlin, A. "Klassische Archäologie Trifft Magnetometrie: Ein Stadtplan von Pompeiopolis (Türkei)." In *Archäometrie Und Denkmalpflege 2009 : Jahrestagung in der Pinakothek der Moderne, München 25.-28. März 2009*, Metalla. Sonderheft 2, edited by A. Hauptmann, and H. Stege, 42–44. Bochum: Deutsches Bergbau-Museum, 2009.

Bettencourt, L.M.A. "The Origins of Scaling in Cities." *Science* 340 (2013): 1438–1441.

Bintliff, J.L. "The archaeological investigation of deserted medieval villages in Greece." In *Rural Settlements in Medieval Europe: Papers of the 'Medieval Europe 1997' conference 7*, edited by G.D. Boe and F. Verhaege, 21-34. Zellik: Archaeological Institute for the Heritage, 1997.

Bintliff, J.L. "Rethinking Early Mediterranean Urbanism." In *Mauerschau : Festschrift Für Manfred Korfmann*, edited by R. Aslan, 153-177. Tübingen: Verlag Bernhard Greiner, 2002.

Blanton, R.E. *Hellenistic, Roman and Byzantine Settlement Patterns of the Coast Lands of Western Rough Cilicia*, BAR International Series 879. Oxford: BAR Publishing, 2000.

Boatwright, M.T. *Hadrian and the Cities of the Roman Empire*. Princeton: Princeton University Press, 2000.

Brandt, H. *Gesellschaft Und Wirtschaft Pamphyliens Und Pisidiens Im Altertum*, Asia Minor Studien 7. Bonn: Habelt R., 1992.

Broughton, T.R.S. "Roman Asia Minor." In *An Economic Survey of Ancient Rome. Volume IV. Africa, Syria, Greece, Asia Minor*, edited by T. Frank, 499–919. New Jersey, New York: Pageant Books, 1938.

Chamberlain, A.T. *Demography in Archaeology*, Cambridge manuals in archaeology. Cambridge: Cambridge University Press, 2006.

Cohen, G.M. *The Hellenistic Settlements in Europe, the Islands, and Asia Minor*, Hellenistic culture and society 17. Berkeley-Los Angeles-Oxford: University of California Press, 1995.

Dalla Rosa, A. "Praktische Lösungen Für Praktische Probleme: Die Gruppierung von Conventus in der Provinz Asia Un Die Bewegungen Des Prokonsuls C. Iulius Severes (Procos. 152/53)." *Zeitschrift für Papyrologie und Epigraphik* 183 (2012): 259–276.

Davies, J.K. "The Well-Balanced Polis: Ephesos." In *The Economies of Hellenistic Societies, Third to First Centuries BC*, edited by V. Gabrielsen, J.K. Davies, and Z. Archibald, 177–206. Oxford-New York: Oxford University Press, 2011.

Drew-Bear, T. "Problèmes de La Géographie Historique En Phrygie: L'exemple d'Alia." In *Aufstieg und Niedergang der römischen Welt Bd. II, 7.2*, edited by H. Temporini, 931-952. Berlin-New York : W. de Gruyter, 1980.

Engels, D.W. *Roman Corinth: An Alternative Model for the Classical City*. Chicago: University of Chicago Press, 1990.

Erdkamp, P. "Structural Determinants of Economic Performance in the Roman world and early-modern Europe. A Comparative Approach." In *Structure and Performance in the Roman Economy: Models, Methods and Case Studies*, Collection Latomus 350, edited by P. Erdkamp and K. Verboven, 17-31. Leuven-Brussels: Latomus, 2015.

Erdkamp, P. "Economic growth in the Roman Mediterranean world: An early good-bye to Malthus?" *Explorations in Economic History* 60 (2016): 1–20.

Faroqhi, S. *Towns and Townsmen of Ottoman Anatolia. Trade, Crafts and Food Production in an Urban Setting, 1520-1650*, Cambridge studies in Islamic civilization. Cambridge: Cambridge University Press, 1984.

Farrington, A. *The Roman Baths of Lycia. An Architectural Study*, British Institute of Archaeology at Ankara monograph 20. Oxford: Oxford University Press, 1995.

Fentress, E., and Badel, J.P. "Cosa in the Republic and Early Empire." In *Cosa V. An Intermittent Town, Excavations 1991-1997*, Memoirs of the American Academy in Rome. Supplementary volume, edited by E. Fentress, 13–62. Ann Arbor: University of Michigan Press, 2003.

Friesen, S.J. *Twice Neokoros. Ephesus, Asia and the Cult of the Flavian Imperial Family*, Religions in the Graeco-Roman world 116. Leiden-New York: Brill, 1993.

Garnsey, P. "Yield of the Land in Ancient Greece." In *Cities, Peasants and Food in Classical Antiquity. Essays in Social and Economic History*, edited by W. Scheidel, 201–213. Cambridge: Cambridge University Press, 2004.

Grainger, J. *The Cities of Pamphylia*. Oxford: Oxbow Books, 2009.

Groh, S. "Neue Forschungen Zur Stadtplanung in Ephesos." *Jahreshefte des Österreichischen Archäologischen Institutes in Wien* 75 (2006): 47–116.

Habicht, C. "New Evidence on the Province of Asia." *Journal of Roman Studies* 65 (1975): 64–91.

Hanfmann, G.M.A., and Waldbaum, J.C. *A Survey of Sardis and the Major Monuments Outside the Citywalls*, Archaeological Exploration of Sardis 1. Cambridge, Mass.: Harvard University Press, 1975.

Hansen, M.H. *The Shotgun Method. The Demography of the Ancient Greek City-State Culture*. Columbia-London: University of Missouri, 2006.

Hansen, M.H., and Nielsen, T.H. *An Inventory of Archaic and Classical Poleis. An Investigation Conducted by The Copenhagen Polis Centre for the Danish National Research Foundation*. Oxford: Oxford University Press, 2004.

Hanson, J.W. "The Urban System of Roman Asia Minor and Wider Urban Connectivity." In *Settlement, Urbanization, and Population*, Oxford studies on the Roman economy 2, edited by A. Bowman, and A. Wilson, 229–275. Oxford: Oxford University Press, 2011.

Hellenkemper, H., and Hild, F. *Lykien Und Pamphylien*, Tabula Imperii Byzantini 8, Denkschriften (Österreichische Akademie der Wissenschaften, Philosophisch-historische Klasse) 320. Vienna: Verlag der Österreichischen Akademie der Wissenschaften, 2004.

Hoepfner, W. *Herakleia Pontike – Ereğli. Eine Baugeschichtliche Untersuchung*, Forschungen an der Nordküste Kleinasiens-Ergänzungsbände zu den Tituli Asiae minoris II/1, Denkschriften (Österreichische Akademie der Wissenschaften. Philosophisch-historische Klasse) 89. Vienna: Verlag der Österreichischen Akademie der Wissenschaften, 1966.

Höhfeld, V. "Überlegungen Zur Potentiellen Tragfähigkeit Des Agrarraumes Im Zentralen Yavu-Berglland (Lykien, Türkei)." In *Lykische Studien 7: Die Chora von Kyaneai*, Tübinger althistorische Studien 2, edited by F. Kolb, 187–202. Bonn: Habelt R., 2006.

Hopkins, K. "Taxes and trade in the Roman Empire (200 B.C. - A.D. 400)." *Journal of Roman Studies* 70 (1980): 101–125.

Humann, K., and Puchstein, O. *Reisen in Kleinasien Und Nordsyrien Ausgeführt Im Auftrage Der Kgl. Pruessischen Akademie Der Wissenschaften. Band I Textband*. Berlin: Gundholzen, 1890.

Jones, A.H.M. *Cities of the Eastern Roman Provinces*, 2nd edition. Oxford: Oxford Clarendon Press, 1971.

Kessener, P. and Piras, S. "The Aspendos Aqueduct and the Roman-Seljuk Bridge Across the Eurymedon." *Adalya* 3 (1998): 149–168.

Knitter, D., Blum, H., Horejs, B., Nakoinz, O., and Schütt, B. "Integrated Centrality Analysis: A Diachronic Comparison of Selected Western Anatolian Locations." *Quaternary International* 312 (2013): 45–56.

Kolb, F. "Bemerkungen Zur Urbanen Ausstattung von Städten Im Westen Und Im Osten Des Römischen Reiches Anhand von Tacitus, Agricola 21 Und Der Konstantinischen Inschrift von Orkistos." *Klio* 75 (1993): 321–341.

Kolb, F. "Einleitung." In *Lykische Studien 7: Die Chora von Kyaneai*, Tübinger althistorische Studien 2, edited by F. Kolb, 1-3. Bonn: Habelt R., 2006.

Kolb, F. *Burg – Polis – Bischofssitz. Geschichte Der Siedlungskammer von Kyaneai in Der Südwesttürkei*. Mainz am Rhein: Zabern, 2008.

Koparal, E. "Land Use and Agricultural Potential in Klazomenian Khora." In *Regional Studies in Archaeology Symposium Proceedings 12-13 May 2011, Ankara*, Yerleşim arkeolojisi serisi 4, edited by B. Erciyas, and E. Sökmen 125–145. Istanbul: Ege Yayinlari, 2014.

LaBuff, J. *Polis Expansion and Elite Power in Hellenistic Karia*. Alexander the Great and the Hellenistic World. Lanham: Lexington Books, 2016.

De Ligt, L. *Peasants, Citizens and Soldiers. Studies in the Demographic History of Roman Italy 225 BC - AD 100*. Cambridge: Cambridge University Press, 2012.

De Ligt, L., Houten, P. and Willet, R. "An Empire of 2000 Cities: Urban Networks and Economic Integration in the Roman Empire." *Tijdschrift Mediterrane Archeologie* 52 (2014): 64.

Ma, J. *Antiochos III and the Cities of Western Asia Minor*. Oxford: Oxford University Press, 1999.

Magie, D. *Roman Rule in Asia Minor to the End of the Third Century after Christ*. Princeton: Princeton University Press, 1950.

Marek, C. *Stadt, Ära Und Territorium in Pontus Bithynia Und Nord-Galatia*, Istanbuler Forschungen 39. Tübingen: E. Wasmuth, 1993.

Martens, F., Vanhaverbeke, H., and Waelkens, M. "Town and Suburbium at Sagalassos: An Interaction Investigated through Survey." In *Thinking about Space: The Potential of Surface Suvery and Contextual Analysis in the Definition of Space in Roman Times*, Studies in eastern Mediterranean archaeology 8, edited by H. Vanhaverbeke, J. Poblome, F. Vermeulen, and M. Waelkens, 127–149. Turnhout: Brepols, 2008.

Matthews, R. and Glatz, C. *At Empires' Edge. Project Paphlagonia Regional Survey in North-Central Turkey*, British Institute of Archaeology at Ankara monograph 44. Ankara: British Institute of Archaeology at Ankara, 2009.

Mcevedy, C. *Cities of the Classical World. An Atlas and Gazetteer of 120 Centres of Ancient Civilization*. London: Penguin, 2011.

McNicoll, A.W. *Hellenistic Fortifications from the Aegean to the Euphrates (with Revisions and an Additional Chapter by N.P. Milner)*, Oxford monographs on classical archaeology. Oxford: Clarendon Press, 1997.

Meriç, R. *Das Hinterland von Ephesos. Archäologisch-topographische Forschungen im Kaystros-Tal*, Ergänzungshefte zu den Jahresheften des Österreichischen Archäologischen Institutes in Wien 12. Vienna: Österreichisches Archäologisches Institut, 2009.

Mitchell, S. *Anatolia: Land, Men, and Gods in Asia Minor. The Celts in Anatolia and the Impact of Roman Rule*. Oxford: Clarendon Press, 1993.

Mitchell, S. "The Settlement in Pisidia in Late Antiquity and the Byzantine Period: Methodological Problems." In, *Byzanz als Raum : zu Methoden und Inhalten der historischen Geographie des östlichen Mittelmeerraumes*, Tabula Imperii Byzantini 7, Denkschriften (Österreichische Akademie der Wissenschaften. Philosophisch-historische Klasse) 283, edited by K. Belke, F. Hild, J. Koder, and P. Soustal, 139-152.Vienna: Verlag der Österreichischen Akademie der Wissenschaften, 2000.

Nielsen, I. *Thermae et Balnea. The Architecture and Cultural History of Roman Public Baths. I Text*, 2nd edition. Aarhus: Aarhus University Press, 1993.

Ottoni, C., Rasteiro, R., Willet, R., Claeys, J., Talloen, P., Van de Vijver, K., Chikhi, L., Poblome, J. and R. Decorte "Comparing Maternal Genetic Variation across Two Millennia Reveals the Demographic History of an Ancient Human Population in Southwest Turkey." *Royal Society Open Science* 3, vol. 2 (2016).

Pirson, F. "Elaia, Der Maritime Satellit Pergamons." *Istanbuler Mitteilungen* 54 (2004): 74–101.

Pirson, F. "Pergamon – Bericht Über Die Arbeiten in Der Kampagne 2007." *Archäologischer Anzeiger* 2008, vol. 2 (2008): 83–155.

Pirson, F. "Pergamon – Bericht Über Die Arbeiten in Der Kampagne 2010." *Archäologischer Anzeiger* 2011, vol. 2 (2011): 81–212.

Pirson, F. "Pergamon - Bericht Über Die Arbeiten in Der Kampagne 2013." *Archäologischer Anzeiger* 2014, vol. 2 (2014): 101–176.

Pirson, F., and Zimmerman, M. "Das Umland von Pergamon – Wirtschafliche Ressourcen, Ländliche Siedlungen Und Politische Repräsentation." In *Pergamon. Panorama Der Antiken Metropole. Begleitbuch Zur Ausstellung*, edited by R. Grüssinger, V. Kästner, and A. Scholl, 58–65. Berlin: Imhof Verlag, 2012.

De Planhol, X. *De La Plaine Pamphylienne Aux Lacs Pisidiens: Nomadisme et Vie Paysanne*, Bibliothèque archéologique et historique de l'Institut français d'archéologie d'Istanbul 3. Paris: Adrien-Maisonneuve, 1959.

Poblome, J., and M. Zelle "The table ware boom: a socio-economic perspective from western Asia Minor." In *Patris Und Imperium: Kulturelle Und Politische Identität in Den Städten Der Römischen Provinzen Kleinasiens in Der Frühen Kaiserzeit*, Babesch Supplement 8, edited by C. Von Bernes, H. Von Hesberg, L. Vandeput, and M. Waelkens, 75–287. Leuven: Peeters, 2002.

Poblome, J. "Life in the Late Antique Countryside of Sagalassos." In *Pisidian Essays in Honour of Hacı Ali Ekinci*, edited by H. Metin, B.A. Polat Becks, R. Becks, and M. Firat, 99-109. Istanbul: Ege Yayinlari, 2015.

Price, S. "Estimating Ancient Greek Populations. The Evidence of Field Survey." In *Settlement, Urbanization, and Population*, Oxford studies on the Roman economy 2, edited by A. Bowman, and A. Wilson, 17–35. Oxford: Oxford University Press, 2011.

Price, S.R.F. *Rituals and Power. The Roman Imperial Cult in Asia Minor*. Cambridge: Cambridge University Press, 1984.

Radt, W. *Pergamon : Geschichte Und Bauten Einer Antiken Metropole*. Darmstadt: Primus Verlag, 1999.

Rauh, N.K., Townsend, R.F., Hoff, M.C., Dillon, M., Doyle, M.W., Ward, C.A., and Rothaus, R.M.. "Life in the Truck Lane: Urban Development in Western

Rough Cilicia." *Jahreshefte des Österreichischen Archäologischen Institutes in Wien* 78 (2009): 253–312.

Willet, R. and Poblome, J. "The Scale of Sagalassos Red Sip Ware Production – Reconstructions of Local Need and Production Output of Roman Imperial Tableware." *Adalya* 18 (2015): 133-157.

Willet, R., and Poblome, J. "Archaeological Surveying in the Western Part of the Ağlasun Valley and within Ağlasun, 2018." *ANMED. News of Archaeology from Anatolia's Mediterranean Areas* 17 (2019): 255–261.

Willet, R. *The Geography of Urbanism in Roman Asia Minor*. Sheffield: Equinox Publishing, 2020.

www.ingramcontent.com/pod-product-compliance
Ingram Content Group UK Ltd.
Pitfield, Milton Keynes, MK11 3LW, UK
UKHW061332200426
11947UKWH00044B/2012